A Member of the International Code Family™

International Energy Conservation Code®

Fort Morgan Public Library
414 Main Street
Fort Morgan, CO
(970) 867-9456

WITHDRAWN

For Reference

Not to be taken from this room

2003

2003 International Energy Conservation Code®

First Printing: January 2003
Second Printing: April 2003

ISBN # 1-892395-68-1 (soft)
ISBN # 1-892395-67-3 (loose-leaf)
ISBN # 1-892395-85-1 (e-document)

COPYRIGHT © 2003
by
INTERNATIONAL CODE COUNCIL, INC.

ALL RIGHTS RESERVED. This 2003 International Energy Conservation Code® is a copyrighted work owned by the International Code Council, Inc. Without advance written permission from the copyright owner, no part of this book may be reproduced, distributed, or transmitted in any form or by any means, including, without limitation, electronic, optical or mechanical means (by way of example and not limitation, photocopying, or recording by or in an information storage retrieval system). For information on permission to copy material exceeding fair use, please contact: Publications, 4051 West Flossmoor Road, Country Club Hills, IL 60478-5795 (Phone 800-214-4321).

Trademarks: "International Code Council," the "International Code Council" logo and the "International Energy Conservation Code" are trademarks of the International Code Council, Inc.

PRINTED IN THE U.S.A.

PREFACE

Introduction

Internationally, code officials recognize the need for a modern, up-to-date energy conservation code addressing the design of energy-efficient building envelopes and installation of energy efficient mechanical, lighting and power systems through requirements emphasizing performance. The *International Energy Conservation Code®*, in this 2003 edition, is designed to meet these needs through model code regulations that will result in the optimal utilization of fossil fuel and nondepletable resources in all communities, large and small.

This comprehensive energy conservation code establishes minimum regulations for energy efficient buildings using prescriptive and performance-related provisions. It is founded on broad-based principles that make possible the use of new materials and new energy efficient designs. This 2003 edition is fully compatible with all the *International Codes* ("I-Codes") published by the International Code Council (ICC), including the *International Building Code, ICC Electrical Code, International Existing Building Code, International Fire Code, International Fuel Gas Code, International Mechanical Code, ICC Performance Code, International Plumbing Code, International Private Sewage Disposal Code, International Property Maintenance Code, International Residential Code, International Urban-Wildland Interface Code* and *International Zoning Code*.

The *International Energy Conservation Code* provisions provide many benefits, among which is the model code development process that offers an international forum for energy professionals to discuss performance and prescriptive code requirements. This forum provides an excellent arena to debate proposed revisions. This model code also encourages international consistency in the application of provisions.

Development

The first edition of the *International Energy Conservation Code* (1998) was based on the 1995 edition of the *Model Energy Code* promulgated by the Council of American Building Officials (CABO) and included changes approved through the CABO Code Development Procedures through 1997. CABO assigned all rights and responsibilities to the International Code Council and its three statutory members: Building Officials and Code Administrators International, Inc. (BOCA), International Conference of Building Officials (ICBO) and Southern Building Code Congress International (SBCCI). This 2003 edition presents the code as originally issued, with changes approved through the ICC Code Development Process through 2002. A new edition such as this is promulgated every three years.

With the development and publication of the family of *International Codes* in 2000, the continued development and maintenance of the model codes individually promulgated by BOCA ("BOCA National Codes"), ICBO ("Uniform Codes") and SBCCI ("Standard Codes") was discontinued. This 2003 *International Energy Conservation Code*, as well as its predecessor — the 2000 edition— is intended to be the successor energy conservation code to those codes previously developed by BOCA, ICBO and SBCCI.

The development of a single set of comprehensive and coordinated *International Codes* was a significant milestone in the development of regulations for the built environment. The timing of this publication mirrors a milestone in the change in structure of the model codes, namely, the pending Consolidation of BOCA, ICBO and SBCCI into the ICC. The activities and services previously provided by the individual model code organizations will be the responsibility of the Consolidated ICC.

This code is founded on principles intended to establish provisions consistent with the scope of an energy conservation code that adequately conserves energy; provisions that do not unnecessarily increase construction costs; provisions that do not restrict the use of new materials, products or methods of construction; and provisions that do not give preferential treatment to particular types or classes of materials, products or methods of construction.

Adoption

The *International Energy Conservation Code* is available for adoption and use by jurisdictions internationally. Its use within a governmental jurisdiction is intended to be accomplished through adoption by reference in accordance with proceedings establishing the jurisdiction's laws. At the time of adoption, jurisdictions should insert the appropriate information in provisions requiring specific local information, such as the name of the adopting jurisdiction. These locations are shown in bracketed words in small capital letters in the code and in the sample ordinance. The sample adoption ordinance on page v addresses several key elements of a code adoption ordinance, including the information required for insertion into the code text.

Maintenance

The *International Energy Conservation Code* is kept up to date through the review of proposed changes submitted by code enforcing officials, industry representatives, design professionals and other interested parties. Proposed changes are carefully considered through an open code development process in which all interested and affected parties may participate.

The contents of this work are subject to change both through the Code Development Cycles and the governmental body that enacts the code into law. For more information regarding the code development process, contact the Code and Standard Development Department of the International Code Council.

While the development procedure of the *International Energy Conservation Code* assures the highest degree of care, ICC and the founding members of ICC—BOCA, ICBO, SBCCI—their members, and those participating in the development of this code do not accept any liability resulting from compliance or noncompliance with the provisions because ICC and its founding members do not have the power or authority to police or enforce compliance with the contents of this code. Only the governmental body that enacts the code into law has such authority.

Letter Designations in Front of Section Numbers

In each code development cycle, proposed changes to this code are considered at the Code Development Hearing by the International Energy Conservation Code Development Committee, whose action constitutes a recommendation to the voting membership for final action on the proposed change. Proposed changes to a code section whose number begins with a letter in brackets are considered by a different code development committee. For instance, proposed changes to code sections which have the letter [EB] in front (e.g., [EB] 101.2.2.1), are considered by the International Existing Building Code Development Committee at the Code Development Hearing. Where this designation is applicable to the entire content of a main section of the code, the designation appears at the main section number and title and is not repeated at every subsection in that section.

The content of sections in this code which begin with a letter designation is maintained by another code development committee in accordance with the following: [B] = International Building Code Development Committee; [EB] = International Existing Building Code Development Committee and [M] = International Mechanical Code Development Committee.

Marginal Markings

Solid vertical lines in the margins within the body of the code indicate a technical change from the requirements of the 2000 edition. Deletion indicators (➡) are provided in the margin where a paragraph or item has been deleted.

ORDINANCE

The *International Codes* are designed and promulgated to be adopted by reference by ordinance. Jurisdictions wishing to adopt the 2003 *International Energy Conservation Code* as an enforceable regulation governing energy efficient building envelopes and installation of energy efficient mechanical, lighting and power systems should ensure that certain factual information is included in the adopting ordinance at the time adoption is being considered by the appropriate governmental body. The following sample adoption ordinance addresses several key elements of a code adoption ordinance, including the information required for insertion into the code text.

SAMPLE ORDINANCE FOR ADOPTION OF THE *INTERNATIONAL ENERGY CONSERVATION CODE*
ORDINANCE NO._____

An ordinance of the **[JURISDICTION]** adopting the 2003 edition of the *International Energy Conservation Code*, regulating and governing energy efficient building envelopes and installation of energy efficient mechanical, lighting and power systems in the **[JURISDICTION]**; providing for the issuance of permits and collection of fees therefor; repealing Ordinance No. _____ of the **[JURISDICTION]** and all other ordinances and parts of the ordinances in conflict therewith.

The **[GOVERNING BODY]** of the **[JURISDICTION]** does ordain as follows:

Section 1. That a certain document, three (3) copies of which are on file in the office of the **[TITLE OF JURISDICTION'S KEEPER OF RECORDS]** of **[NAME OF JURISDICTION]**, being marked and designated as the *International Energy Conservation Code*, 2003 edition, including the Appendix **[FILL IN THE APPENDIX DETAILS BEING ADOPTED]**, as published by the International Code Council, be and is hereby adopted as the Energy Conservation Code of the **[JURISDICTION]**, in the State of **[STATE NAME]** for regulating and governing energy efficient building envelopes and installation of energy efficient mechanical, lighting and power systems as herein provided; providing for the issuance of permits and collection of fees therefor; and each and all of the regulations, provisions, penalties, conditions and terms of said Energy Conservation Code on file in the office of the **[JURISDICTION]** are hereby referred to, adopted, and made a part hereof, as if fully set out in this ordinance, with the additions, insertions, deletions and changes, if any, prescribed in Section 2 of this ordinance.

Section 2. The following sections are hereby revised:

Section 101.1. Insert: **[NAME OF JURISDICTION]**.

Section 3. That Ordinance No. _____ of **[JURISDICTION]** entitled **[FILL IN HERE THE COMPLETE TITLE OF THE ORDINANCE OR ORDINANCES IN EFFECT AT THE PRESENT TIME SO THAT THEY WILL BE REPEALED BY DEFINITE MENTION]** and all other ordinances or parts of ordinances in conflict herewith are hereby repealed.

Section 4. That if any section, subsection, sentence, clause or phrase of this ordinance is, for any reason, held to be unconstitutional, such decision shall not affect the validity of the remaining portions of this ordinance. The **[GOVERNING BODY]** hereby declares that it would have passed this ordinance, and each section, subsection, clause or phrase thereof, irrespective of the fact that any one or more sections, subsections, sentences, clauses and phrases be declared unconstitutional.

Section 5. That nothing in this ordinance or in the Energy Conservation Code hereby adopted shall be construed to affect any suit or proceeding impending in any court, or any rights acquired, or liability incurred, or any cause or causes of action acquired or existing, under any act or ordinance hereby repealed as cited in Section 2 of this ordinance; nor shall any just or legal right or remedy of any character be lost, impaired or affected by this ordinance.

Section 6. That the **[JURISDICTION'S KEEPER OF RECORDS]** is hereby ordered and directed to cause this ordinance to be published. (An additional provision may be required to direct the number of times the ordinance is to be published and to specify that it is to be in a newspaper in general circulation. Posting may also be required.)

Section 7. That this ordinance and the rules, regulations, provisions, requirements, orders and matters established and adopted hereby shall take effect and be in full force and effect **[TIME PERIOD]** from and after the date of its final passage and adoption.

TABLE OF CONTENTS

CHAPTER 1 ADMINISTRATION 1
Section
101 General 1
102 Materials, Systems and Equipment 2
103 Alternate Materials—Method of Construction, Design or Insulating Systems 3
104 Construction Documents 4
105 Inspections 4
106 Validity 4
107 Referenced Standards 4

CHAPTER 2 DEFINITIONS 5
Section
201 General 5
202 General Definitions 5

CHAPTER 3 DESIGN CONDITIONS 11
Section
301 General 11
302 Thermal Design Parameters 11

CHAPTER 4 RESIDENTIAL BUILDING DESIGN BY SYSTEMS ANALYSIS AND DESIGN OF BUILDINGS UTILIZING RENEWABLE ENERGY SOURCES 13
Section
401 General 13
402 Systems Analysis 13

CHAPTER 5 RESIDENTIAL BUILDING DESIGN BY COMPONENT PERFORMANCE APPROACH 17
Section
501 General 17
502 Building Envelope Requirements 17
503 Building Mechanical Systems and Equipment 37
504 Service Water Heating 42
505 Electrical Power and Lighting 43

CHAPTER 6 SIMPLIFIED PRESCRIPTIVE REQUIREMENTS FOR DETACHED ONE- AND TWO-FAMILY DWELLINGS AND GROUP R-2, R-4 OR TOWNHOUSE RESIDENTIAL BUILDINGS 45
Section
601 General 45
602 Building Envelope 45
603 Mechanical Systems 50
604 Service Water Heating 50
605 Electrical Power and Lighting 50

CHAPTER 7 BUILDING DESIGN FOR ALL COMMERCIAL BUILDINGS 51
Section
701 General 51

CHAPTER 8 DESIGN BY ACCEPTABLE PRACTICE FOR COMMERCIAL BUILDINGS 53
Section
801 General 53
802 Building Envelope Requirements 53
803 Building Mechanical Systems 59
804 Service Water Heating 72
805 Electrical Power and Lighting Systems 74
806 Total Building Performance 75

CHAPTER 9 CLIMATE MAPS 145
Section
901 General 145
902 Climate Zones 145

CHAPTER 10 REFERENCED STANDARDS 197

APPENDIX 201

INDEX 209

CHAPTER 1
ADMINISTRATION

SECTION 101
GENERAL

101.1 Title. These regulations shall be known as the *Energy Conservation Code* of [NAME OF JURISDICTION], and shall be cited as such. It is referred to herein as "this code."

101.2 Scope. This code establishes minimum prescriptive and performance-related regulations for the design of energy-efficient buildings and structures or portions thereof that provide facilities or shelter for public assembly, educational, business, mercantile, institutional, storage and residential occupancies, as well as those portions of factory and industrial occupancies designed primarily for human occupancy. This code thereby addresses the design of energy-efficient building envelopes and the selection and installation of energy-efficient mechanical, service water-heating, electrical distribution and illumination systems and equipment for the effective use of energy in these buildings and structures.

> **Exception:** Energy conservation systems and components in existing buildings undergoing repair, alteration or additions, and change of occupancy, shall be permitted to comply with the *International Existing Building Code*.

101.2.1 Exempt buildings. Buildings and structures indicated in Sections 101.2.1.1 and 101.2.1.2 shall be exempt from the building envelope provisions of this code, but shall comply with the provisions for building, mechanical, service water heating and lighting systems.

> **101.2.1.1 Separated buildings.** Buildings and structures, or portions thereof separated by building envelope assemblies from the remainder of the building, that have a peak design rate of energy usage less than 3.4 Btu/h per square foot (10.7 W/m^2) or 1.0 watt per square foot (10.7 W/m^2) of floor area for space conditioning purposes.

> **101.2.1.2 Unconditioned buildings.** Buildings and structures or portions thereof which are neither heated nor cooled.

101.2.2 Applicability. The provisions of this code shall apply to all matters affecting or relating to structures and premises, as set forth in Section 101. Where, in a specific case, different sections of this code specify different materials, methods of construction or other requirements, the most restrictive shall govern.

> **[EB] 101.2.2.1 Existing installations.** Except as otherwise provided for in this chapter, a provision in this code shall not require the removal, alteration or abandonment of, nor prevent the continued utilization and maintenance of, an existing building envelope, mechanical, service water-heating, electrical distribution or illumination system lawfully in existence at the time of the adoption of this code.

> **[EB] 101.2.2.2 Additions, alterations or repairs.** Additions, alterations, renovations or repairs to a building envelope, mechanical, service water-heating, electrical distribution or illumination system or portion thereof shall conform to the provisions of this code as they relate to new construction without requiring the unaltered portion(s) of the existing system to comply with all of the requirements of this code. Additions, alterations or repairs shall not cause any one of the aforementioned and existing systems to become unsafe, hazardous or overloaded.

> **[EB] 101.2.2.3 Historic buildings.** The provisions of this code relating to the construction, alteration, repair, enlargement, restoration, relocation or movement of buildings or structures shall not be mandatory for existing buildings or structures specifically identified and classified as historically significant by the state or local jurisdiction, listed in *The National Register of Historic Places* or which have been determined to be eligible for such listing.

> **[EB] 101.2.2.4 Change in occupancy.** It shall be unlawful to make a change in the occupancy of any building or structure which would result in an increase in demand for either fossil fuel or electrical energy supply unless such building or structure is made to comply with the requirements of this code or otherwise approved by the authority having jurisdiction. The code official shall certify that such building or structure meets the intent of the provisions of law governing building construction for the proposed new occupancy and that such change of occupancy does not result in any increase in demand for either fossil fuel or electrical energy supply or any hazard to the public health, safety or welfare.

101.2.3 Mixed occupancy. When a building houses more than one occupancy, each portion of the building shall conform to the requirements for the occupancy housed therein. Where minor accessory uses do not occupy more than 10 percent of the area of any floor of a building, the major use shall be considered the building occupancy. Buildings, other than detached one- and two-family dwellings and townhouses, with a height of four or more stories above grade shall be considered commercial buildings for purposes of this code, regardless of the number of floors that are classified as residential occupancy.

101.3 Intent. The provisions of this code shall regulate the design of building envelopes for adequate thermal resistance and low air leakage and the design and selection of mechanical, electrical, service water-heating and illumination systems and equipment which will enable effective use of energy in new building construction. It is intended that these provisions provide flexibility to permit the use of innovative approaches and techniques to achieve effective utilization of energy. This code is not intended to abridge safety, health or environmental requirements under other applicable codes or ordinances.

101.4 Compliance. Compliance with this code shall be determined in accordance with Sections 101.4.1 and 101.4.2.

101.4.1 Residential buildings. For residential buildings the following shall be used as the basis for compliance assessment: a systems approach for the entire building (Chapter 4), an approach based on performance of individual components of the building envelope (Chapter 5), an approach based on performance of the total building envelope (Chapter 5), an approach based on acceptable practice for each envelope component (Chapter 5), an approach by prescriptive specification for individual components of the building envelope (Chapter 5), or an approach based on simplified, prescriptive specification (Chapter 6) where the conditions set forth in Section 101.4.1.1 or 101.4.1.2 are satisfied.

101.4.1.1 Detached one- and two-family dwellings. When the glazing area does not exceed 15 percent of the gross area of exterior walls.

101.4.1.2 Residential buildings, Group R-2, R-4 or townhouses. When the glazing area does not exceed 25 percent of the gross area of exterior walls.

101.4.2 Commercial buildings. For commercial buildings, a prescriptive or performance-based approach (Chapter 7) or as specified by acceptable practice (Chapter 8) shall be used as the basis for compliance assessment.

SECTION 102
MATERIALS, SYSTEMS AND EQUIPMENT

102.1 General. Materials, equipment and systems shall be identified in a manner that will allow a determination of their compliance with the applicable provisions of this code.

102.2 Materials, equipment and systems installation. All insulation materials, caulking and weatherstripping, fenestration assemblies, mechanical equipment and systems components, and water-heating equipment and system components shall be installed in accordance with the manufacturer's installation instructions.

102.3 Maintenance information. Required regular maintenance actions shall be clearly stated and incorporated on a readily accessible label. Such label shall include the title or publication number, the operation and maintenance manual for that particular model and type of product. Maintenance instructions shall be furnished for equipment that requires preventive maintenance for efficient operation.

102.4 Insulation installation. Roof/ceiling, floor, wall cavity and duct distribution systems insulation shall be installed in a manner that permits inspection of the manufacturer's R-value identification mark.

102.4.1 Protection of exposed foundation insulation. Insulation applied to the exterior of foundation walls and around the perimeter of slab-on-grade floors shall have a rigid, opaque and weather-resistant protective covering to prevent the degradation of the insulation's thermal performance. The protective covering shall cover the exposed area of the exterior insulation and extend a minimum of 6 inches (153 mm) below grade.

102.5 Identification. Materials, equipment and systems shall be identified in accordance with Sections 102.5.1, 102.5.2 and 102.5.3.

102.5.1 Building envelope insulation. A thermal resistance (R) identification mark shall be applied by the manufacturer to each piece of building envelope insulation 12 inches (305 mm) or greater in width.

Alternatively, the insulation installer shall provide a signed and dated certification for the insulation installed in each element of the building envelope, listing the type of insulation installations in roof/ceilings, the manufacturer and the R-value. For blown-in or sprayed insulation, the installer shall also provide the initial installed thickness, the settled thickness, the coverage area and the number of bags installed. Where blown-in or sprayed insulation is installed in walls, floors and cathedral ceilings, the installer shall provide a certification of the installed density and R-value. The installer shall post the certification in a conspicuous place on the job site.

102.5.1.1 Roof/ceiling insulation. The thickness of roof/ceiling insulation that is either blown in or sprayed shall be identified by thickness markers that are labeled in inches or millimeters installed at least one for every 300 square feet (28 m^2) throughout the attic space. The markers shall be affixed to the trusses or joists and marked with the minimum initial installed thickness and minimum settled thickness with numbers a minimum of 1 inch (25 mm) in height. Each marker shall face the attic access. The thickness of installed insulation shall meet or exceed the minimum initial installed thickness shown by the marker.

102.5.2 Fenestration product rating, certification and labeling. U-factors of fenestration products (windows, doors and skylights) shall be determined in accordance with NFRC 100 by an accredited, independent laboratory, and labeled and certified by the manufacturer. The solar heat gain coefficient (SHGC) of glazed fenestration products (windows, glazed doors and skylights) shall be determined in accordance with NFRC 200 by an accredited, independent laboratory, and labeled and certified by the manufacturer. Where a shading coefficient for a fenestration product is used, it shall be determined by converting the product's SHGC, as determined in accordance with NFRC 200, to a shading coefficient, by dividing the SHGC by 0.87. Such certified and labeled U-factors and SHGCs shall be accepted for purposes of determining compliance with the building envelope requirements of this code.

When a manufacturer has not determined product U-factor in accordance with NFRC 100 for a particular product line, compliance with the building envelope requirements of this code shall be determined by assigning such products a default U-factor in accordance with Tables 102.5.2(1) and 102.5.2(2). When a SHGC or shading coefficient is used for code compliance and a manufacturer has not determined product SHGC in accordance with NFRC 200 for a particular product line, compliance with the building envelope requirements of this code shall be determined by assigning such products a default SHGC in accordance with Table

102.5.2(3). Product features must be verifiable for the product to qualify for the default value associated with those features. Where the existence of a particular feature cannot be determined with reasonable certainty, the product shall not receive credit for that feature. Where a composite of materials from two different product types is used, the product shall be assigned the higher *U*-factor.

TABLE 102.5.2(1)
***U*-FACTOR DEFAULT TABLE FOR WINDOWS, GLAZED DOORS AND SKYLIGHTS**

FRAME MATERIAL AND PRODUCT TYPE[a]	SINGLE GLAZED	DOUBLE GLAZED
Metal without thermal break:		
Curtain wall	1.22	0.79
Fixed	1.13	0.69
Garden window	2.60	1.81
Operable (including sliding and swinging glass doors)	1.27	0.87
Site-assembled sloped/overhead glazing	1.36	0.82
Skylight	1.98	1.31
Metal with thermal break:		
Curtain wall	1.11	0.68
Fixed	1.07	0.63
Operable (including sliding and swinging glass doors)	1.08	0.65
Site-assembled sloped/overhead glazing	1.25	0.70
Skylight	1.89	1.11
Reinforced vinyl/metal clad wood:		
Fixed	0.98	0.56
Operable (including sliding and swinging glass doors)	0.90	0.57
Skylight	1.75	1.05
Wood/vinyl/fiberglass:		
Fixed	0.98	0.56
Garden window	2.31	1.61
Operable (including sliding and swinging glass doors)	0.89	0.55
Skylight	1.47	0.84

a. Glass-block assemblies with mortar but without reinforcing or framing shall have a *U*-factor of 0.60.

TABLE 102.5.2(2)
***U*-FACTOR DEFAULT TABLE FOR NONGLAZED DOORS**

DOOR TYPE	WITH FOAM CORE	WITHOUT FOAM CORE
Steel doors (1.75 inches thick)	0.35	0.60
	WITH STORM DOOR	WITHOUT STORM DOOR
Wood doors (1.75 inches thick)		
Hollow core flush	0.32	0.46
Panel with 0.438-inch panels	0.36	0.54
Panel with 1.125-inch panels	0.28	0.39
Solid core flush	0.26	0.40

For SI: 1 inch = 25.4 mm.

102.5.3 Duct distribution systems insulation. A thermal resistance (*R*) identification mark shall be applied by the manufacturer in maximum intervals of no greater than 10 feet (3048 mm) to insulated flexible duct products showing the thermal performance *R*-value for the duct insulation itself (excluding air films, vapor retarders or other duct components).

SECTION 103
ALTERNATE MATERIALS—METHOD OF CONSTRUCTION, DESIGN OR INSULATING SYSTEMS

103.1 General. The provisions of this code are not intended to prevent the use of any material, method of construction, design or insulating system not specifically prescribed herein, provided that such construction, design or insulating system has been approved by the code official as meeting the intent of the code.

Compliance with specific provisions of this code shall be determined through the use of computer software, worksheets, compliance manuals and other similar materials when they have been approved by the code official as meeting the intent of this code.

TABLE 102.5.2(3)
SHGC DEFAULT TABLE FOR FENESTRATION

PRODUCT DESCRIPTION	SINGLE GLAZED				DOUBLE GLAZED			
	Clear	Bronze	Green	Gray	Clear + Clear	Bronze + Clear	Green + Clear	Gray + Clear
Metal frames								
Fixed	0.78	0.67	0.65	0.64	0.68	0.57	0.55	0.54
Operable	0.75	0.64	0.62	0.61	0.66	0.55	0.53	0.52
Nonmetal frames								
Fixed	0.75	0.64	0.62	0.61	0.66	0.54	0.53	0.52
Operable	0.63	0.54	0.53	0.52	0.55	0.46	0.45	0.44

SECTION 104
CONSTRUCTION DOCUMENTS

104.1 General. Construction documents and other supporting data shall be submitted in one or more sets with each application for a permit. The construction documents and designs submitted under the provisions of Chapter 4 shall be prepared by a registered design professional where required by the statutes of the jurisdiction in which the project is to be constructed. Where special conditions exist, the code official is authorized to require additional construction documents to be prepared by a registered design professional.

Exceptions:

1. The code official is authorized to waive the submission of construction documents and other supporting data not required to be prepared by a registered design professional if it is found that the nature of the work applied for is such that reviewing of construction documents is not necessary to obtain compliance with this code.

2. For residential buildings having a conditioned floor area of 5,000 square feet (465 m^2) or less, designs submitted under the provisions of Chapter 4 shall be prepared by anyone having qualifications acceptable to the code official.

104.2 Information on construction documents. Construction documents shall be drawn to scale upon suitable material. Electronic media documents are permitted to be submitted when approved by the code official. Construction documents shall be of sufficient clarity to indicate the location, nature and extent of the work proposed and show in sufficient detail pertinent data and features of the building and the equipment and systems as herein governed, including, but not limited to, design criteria, exterior envelope component materials, *U*-factors of the envelope systems, *U*-factors of fenestration products, *R*-values of insulating materials, size and type of apparatus and equipment, equipment and systems controls and other pertinent data to indicate compliance with the requirements of this code and relevant laws, ordinances, rules and regulations, as determined by the code official.

SECTION 105
INSPECTIONS

105.1 General. Construction or work for which a permit is required shall be subject to inspection by the code official.

105.2 Approvals required. No work shall be done on any part of the building or structure beyond the point indicated in each successive inspection without first obtaining the written approval of the code official. No construction shall be concealed without inspection approval.

105.3 Final inspection. There shall be a final inspection and approval for buildings when completed and ready for occupancy.

105.4 Reinspection. A structure shall be reinspected when determined necessary by the code official.

SECTION 106
VALIDITY

106.1 General. If a section, subsection, sentence, clause or phrase of this code is, for any reason, held to be unconstitutional, such decision shall not affect the validity of the remaining portions of this code.

SECTION 107
REFERENCED STANDARDS

107.1 General. The standards, and portions thereof, which are referred to in this code and listed in Chapter 10, shall be considered part of the requirements of this code to the extent of such reference.

107.2 Conflicting requirements. When a section of this code and a section of a referenced standard from Chapter 10 specify different materials, methods of construction or other requirements, the provisions of this code shall apply.

CHAPTER 2
DEFINITIONS

SECTION 201
GENERAL

201.1 Scope. Unless otherwise expressly stated, the following words and terms shall, for the purposes of this code, have the meanings indicated in this chapter.

201.2 Interchangeability. Words used in the present tense include the future; words in the masculine gender include the feminine and neuter; the singular number includes the plural and the plural, the singular.

201.3 Terms defined in other codes. Where terms are not defined in this code and are defined in the *International Building Code*, ICC *Electrical Code, International Fire Code, International Fuel Gas Code, International Mechanical Code* or the *International Plumbing Code*, such terms shall have meanings ascribed to them as in those codes.

201.4 Terms not defined. Where terms are not defined through the methods authorized by this section, such terms shall have ordinarily accepted meanings such as the context implies.

SECTION 202
GENERAL DEFINITIONS

ACCESSIBLE (AS APPLIED TO EQUIPMENT). Admitting close approach because not guarded by locked doors, elevation or other effective means (see "Readily accessible").

ADDITION. An extension or increase in the height, conditioned floor area or conditioned volume of a building or structure.

AIR TRANSPORT FACTOR. The ratio of the rate of useful sensible heat removal from the conditioned space to the energy input to the supply and return fan motor(s), expressed in consistent units and under the designated operating conditions.

ALTERATION. Any construction, renovation or change in a mechanical system that involves an extension, addition or change to the arrangement, type or purpose of the original installation.

ANNUAL FUEL UTILIZATION EFFICIENCY (AFUE). The ratio of annual output energy to annual input energy which includes any nonheating season pilot input loss, and for gas or oil-fired furnaces or boilers, does not include electrical energy.

APPROVED. Approved by the code official or other authority having jurisdiction as the result of investigation and tests conducted by said official or authority, or by reason of accepted principles or tests by nationally recognized organizations.

AUTOMATIC. Self-acting, operating by its own mechanism when actuated by some impersonal influence, as, for example, a change in current strength, pressure, temperature or mechanical configuration (see "Manual").

BASEMENT WALL. The opaque portion of a wall which encloses one side of a basement and having an average below-grade area greater than or equal to 50 percent of its total wall area, including openings (see "Gross area of exterior walls").

BTU. Abbreviation for British thermal unit, which is the quantity of heat required to raise the temperature of 1 pound (0.454 kg) of water 1°F (D0.56°C), (1 Btu = 1,055 J).

BUILDING. Any structure occupied or intended for supporting or sheltering any use or occupancy.

BUILDING ENVELOPE. The elements of a building which enclose conditioned spaces through which thermal energy is capable of being transferred to or from the exterior or to or from spaces exempted by the provisions of Section 101.2.1.

CODE OFFICIAL. The officer or other designated authority charged with the administration and enforcement of this code, or a duly authorized representative.

COEFFICIENT OF PERFORMANCE (COP)—COOLING. The ratio of the rate of heat removal to the rate of energy input in consistent units, for a complete cooling system or factory-assembled equipment, as tested under a nationally recognized standard or designated operating conditions.

COEFFICIENT OF PERFORMANCE (COP)—HEAT PUMP—HEATING. The ratio of the rate of heat delivered to the rate of energy input, in consistent units, for a complete heat pump system under designated operating conditions. Supplemental heat shall not be considered when checking compliance with the heat pump equipment (COPs listed in the tables in Sections 503 and 803).

COMFORT ENVELOPE. The areas defined on a psychrometric chart and enclosing the range of operative temperatures and humidities for both the winter and summer comfort zones as depicted in Figure 2 of ASHRAE 55.

COMMERCIAL BUILDING. All buildings other than detached one- and two-family dwellings, townhouses and residential buildings, Groups R-2 and R-4.

CONDENSER. A heat exchanger designed to liquefy refrigerant vapor by removal of heat.

CONDENSING UNIT. A specific refrigerating machine combination for a given refrigerant, consisting of one or more power-driven compressors, condensers, liquid receivers (when required), and the regularly furnished accessories.

CONDITIONED FLOOR AREA. The horizontal projection of that portion of interior space which is contained within exterior walls and which is conditioned directly or indirectly by an energy-using system.

CONDITIONED SPACE. A heated or cooled space, or both, within a building and, where required, provided with humidification or dehumidification means so as to be capable of maintaining a space condition falling within the comfort envelope set forth in ASHRAE 55.

DEFINITIONS

COOLED SPACE. Space within a building which is provided with a positive cooling supply (see "Positive cooling supply").

CRAWL SPACE WALL. The opaque portion of a wall which encloses a crawl space and is partially or totally below grade.

DEADBAND. The temperature range in which no heating or cooling is used.

DEGREE DAY, COOLING. A unit, based on temperature difference and time, used in estimating cooling energy consumption and specifying nominal cooling load of a building in summer. For any one day, when the mean temperature is more than 65°F (18°C), there are as many degree days as there are degrees Fahrenheit (Celsius) difference in temperature between the mean temperature for the day and 65°F (18°C). Annual cooling degree days (CDD) are the sum of the degree days over a calendar year.

DEGREE DAY, HEATING. A unit, based on temperature difference and time, used in estimating heating energy consumption and specifying nominal heating load of a building in winter. For any one day, when the mean temperature is less than 65°F (18°C), there are as many degree days as there are degrees Fahrenheit (Celsius) difference in temperature between the mean temperature for the day and 65°F (18°C). Annual heating degree days (HDD) are the sum of the degree days over a calendar year.

DUCT. A tube or conduit utilized for conveying air. The air passages of self-contained systems are not to be construed as air ducts.

DUCT SYSTEM. A continuous passageway for the transmission of air that, in addition to ducts, includes duct fittings, dampers, plenums, fans and accessory air-handling equipment and appliances.

DWELLING UNIT. A single housekeeping unit comprised of one or more rooms providing complete independent living facilities for one or more persons, including permanent provisions for living, sleeping, eating, cooking and sanitation.

ECONOMIZER. A ducting arrangement and automatic control system that allows a cooling supply fan system to supply outdoor air to reduce or eliminate the need for mechanical refrigeration during mild or cold weather.

ENERGY. The capacity for doing work (taking a number of forms) which is capable of being transformed from one into another, such as thermal (heat), mechanical (work), electrical and chemical in customary units, measured in joules (J), kilowatt-hours (kWh) or British thermal units (Btu).

ENERGY ANALYSIS. A method for determining the annual (8,760 hours) energy use of the proposed design and standard design based on hour-by-hour estimates of energy use.

ENERGY COST. The total estimated annual cost for purchased energy for the building, including any demand charges, fuel adjustment factors and delivery charges applicable to the building.

ENERGY EFFICIENCY RATIO (EER). The ratio of net equipment cooling capacity in British thermal units per hour (Btu/h) (W) to total rate of electric input in watts under designated operating conditions. When consistent units are used, this ratio becomes equal to COP (see also "Coefficient of performance").

EVAPORATOR. That part of the system in which liquid refrigerant is vaporized to produce refrigeration.

EXTERIOR ENVELOPE. See "Building envelope."

EXTERIOR WALL. An above-grade wall enclosing conditioned space which is vertical or sloped at an angle 60 degrees (1.1 rad) or greater from the horizontal (see "Roof assembly"). Includes between-floor spandrels, peripheral edges of floors, roof and basement knee walls, dormer walls, gable end walls, walls enclosing a mansard roof, and basement walls with an average below-grade wall area which is less than 50 percent of the total opaque and nonopaque area of that enclosing side.

FENESTRATION. Skylights, roof windows, vertical windows (whether fixed or moveable), opaque doors, glazed doors, glass block, and combination opaque/glazed doors.

FURNACE, DUCT. A furnace normally installed in distribution ducts of air-conditioning systems to supply warm air for heating and which depends on a blower not furnished as part of the duct furnace for air circulation.

FURNACE, WARM AIR. A self-contained, indirect-fired or electrically heated furnace that supplies heated air through ducts to spaces that require it.

GLAZING AREA. Total area of the glazed fenestration measured using the rough opening and including sash, curbing or other framing elements that enclose conditioned space. Glazing area includes the area of glazed fenestration assemblies in walls bounding conditioned basements. For doors where the daylight opening area is less than 50 percent of the door area, the glazing area is the daylight opening area. For all other doors, the glazing area is the rough opening area for the door including the door and the frame.

GROSS AREA OF EXTERIOR WALLS. The normal projection of all exterior walls, including the area of all windows and doors installed therein (see "Exterior wall").

GROSS FLOOR AREA. The sum of the areas of several floors of the building, including basements, cellars, mezzanine and intermediate floored tiers and penthouses of headroom height, measured from the exterior faces of exterior walls or from the centerline of walls separating buildings, but excluding:

1. Covered walkways, open roofed-over areas, porches and similar spaces.
2. Pipe trenches, exterior terraces or steps, chimneys, roof overhangs and similar features.

HEAT. The form of energy that is transferred by virtue of a temperature difference or a change in state of a material.

HEAT CAPACITY (HC). The amount of heat necessary to raise the temperature of a given mass by one degree. The heat capacity of a building element is the sum of the heat capacities of each of its components.

HEAT PUMP. A refrigeration system that extracts heat from one substance and transfers it to another portion of the same substance or to a second substance at a higher temperature for a beneficial purpose.

HEAT REJECTION EQUIPMENT. Equipment used in comfort cooling systems such as air-cooled condensers, open cooling towers, closed-circuit cooling towers and evaporative condensers.

HEAT TRAP. An arrangement of piping and fittings, such as elbows, or a commercially available heat trap, that prevents thermosyphoning of hot water during standby periods.

HEATED SLAB. Slab-on-grade construction in which the heating elements or hot air distribution system is in contact with or placed within the slab or the subgrade.

HEATED SPACE. Space within a building which is provided with a positive heat supply (see "Positive heat supply"). Finished living space within a basement with registers or heating devices designed to supply heat to a basement space shall automatically define that space as heated space.

HEATING SEASONAL PERFORMANCE FACTOR (HSPF). The total heating output of a heat pump during its normal annual usage period for heating, in Btu, divided by the total electric energy input during the same period, in watt hours, as determined by DOE 10 CFR Part 430, Subpart B, Test Procedures and based on Region 4.

HUMIDISTAT. A regulatory device, actuated by changes in humidity, used for automatic control of relative humidity.

HVAC. Heating, ventilating and air conditioning.

HVAC SYSTEM. The equipment, distribution network, and terminals that provide either collectively or individually the processes of heating, ventilating, or air conditioning to a building.

HVAC SYSTEM COMPONENTS. HVAC system components provide, in one or more factory-assembled packages, means for chilling or heating water, or both, with controlled temperature for delivery to terminal units serving the conditioned spaces of the building. Types of HVAC system components include, but are not limited to, water chiller packages, reciprocating condensing units and water source (hydronic) heat pumps (see "HVAC system equipment").

HVAC SYSTEM EQUIPMENT. HVAC system equipment provides, in one (single package) or more (split system) factory-assembled packages, means for air circulation, air cleaning, air cooling with controlled temperature and dehumidification and, optionally, either alone or in combination with a heating plant, the functions of heating and humidifying. The cooling function is either electrically or heat operated and the refrigerant condenser is air, water or evaporatively cooled. Where the equipment is provided in more than one package, the separate packages shall be designed by the manufacturer to be used together. The equipment shall be permitted to provide the heating function as a heat pump or by the use of electric or fossil-fuel-fired elements. (The word "equipment" used without a modifying adjective, in accordance with common industry usage, applies either to HVAC system equipment or HVAC system components.)

INFILTRATION. The uncontrolled inward air leakage through cracks and interstices in any building element and around windows and doors of a building caused by the pressure effects of wind or the effect of differences in the indoor and outdoor air density or both.

INSULATING SHEATHING. An insulating board having a minimum thermal resistance of R-2 of the core material.

INTEGRATED PART-LOAD VALUE (IPLV). A single measure of merit, based on part-load EER or COP expressing part-load efficiency for air-conditioning and heat pump equipment on the basis of weighted operation at various load capacities for the equipment.

LABELED. Devices, equipment, appliances, assemblies or materials to which have been affixed a label, seal, symbol or other identifying mark of a nationally recognized testing laboratory, inspection agency or other organization concerned with product evaluation that maintains periodic inspection of the production of the above-labeled items and by whose label the manufacturer attests to compliance with applicable nationally recognized standards.

LISTED. Equipment, appliances, assemblies or materials included in a list published by a nationally recognized testing laboratory, inspection agency or other organization concerned with product evaluation that maintains periodic inspection of production of listed equipment, appliances, assemblies or material, and whose listing states either that the equipment, appliances, assemblies, or material meets nationally recognized standards or has been tested and found suitable for use in a specified manner.

LOW-VOLTAGE LIGHTING. Lighting equipment that is powered through a transformer such as cable conductor, rail conductor and track lighting.

LUMINAIRE. A complete lighting unit consisting of at least one lamp and the parts designed to distribute the light, to position and protect the lamp, to connect the lamp to the power supply and ballasting, when applicable. Luminaires are commonly referred to as "lighting fixtures."

MANUAL. Capable of being operated by personal intervention (see "Automatic").

MULTIFAMILY DWELLING. A building containing three or more dwelling units.

MULTIPLE SINGLE-FAMILY DWELLING (TOWNHOUSE). A building not more than three stories in height consisting of multiple single-family dwelling units, constructed in a group of three or more attached units in which each unit extends from foundation to roof and with open space on at least two sides.

OCCUPANCY. The purpose for which a building, or portion thereof, is utilized or occupied.

OPAQUE AREAS. All exposed areas of a building envelope which enclose conditioned space, except openings for windows, skylights, doors and building service systems.

OUTDOOR AIR. Air taken from the outdoors and, therefore, not previously circulated through the system.

OZONE DEPLETION FACTOR. A relative measure of the potency of chemicals in depleting stratospheric ozone. The ozone depletion factor potential depends on the chlorine and bromine content and the atmospheric lifetime of the chemical. The depletion factor potential is normalized such that the factor for CFC-11 is set equal to unity and the factors for the other chemicals indicate their potential relative to CFC-11.

DEFINITIONS

PACKAGED TERMINAL AIR CONDITIONER (PTAC). A factory-selected wall sleeve and separate unencased combination of heating and cooling components, assemblies or sections (intended for mounting through the wall to serve a single room or zone). It includes heating capability by hot water, steam or electricity. (For the complete technical definition, see ARI 310/380.)

PACKAGED TERMINAL HEAT PUMP. A PTAC capable of using the refrigeration system in a reverse cycle or heat pump mode to provide heat. (For the complete technical definition, see ARI 310/380.)

POSITIVE COOLING SUPPLY. Mechanical cooling deliberately supplied to a space, such as through a supply register. Also, mechanical cooling indirectly supplied to a space through uninsulated surfaces of space-cooling components, such as evaporator coil cases and cooling distribution systems which continually maintain air temperatures within the space of 85°F (29°C) or lower during normal operation. To be considered exempt from inclusion in this definition, such surfaces shall comply with the insulation requirements of this code.

POSITIVE HEAT SUPPLY. Heat deliberately supplied to a space by design, such as a supply register, radiator or heating element. Also, heat indirectly supplied to a space through uninsulated surfaces of service water heaters and space-heating components, such as furnaces, boilers and heating and cooling distribution systems which continually maintain air temperature within the space of 50°F (10°C) or higher during normal operation. To be considered exempt from inclusion in this definition, such surfaces shall comply with the insulation requirements of this code.

PROPOSED DESIGN. A description of the proposed building design used to estimate annual energy costs for determining compliance based on total building performance.

READILY ACCESSIBLE. Capable of being reached quickly for operation, renewal or inspections, without requiring those to whom ready access is requisite to climb over or remove obstacles or to resort to portable ladders or access equipment (see "Accessible").

REFRIGERANT. A substance utilized to produce refrigeration by its expansion or vaporization or absorption.

RENEWABLE ENERGY SOURCES. Sources of energy (excluding minerals) derived from incoming solar radiation, including natural daylighting and photosynthetic processes; from phenomena resulting therefrom, including wind, waves and tides, lake or pond thermal differences; and from the internal heat of the earth, including nocturnal thermal exchanges.

REPAIR. The reconstruction or renewal of any part of an existing building for the purpose of its maintenance.

RESIDENTIAL BUILDING, GROUP R-2. Residential occupancies containing more than two dwelling units where the occupants are primarily permanent in nature such as apartment houses, boarding houses (not transient), convents, monasteries, rectories, fraternities and sororities, dormitories and rooming houses. For the purpose of this code, reference to Group R-2 occupancies shall refer to buildings that are three stories or less in height above grade.

RESIDENTIAL BUILDING, GROUP R-4. Residential occupancies shall include buildings arranged for occupancies as Residential Care/Assisted Living Facilities including more than five but not more than 16 occupants, excluding staff. For the purpose of this code, reference to Group R-4 occupancies shall refer to buildings which are three stories or less in height above grade.

ROOF ASSEMBLY. A roof assembly shall be considered as all roof/ceiling components of the building envelope through which heat flows, thus creating a building transmission heat loss or gain, where such assembly is exposed to outdoor air and encloses conditioned space.

The gross area of a roof assembly consists of the total interior surface of all roof/ceiling components, including opaque surfaces, dormer and bay window roofs, trey ceilings, overhead portions of an interior stairway to an unconditioned attic, doors and hatches, glazing and skylights exposed to conditioned space, that are horizontal or sloped at an angle less than 60 degrees (1.1 rad) from the horizontal (see "Exterior wall"). A roof assembly, or portions thereof, having a slope of 60 degrees (1.1 rad) or greater from horizontal shall be considered in the gross area of exterior walls and thereby excluded from consideration in the roof assembly. Skylight shaft walls 12 inches (305 mm) in depth or greater (as measured from the ceiling plane to the roof deck) shall be considered in the gross area of exterior walls and are thereby excluded from consideration in the roof assembly.

ROOM AIR CONDITIONER. An encased assembly designed as a unit for mounting in a window or through a wall, or as a console. It is designed primarily to provide free delivery of conditioned air to an enclosed space, room or zone. It includes a prime source of refrigeration for cooling and dehumidification and means for circulating and cleaning air, and shall be permitted to also include means for ventilating and heating.

SASH CRACK. The sum of all perimeters of all window sashes, based on overall dimensions of such parts, expressed in feet. If a portion of one sash perimeter overlaps a portion of another sash perimeter, the overlapping portions are only counted once.

SCREW LAMP HOLDERS. A lamp base that requires a screw-in-type lamp such as an incandescent or tungsten-halogen bulb.

SEASONAL ENERGY EFFICIENCY RATIO (SEER). The total cooling output of an air conditioner during its normal annual usage period for cooling, in Btu/h (W), divided by the total electric energy input during the same period, in watt-hours, as determined by DOE 10 CFR Part 430, Subpart B, Test Procedures.

SERVICE SYSTEMS. All energy-using systems in a building that are operated to provide services for the occupants or processes housed therein, including HVAC, service water heating, illumination, transportation, cooking or food preparation, laundering and similar functions.

SERVICE WATER HEATING. Supply of hot water for purposes other than comfort heating.

DEFINITIONS

SIMULATION TOOL. An approved software program or calculation-based methodology that projects the hour-by-hour loads and annual energy use of a building.

SKYLIGHT. Glazing that is horizontal or sloped at an angle less than 60 degrees (1.1 rad) from the horizontal (see "Glazing area").

SLAB-ON-GRADE FLOOR INSULATION. Insulation around the perimeter of the floor slab or its supporting foundation when the top edge of the floor perimeter slab is above the finished grade or 12 inches (305 mm) or less below the finished grade.

SOLAR ENERGY SOURCE. Source of natural daylighting and of thermal, chemical or electrical energy derived directly from conversion of incident solar radiation.

STANDARD DESIGN. A version of the proposed design that meets the minimum requirements of this code and is used to determine the maximum annual energy cost requirement for compliance based on total building performance.

STANDARD TRUSS. Any construction that does not permit the roof/ceiling insulation to achieve the required R-value over the exterior walls.

SUNROOM ADDITION. A one-story structure added to an existing dwelling with a glazing area in excess of 40 percent of the gross area of the structure's exterior walls and roof.

SYSTEM. A combination of central or terminal equipment or components or controls, accessories, interconnecting means, and terminal devices by which energy is transformed so as to perform a specific function, such as HVAC, service water heating or illumination.

THERMAL CONDUCTANCE. Time rate of heat flow through a body (frequently per unit area) from one of its bounding surfaces to the other for a unit temperature difference between the two surfaces, under steady conditions (Btu/h · ft² · °F) [W/(m² · K)].

THERMAL ISOLATION. A separation of conditioned spaces, between a sunroom addition and a dwelling unit, consisting of existing or new wall(s), doors and/or windows. New wall(s), doors and/or windows shall meet the prescriptive envelope component criteria in Table 502.2.5.

THERMAL RESISTANCE (R). The reciprocal of thermal conductance (h · ft² · °F/Btu) [(m² · K)/W].

THERMAL RESISTANCE, OVERALL (R_o). The reciprocal of overall thermal conductance (h · ft² · °F/Btu)[(m² · K)/W]. The overall thermal resistance of the gross area or individual component of the exterior building envelope (such as roof/ceiling, exterior wall, floor, crawl space wall, foundation, window, skylight, door, opaque wall, etc.), which includes the area weighted R-values of the specific component assemblies (such as air film, insulation, drywall, framing, glazing, etc.).

THERMAL TRANSMITTANCE (U). The coefficient of heat transmission (air to air). It is the time rate of heat flow per unit area and unit temperature difference between the warm-side and cold-side air films (Btu/h · ft² · °F) [W/(m² · K)]. The U-factor applies to combinations of different materials used in series along the heat flow path, single materials that comprise a building section, cavity airspaces and surface air films on both sides of a building element.

THERMAL TRANSMITTANCE, OVERALL (U_o). The overall (average) heat transmission of a gross area of the exterior building envelope (Btu/h · ft² · °F) [W/(m² · K)]. The U_o-factor applies to the combined effect of the time rate of heat flow through the various parallel paths, such as windows, doors and opaque construction areas, comprising the gross area of one or more exterior building components, such as walls, floors or roof/ceilings.

THERMOSTAT. An automatic control device actuated by temperature and designed to be responsive to temperature.

TOWNHOUSE. See "Multiple single-family dwelling."

UNITARY COOLING AND HEATING EQUIPMENT. One or more factory-made assemblies which include an evaporator or cooling coil, a compressor and condenser combination, and which shall be permitted to include a heating function as well. When heating and cooling equipment is provided in more than one assembly, the separate assemblies shall be designed to be used together.

UNITARY HEAT PUMP. One or more factory-made assemblies which include an indoor conditioning coil, compressor(s) and outdoor coil or refrigerant-to-water heat exchanger, including means to provide both heating and cooling functions. When heat pump equipment is provided in more than one assembly, the separate assemblies shall be designed to be used together.

VENTILATION. The process of supplying or removing air by natural or mechanical means to or from any space. Such air shall be permitted to be conditioned or unconditioned.

VENTILATION AIR. That portion of supply air which comes from outside (outdoors) plus any recirculated air that has been treated to maintain the desired quality of air within a designated space.

WATER HEATER, INSTANTANEOUS. A water heater with an input rating of at least 4,000 Btu/h per gallon (310 W/L) stored water and a storage capacity of less than 10 gallons (38 L).

WATER HEATER, STORAGE. A water heater with an input rating less than 4,000 Btu/h per gallon (310 W/L) of stored water or storage capacity of at least 10 gallons (38 L).

WINDOW PROJECTION FACTOR. A measure of the portion of glazing that is shaded by an eave or overhang.

ZONE. A space or group of spaces within a building with heating or cooling requirements, or both, sufficiently similar so that comfort conditions can be maintained throughout by a single controlling device.

CHAPTER 3
DESIGN CONDITIONS

SECTION 301
GENERAL

301.1 Design criteria. The criteria of this chapter establish the design conditions for use with Chapters 4, 5, 6 and 8.

SECTION 302
THERMAL DESIGN PARAMETERS

302.1 Exterior design conditions. The following design parameters in Table 302.1 shall be used for calculations required under this code.

TABLE 302.1
EXTERIOR DESIGN CONDITIONS

CONDITION	VALUE
Winter[a], Design Dry-bulb (°F)	
Summer[a], Design Dry-bulb (°F)	
Summer[a], Design Wet-bulb (°F)	
Degree days heating[b]	
Degree days cooling[b]	
Climate zone[c]	

For SI: °C = [(°F)-32]/1.8.

a. The outdoor design temperature shall be selected from the columns of $97^1/_2$ percent values for winter and $2^1/_2$ percent values for summer from tables in the ASHRAE *Fundamentals Handbook*. Adjustments shall be permitted to reflect local climates which differ from the tabulated temperatures, or local weather experience determined by the code official.

b. The degree days heating (base 65°F) and cooling (base 65°F) shall be selected from NOAA "Annual Degree Days to Selected Bases Derived from the 1961-1990 Normals," data available from adjacent military installations, or other source of local weather data acceptable to the code official.

c. The climate zone shall be selected from the applicable map provided in Figures 902.1(1) through 902.1(51) in Chapter 9 of this code.

CHAPTER 4
RESIDENTIAL BUILDING DESIGN BY SYSTEMS ANALYSIS AND DESIGN OF BUILDINGS UTILIZING RENEWABLE ENERGY SOURCES

SECTION 401
GENERAL

401.1 Scope. This chapter establishes design criteria in terms of total energy use by a residential building, including all of its systems.

SECTION 402
SYSTEMS ANALYSIS

402.1 Analysis procedure. Except as explicitly specified by this chapter, the standard design home shall be configured and simulated using identical methods and techniques as are used in the configuration and simulation of the proposed design home.

402.2 Energy analysis. Compliance with this chapter will require an analysis of the annual energy usage, hereinafter called an "annual energy analysis."

Exception: Chapters 5 and 6 establish criteria for different energy-consuming and enclosure elements of the building which, if followed, will eliminate the requirement for an annual energy analysis while meeting the intent of this code.

402.2.1 Standard design. A building designed in accordance with this chapter will be deemed as complying with this code if the calculated annual energy consumption is not greater than a similar building (defined as a "standard design") whose enclosure elements and energy-consuming systems are designed in accordance with Chapter 5. Specific building envelope elements of the standard design shall comply with Sections 402.2.1.1 through 402.2.1.4.

402.2.1.1 Exterior walls. The exterior wall assembly U-Factors for the standard design shall be selected by climate in accordance with Table 402.2.1.1.

402.2.1.2 Fenestration U-factor. The fenestration system U-Factor used in the standard design shall be selected by climate in accordance with Table 402.2.1.2.

402.2.1.3 Window area. The window area of the standard design, inclusive of the framed sash and glazing area, shall be equal to 18 percent of the conditioned floor area of the proposed design.

402.2.1.4 Skylights. Skylights and other nonvertical roof glazing elements shall not be included in the standard design, and ceiling U-factors used in the standard design shall not include such elements in their computation.

402.2.2 Proposed design. For a proposed alternative building design (defined as a "proposed design") to be considered similar to a "standard design," it shall utilize the same nonrenewable energy source(s) for the same functions and have equal conditioned floor area and the same ratio of thermal envelope area to floor area (i.e., the same geometry), exterior design conditions, occupancy, climate data, and usage operational schedule as the standard design. Where an energy end use (such as space heating or domestic water) is to be provided entirely from renewable energy sources in a proposed design, the standard design shall assume an equipment type using a nonrenewable energy source common to that region for that end use as approved by the code official.

402.2.2.1 Orientation for groups of buildings. The worst possible orientation of the proposed design, in terms of annual energy use, considering north, northeast, east, southeast, south, southwest, west and northwest orientations, shall be used to represent group of otherwise identical designs.

TABLE 402.2.1.1
STANDARD DESIGN WALL ASSEMBLY U-FACTORS (U_w)

HEATING DEGREE DAYS[a]	U_w (air to air)[b]
≥ 13,000	0.038
9,000 - 12,999	0.046
6,500 - 8,999	0.052
4,500 - 6,499	0.058
3,500 - 4,499	0.064
2,600 - 3,499	0.076
< 2,600	0.085

a. From Table 302.1.
b. Including framing effects.

TABLE 402.2.1.2
STANDARD DESIGN FENESTRATION SYSTEM U-FACTORS (U_g or U_f)

HEATING DEGREE DAYS[a]	U_g FOR SECTION 502.2.1.1 AND U_f FOR SECTION 502.2.3.1 (air to air)[b]
≥ 13,000	0.25
9,000 - 12,999	0.26
6,500 - 8,999	0.28
4,500 - 6,499	0.30
3,500 - 4,499	0.41
2,600 - 3,499	0.44
700 - 2,599	0.47
< 700	0.74

a. From Table 302.1.
b. Entire assembly, including sash.

RESIDENTIAL BUILDING DESIGN

402.2.3 Input values for residential buildings. The input values in Sections 402.2.3.1 through 402.2.3.11 shall be used in calculating annual energy performance. The requirements of this section specifically indicate which variables shall remain constant between the standard design and proposed design calculations. The standard design shall be a base version of the design that directly complies with the provisions of this code. The proposed building shall be permitted to utilize a design methodology that is demonstrated, through calculations satisfactory to the code official, to have equal or lower annual energy use than the standard design.

402.2.3.1 Glazing systems. The input values in Sections 402.2.3.1.1 through 402.2.3.1.4, specific to glazing systems, shall be used in calculating annual energy performance.

402.2.3.1.1 Orientation, standard design. As a minimum, equal areas on north, east, south and west exposures shall be assumed.

402.2.3.1.2 Exterior shading, standard design. Glazing areas in the standard design shall not be provided with exterior shading such as roof overhangs. Energy performance impacts of added exterior shading for glazing areas which are accounted for in the proposed design for a specific building shall be permitted, provided that the code official approves the actual installation of such systems.

402.2.3.1.3 Fenestration system solar heat gain coefficient, standard design. The fenestration system solar heat gain coefficient (SHGC), inclusive of framed sash and glazing area, of the glazing systems in the standard design shall be 0.40 for HDD < 3,500 and 0.68 for HDD ≥ 3,500 during periods of mechanical heating and cooling operation. These fenestration system SHGC values shall be multiplied together with (added in series to) the interior shading values as specified in Section 402.2.3.1.4 to arrive at an overall solar heat gain coefficient for the installed glazing system.

Where the SHGC characteristics of the proposed fenestration products are not known, the default SHGC values given in Table 102.5.2(3) shall be used for the proposed design.

402.2.3.1.4 Interior shading, standard design and proposed design. The same schedule of interior shading values, expressed as the fraction of the solar heat gain admitted by the fenestration system that is also admitted by the interior shading, shall be assumed for the standard and proposed designs.

The values used for interior shading shall be 0.70 in summer, and 0.90 in winter.

402.2.3.2 Heat storage (thermal mass). The following input values, specific to heat storage (thermal mass), shall be used in calculating annual energy performance:

Internal mass: 8 pounds per square foot (39 kg/m^2)

Structural mass: 3.5 pounds per square foot (17 kg/m^2)

402.2.3.3 Building thermal envelope — surface areas and volume. The input values in Sections 402.2.3.3.1 through 402.2.3.3.4, specific to building thermal envelope surface areas, shall be used in calculating annual energy performance.

402.2.3.3.1 Floors, walls, ceiling. The standard and proposed designs shall have equal areas.

402.2.3.3.2 Foundation and floor type. The foundation and floor type for both the standard and proposed designs shall be equal.

402.2.3.3.3 Doors. The opaque door area of the standard design shall equal that of the proposed design and shall have a U-factor of 0.2 Btu/hr · ft^2 · °F [1.14 W/(m^2 · K)].

402.2.3.3.4 Building volume. The volume of both the standard and proposed designs shall be equal.

402.2.3.4 Heating and cooling controls. Unless otherwise specified by local codes, heating and cooling thermostats shall comply with Table 402.2.3.4 for the standard and proposed designs. The input values specific to heating and cooling controls, shall be used in calculating annual energy performance.

TABLE 402.2.3.4
HEATING AND COOLING CONTROLS

PARAMETER	STANDARD DESIGN VALUE	proposed DESIGN VALUE
Heating	68°F	68°F
Cooling	78°F	78°F
Setback/setup	5°F	Maximum of 5°F
Setback/setup duration	6 hours per day	Maximum of 6 hours per day
Number of setback/setup periods per unit[a]	1	Maximum of 1
Maximum number of zones per unit[a]	2	2
Number of thermostats per zone	1	1

For SI: °C = [(°F) -32]/1.8

a. Units = Number of dwelling units in standard and proposed designs.

402.2.3.5 Internal heat gains. Equation 4-1 shall be used to determine the input values, specific to internal heat gains, that shall be used in both the standard design and the proposed design in calculating annual energy performance:

$$I\text{-}Gain = 17,900 + (23.8 \cdot CFA) + (4140 \cdot BR)$$

(Equation 4-1)

where:

$I\text{-}Gain$ = Internal gains in Btu/day (kWh/day) per dwelling unit.

CFA = Conditioned floor area.

BR = Number of bedrooms.

402.2.3.6 Domestic hot water (calculate, then constants). The following input values, specific to domestic hot water, shall be used in calculating annual energy performance.

Temperature set point   120°F (49°C)

Daily hot water consumption   Gallons = (30 × a) + (10 × b)

where:

a = Number of dwelling units in standard and proposed designs.

b = Number of bedrooms in each dwelling unit.

402.2.3.7 Site weather data (constants). The typical meteorological year (TMY2), or its "ersatz" equivalent, from the National Oceanic and Atmospheric Administration (NOAA), or an approved equivalent, for the closest available location shall be used.

402.2.3.8 Forced-air distribution system loss factors (DLF). The heating and cooling system efficiency shall be proportionately adjusted for those portions of the ductwork located outside or inside the conditioned space using the values shown below:

System Operating Mode	Duct Location	
	Outside	Inside
Heating	0.80	1.00
Cooling	0.80	1.00

Note: Ducts located in a space that contains a positive heating supply or cooling supply, or both, shall be considered inside the building envelope.

Impacts from improved distribution loss factors (DLF) shall be accounted for in the proposed design only if the entire air distribution system is specified on the construction documents to be substantially leak free, and is tested after installation to ensure that the installation is substantially leak free. "Substantially leak free" shall be defined as the condition under which the entire air distribution system (including the air handler cabinet) is capable of maintaining a 0.1-inch w.g. (25 Pa) internal pressure at 5 percent or less of the air handler's rated airflow when the return grilles and supply registers are sealed off. This test shall be conducted using methods and procedures as specified in Section 3 of the SMACNA *HVAC Air Duct Leakage Test Manual*, or by using other, similar pressurization test methods and as approved by the code official.

Where test results show that the entire distribution system is substantially leak free, then seasonal DLFs shall be calculated separately for heating and cooling modes using engineering methods capable of considering the net seasonal cooling energy heat gain impacts and the net seasonal heating energy heat loss impacts that result from the portion of the thermal air distribution system that is located outside the conditioned space. Once these heating and cooling season "distribution system energy impacts" are known, then heating and cooling mode DLFs for the proposed design shall be calculated using Equations 4-2 and 4-3:

Total Seasonal Energy = Seasonal Building Energy + Distribution System Energy Impacts

(Equation 4-2)

DLF = Seasonal Building Energy/Total Seasonal Energy

(Equation 4-3)

Once the DLFs for the heating and cooling seasons are known, the total "adjusted system efficiency" is calculated using Equation 4-4:

Adjusted System Efficiency = (Equipment Efficiency × DLF × Percent of Duct Outside) + (Equipment Efficiency × DLF × Percent of Duct Inside)

(Equation 4-4)

Equation 4-4 shall be used to develop adjusted system efficiency for each heating and cooling system included in the standard design. Where a single system provides both heating and cooling, efficiencies shall be calculated separately for heating and cooling modes.

402.2.3.9 Air infiltration. Annual average air changes per hour (ACH) for the standard design shall be determined using the following equation:

ACH = Normalized Leakage × Weather Factor

(Equation 4-5)

where:

Normalized leakage = 0.57

and

Weather factor is determined in accordance with the weather factors (W) given by ASHRAE 136, as taken from the weather station nearest the building site.

Where the proposed design takes credit for reduced ACH levels, documentation of measures providing such reductions, and results of a post-construction blower-door test shall be provided to the code official using ASTM E 779. No energy credit shall be granted for ACH levels below 0.35.

402.2.3.10 Foundation walls. When performing annual energy analyses for buildings with insulated basement or crawl space walls, the design U-factors taken from Table 502.2 for these walls of the standard building shall be permitted to be decreased by accounting for the R-values of the adjacent soil, provided that the foundation wall U-factor of the proposed building also accounts for the R-value of the adjacent soil.

402.2.3.11 Heating and cooling system equipment efficiency, standard design. The efficiency of the heating and cooling equipment shall meet, but not exceed the minimum efficiency requirement in Section 503.2. Where the proposed design utilizes an electric resistance space heating system as the primary heating source, the standard design shall utilize an air-cooled heat pump that meets but does not exceed the minimum efficiency requirements in Section 503.2.

Exception: Zonal electric-resistance space heating equipment in buildings in Climate Zones 1a through 4b as indicated in Table 302.1

402.3 Design. The standard design, conforming to the criteria of Chapter 5 and the proposed design, shall be designed on a common basis as specified in Sections 402.3.1 through 402.3.3.

402.3.1 Units of energy. The comparison shall be expressed as Btu input per square foot (W/m^2) of gross floor area per year at the building site.

402.3.2 Equivalent energy units. If the proposed design results in an increase in consumption of one energy source and a decrease in another energy source, even though similar sources are used for similar purposes, the difference in each energy source shall be converted to equivalent energy units for purposes of comparing the total energy used.

402.3.3 Site energy. The different energy sources shall be compared on the basis of energy use at the site where: 1 kWh = 3,413 Btu.

402.4 Analysis procedure. The analysis of the annual energy usage of the standard and the proposed alternative building and system designs shall meet the criteria specified in Sections 402.4.1 and 402.4.2.

402.4.1 Load calculations. The building heating and cooling load calculation procedures used for annual energy consumption analysis shall be detailed to permit the evaluation of effect of factors specified in Section 402.4.

402.4.2 Simulation details. The calculation procedure used to simulate the operation of the building and its service systems through a full-year operating period shall be detailed to permit the evaluation of the effect of system design, climatic factors, operational characteristics, and mechanical equipment on annual energy usage. Manufacturer's data or comparable field test data shall be used when available in the simulation of systems and equipment. The calculation procedure shall be based on 8,760 hours of operation of the building and its service systems and shall utilize the design methods specified in the ASHRAE *Fundamentals Handbook*.

402.5 Calculation procedure. The calculation procedure shall include the items specified in Sections 402.5.1 through 402.5.7.

402.5.1 Design requirements. Environmental requirements as required in Chapter 3.

402.5.2 Climatic data. Coincident hourly data for temperatures, solar radiation, wind and humidity of typical days in the year representing seasonal variation.

402.5.3 Building data. Orientation, size, shape, framing, mass, air, moisture and heat transfer characteristics.

402.5.4 Operational characteristics. Temperature, humidity, ventilation, illumination and control mode for occupied and unoccupied hours.

402.5.5 Mechanical equipment. Design capacity and part-load profile.

402.5.6 Building loads. Internal heat generation, lighting, equipment and number of people during occupied and unoccupied periods.

402.5.7 Use of approved calculation tool. The same calculation tool shall be used to estimate the annual energy usage for space heating and cooling of the standard design and the proposed design. The calculation tool shall be approved by the code official.

402.6 Documentation. Proposed alternative designs, submitted as requests for exception to the standard design criteria, shall be accompanied by an energy analysis comparison report. The report shall provide technical detail on the standard and proposed designs and on the data used in and resulting from the comparative analysis to verify that both the analysis and the designs meet the criteria of Chapter 4.

Exception: Proposed alternative designs for residential buildings having a conditioned floor area of 5,000 square feet (464 m^2) or less are exempted from the hourly analysis described in Sections 402.4 and 402.5. However, a comparison of energy consumption using correlation methods based on full-year hourly simulation analysis or other engineering methods that are capable of estimating the annual heating, cooling and hot water use between the proposed alternative design and the standard design shall be provided.

CHAPTER 5

RESIDENTIAL BUILDING DESIGN BY COMPONENT PERFORMANCE APPROACH

SECTION 501
GENERAL

501.1 Scope. Residential buildings or portions thereof that enclose conditioned space shall be constructed to meet the requirements of this chapter.

SECTION 502
BUILDING ENVELOPE REQUIREMENTS

502.1 General requirements. The building envelope shall comply with the applicable provisions of Sections 502.1.1 through 502.1.5 regardless of the means of demonstrating envelope compliance as set forth in Section 502.2.

502.1.1 Moisture control. The design shall not create conditions of accelerated deterioration from moisture condensation. Frame walls, floors and ceilings not ventilated to allow moisture to escape shall be provided with an approved vapor retarder having a permeance rating of 1 perm (5.7×10^{-11} kg/Pa · s · m²) or less, when tested in accordance with the dessicant method using Procedure A of ASTM E 96. The vapor retarder shall be installed on the warm-in-winter side of the thermal insulation.

Exceptions:

1. In construction where moisture or its freezing will not damage the materials.
2. Where the county in which the building is being constructed is considered a hot and humid climate area and identified as such in Figures 902.1(1) through 902.1(51) in Chapter 9 of this code.
3. Where other approved means to avoid condensation in unventilated framed wall, floor, roof and ceiling cavities are provided.

502.1.2 Masonry veneer. When insulation is placed on the exterior of a foundation supporting a masonry veneer exterior, the horizontal foundation surface supporting the veneer is not required to be insulated to satisfy any foundation insulation requirement.

502.1.3 Recessed lighting fixtures. When installed in the building envelope, recessed lighting fixtures shall meet one of the following requirements:

1. Type IC rated, manufactured with no penetrations between the inside of the recessed fixture and ceiling cavity and sealed or gasketed to prevent air leakage into the unconditioned space.
2. Type IC or non-IC rated, installed inside a sealed box constructed from a minimum 0.5-inch-thick (12.7 mm) gypsum wallboard or constructed from a preformed polymeric vapor barrier, or other air-tight assembly manufactured for this purpose, while maintaining required clearances of not less than 0.5 inch (12.7 mm) from combustible material and not less than 3 inches (76 mm) from insulation material.
3. Type IC rated, in accordance with ASTM E 283 admitting no more than 2.0 cubic feet per minute (cfm) (0.944 L/s) of air movement from the conditioned space to the ceiling cavity. The lighting fixture shall be tested at 1.57 pounds per square inch (psi) (75 Pa) pressure difference and shall be labeled.

502.1.4 Air leakage. Provisions for air leakage shall be in accordance with Sections 502.1.4.1 and 502.1.4.2.

502.1.4.1 Window and door assemblies. Window and door assemblies installed in the building envelope shall comply with the maximum allowable air leakage rates in Table 502.1.4.1.

Exception: Site-constructed windows and doors sealed in accordance with Section 502.1.4.2.

TABLE 502.1.4.1
ALLOWABLE AIR LEAKAGE RATES[a, b]

WINDOWS (cfm per square foot of window area)	DOORS (cfm per square foot of door area)	
	Sliding	Swinging
0.3[b, d]	0.3[d]	0.5[c, d]

For SI: 1 cfm/ft² = 5 L/S · m².
a. When tested in accordance with ASTM E 283.
b. See AAMA/WDMA 101/I.S.2.
c. Requirement based on assembly area.
d. See NFRC 400.

502.1.4.2 Caulking and sealants. Exterior joints, seams or penetrations in the building envelope, that are sources of air leakage, shall be sealed with durable caulking materials, closed with gasketing systems, taped or covered with moisture vapor-permeable housewrap. Sealing materials spanning joints between dissimilar construction materials shall allow for differential expansion and contraction of the construction materials.

This includes sealing around tubs and showers, at the attic and crawl space panels, at recessed lights and around all plumbing and electrical penetrations. These are openings located in the building envelope between conditioned space and unconditioned space or between the conditioned space and the outside.

502.1.5 Fenestration solar heat gain coefficient. In locations with heating degree days (HDD) less than 3,500, the combined solar heat gain coefficient (the area-weighted average) of all glazed fenestration products (including the effects of any permanent exterior solar shading devices) in the building shall not exceed 0.4.

502.2 Heating and cooling criteria. The building envelope shall meet the provisions of Table 502.2. Compliance shall be demonstrated in accordance with Section 502.2.1, 502.2.2, 502.2.3, 502.2.4 or 502.2.5, as applicable.

Energy measure tradeoffs utilizing equipment exceeding the requirements of Section 503, 504 or 505 shall only use the compliance methods described in Chapter 4.

TABLE 502.2
HEATING AND COOLING CRITERIA[a]

ELEMENT	MODE	TYPE A-1 RESIDENTIAL BUILDINGS U_o	TYPE-2 RESIDENTIAL BUILDINGS U_o
Walls	Heating or cooling	—	—
Roof/ceiling	Heating or cooling	—	—
Floors over unheated spaces	Heating or cooling	—	—
Heated slab on grade[b,f]	Heating	R-value =	R-value =
Unheated slab on grade[c,d,f]	Heating	R-value =	R-value =
Basement wall[e,f]	Heating or cooling	U-factor =	U-factor =
Crawl space wall[e,f]	Heating or cooling	U-factor =	U-factor =

For SI: 1 Btu/h · ft^2 · °F = 5.678 W/(m^2 · K), °C = [(°F)-32]/1.8.

a. Values shall be determined by using the graphs [Figures 502.2(1), 502.2(2), 502.2(3), 502.2(4), 502.2(5) and 502.2(6)] using HDD as specified in Section 302.
b. There are no insulation requirements for heated slabs in locations having less than 500 HDD.
c. There are no insulation requirements for unheated slabs in locations having less than 2,500 HDD.
d. Slab edge insulation is not required for unheated slabs in areas of very heavy termite infestation probability in accordance with Section 502.2.1.4, and as shown in Figure 502.2(7).
e. Basement and crawl space wall U-factors shall be based on the wall components and surface air films. Adjacent soil shall not be considered in the determination of the U-factor.
f. Typical foundation insulation techniques can be found in the DOE *Building Foundation Design Handbook*.

502.2.1 Compliance by performance on an individual component basis. Each component of the building envelope shall meet the provisions of Table 502.2 as provided in Sections 502.2.1.1 through 502.2.1.6.

502.2.1.1 Walls. The combined thermal transmittance value (U_o) of the gross area of exterior walls shall not exceed the value given in Table 502.2. Equation 5-1 shall be used to determine acceptable combinations to meet this requirement:

$$U_o = \frac{(U_w \times A_w) + (U_g \times A_g) + (U_d \times A_d)}{A_o}$$

(Equation 5-1)

where:

U_o = The average thermal transmittance of the gross area of the exterior walls.

A_o = The gross area of exterior walls.

U_w = The combined thermal transmittance of the various paths of heat transfer through the opaque exterior wall area.

A_w = Area of exterior walls that are opaque.

U_g = The combined thermal transmittance of all glazing within the gross area of exterior walls.

A_g = The area of all glazing within the gross area of exterior walls.

U_d = The combined thermal transmittance of all opaque doors within the gross area of exterior walls.

A_d = The area of all opaque doors within the gross area of exterior walls.

Notes: (1) When more than one type of wall, window or door is used, the U and A terms for those items shall be expanded into subelements as:

$(U_{w1} A_{w1}) + (U_{w2} A_{w2}) + (U_{w3} A_{w3}) +...$ (etc.)

(Equation 5-2)

(2) Access doors or hatches in a wall assembly shall be included as a subelement of the wall assembly.

502.2.1.1.1 Steel stud framed walls. When the walls contain steel stud framing, the value of U_w used in Equation 5-1 shall be recalculated using a series path procedure to correct for parallel path thermal bridging. The U_w for purposes of Equation 5-1 of steel stud walls shall be determined as follows:

$$U_w = \frac{1}{[R_s + (R_{ins} \times F_c)]}$$

(Equation 5-3)

where:

R_s = The total thermal resistance of the elements comprising the wall assembly along the path of heat transfer, excluding the cavity insulation and the steel stud.

R_{ins} = The R-value of the cavity insulation.

F_c = The correction factor listed in Table 502.2.1.1.1.

Exception: Overall system tested U_w values for steel stud framed walls from approved laboratories, when such data are acceptable to the code official.

RESIDENTIAL — COMPONENT PERFORMANCE APPROACH

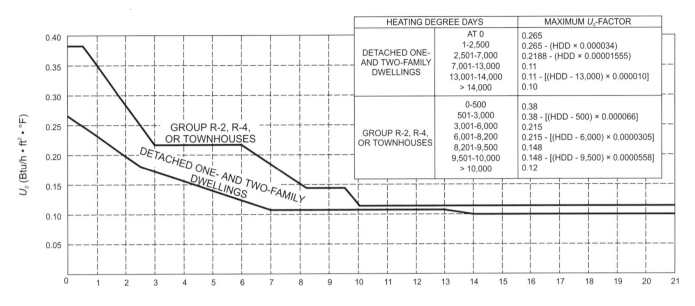

For SI: 1 Btu/h · ft² · °F = 5.678W/(m² · K), °C = [(°F)-32]/1.8.

FIGURE 502.2(1)
U_o-FACTORS—WALLS: RESIDENTIAL BUILDINGS

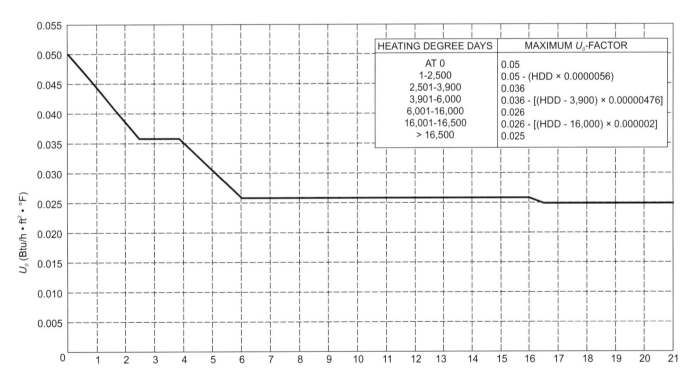

For SI: 1 Btu/h · ft² · °F = 5.678W/(m² · K), °C = [(°F)-32]/1.8.

FIGURE 502.2(2)
U_o-FACTORS—ROOF/CEILINGS

RESIDENTIAL — COMPONENT PERFORMANCE APPROACH

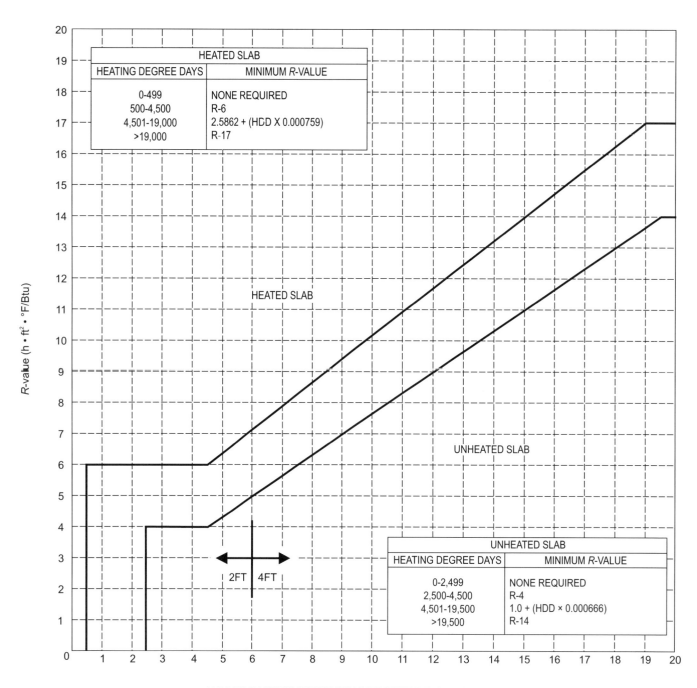

FIGURE 502.2(3)
R-VALUES—SLAB ON GRADE

RESIDENTIAL — COMPONENT PERFORMANCE APPROACH

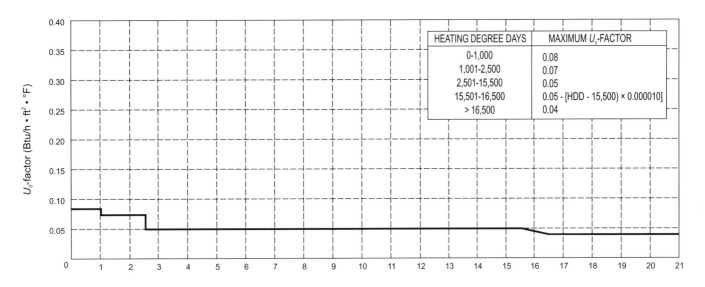

ANNUAL FAHRENHEIT HEATING DEGREE DAYS, BASE 65°F (in thousands)

For SI: 1 Btu/h · ft² · °F = 5.678W/(m² · K), °C = [(°F)-32]/1.8.

FIGURE 502.2(4)
U_o-FACTORS—FLOOR OVER UNHEATED SPACES

RESIDENTIAL — COMPONENT PERFORMANCE APPROACH

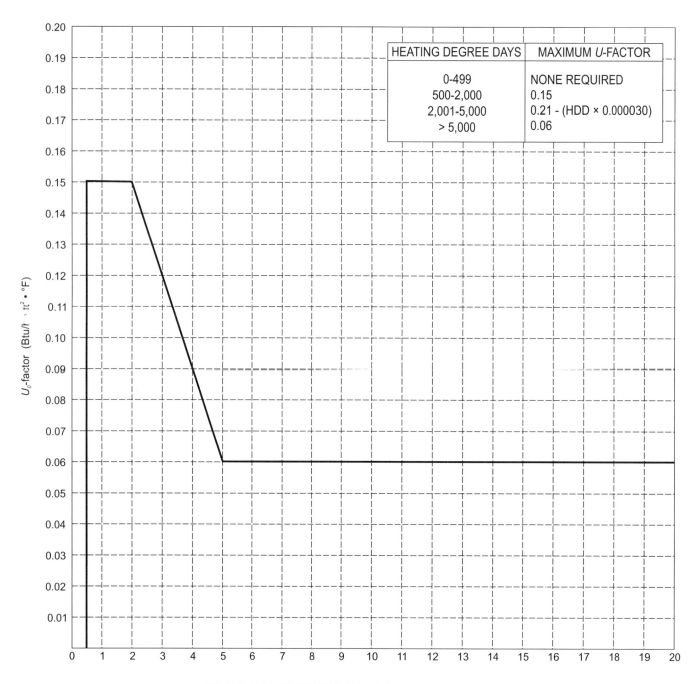

For SI: 1 Btu/h · ft² · °F = 5.678W/(m² · K), °C = [(°F)-32]/1.8.

FIGURE 502.2(5)
***U*-FACTORS—CRAWL SPACE WALLS**

RESIDENTIAL — COMPONENT PERFORMANCE APPROACH

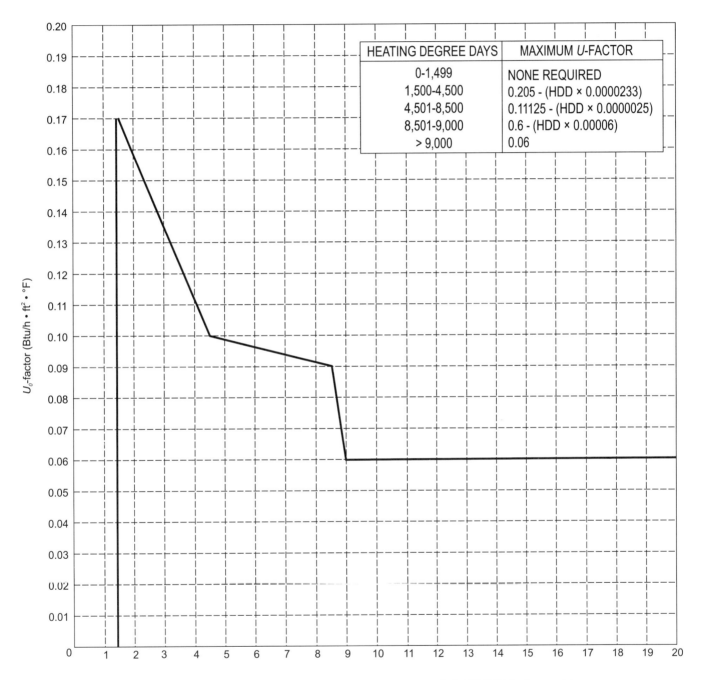

For SI: 1 Btu/h · ft² · °F = 5.678W/(m² · K), °C = [(°F)-32]/1.8.

FIGURE 502.2(6)
U-FACTORS—BASEMENT WALLS

RESIDENTIAL — COMPONENT PERFORMANCE APPROACH

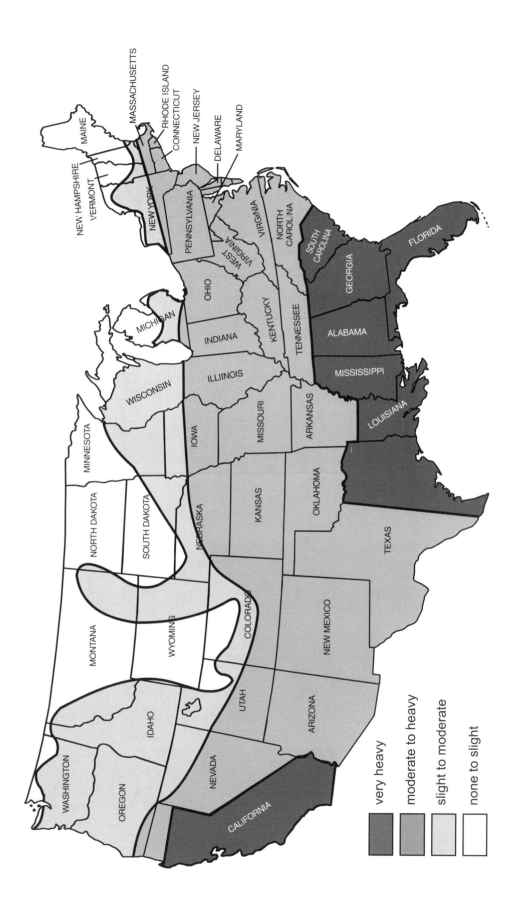

FIGURE 502.2(7)
TERMITE INFESTATION PROBABILITY MAP

TABLE 502.2.1.1.1
F_c VALUES FOR WALL SECTIONS WITH STEEL STUDS PARALLEL PATH CORRECTION FACTORS

NOMINAL STUD SIZE[a]	SPACING OF FRAMING (inches)	CAVITY INSULATION R-VALUE	CORRECTION FACTOR
2 × 4	16 o.c.	R-11 R-13 R-15	0.50 0.46 0.43
2 × 4	24 o.c.	R-11 R-13 R-15	0.60 0.55 0.52
2 × 6	16 o.c.	R-19 R-21	0.37 0.35
2 × 6	24 o.c.	R-19 R-21	0.45 0.43
2 × 8	16 o.c.	R-25	0.31
2 × 8	24 o.c.	R-25	0.38

For SI: 1 inch = 25.4 mm.

a. Applies to steel studs up to a maximum thickness of 0.064 inches (16 gage).

502.2.1.1.2 Mass walls. When thermal mass credit is desired for an exterior wall having a heat capacity greater than or equal to 6 Btu/ft² · °F [1.06 kJ/(m² · K)], the U_w for such a wall shall be less than or equal to the applicable value in Table 502.2.1.1.2(1), 502.2.1.1.2(2) or 502.2.1.1.2(3) based on the U_w required for an exterior wall having a heat capacity less than 6 Btu/ft² · °F [1.06 kJ/(m² · K)] as determined by Section 502.2.1.1, Equation 5-1 and Figure 502.2(1).

Note: Masonry or concrete walls having a mass greater than or equal to 30 lb/ft² (146 kg/m²) of exterior wall area and solid wood walls having a mass greater than or equal to 20 lb/ft² (98 kg/m²) of exterior wall area have heat capacities equal to or exceeding 6 Btu/ft² · °F [1.06 kJ/(m² · K)] of exterior wall area.

The heat capacity of the wall shall be determined using Equation 5-4 as follows:

$HC = w \times c$ **(Equation 5-4)**

where:

HC = Heat capacity of the exterior wall, Btu/ft² · °F [kJ/(m² · K)] of exterior wall area.

w = Mass of the exterior wall, lb/ft² (kg/m²) of exterior wall area is the density of the exterior wall material, lb/ft³ (kg/m³) multiplied by the thickness of the exterior wall, ft (m).

c = Specific heat of the exterior wall material, Btu/lb · °F [kJ/(kg · K)] of exterior wall area as determined from Chapter 24 of the ASHRAE *Fundamentals Handbook*.

502.2.1.2 Roof/ceiling. The combined thermal transmittance value (U_o) of the gross area of the roof or ceiling assembly shall not exceed the value given in Table 502.2. Equation 5-5 shall be used to determine acceptable combinations to meet this requirement.

$$U_o = \frac{(U_R \times A_R) + (U_S \times A_S)}{A_o}$$ **(Equation 5-5)**

where:

U_o = The average thermal transmittance of the gross roof/ceiling area, Btu/h · ft² · °F [W/(m² · K)].

A_o = The gross area of the roof/ceiling assembly, square feet (m²).

U_R = The combined thermal transmittance of the various paths of heat transfer through the opaque roof/ceiling area, Btu/h · ft² · °F [W/(m² · K)].

A_R = Opaque roof/ceiling assembly area, square feet (m²).

U_s = The combined thermal transmittance of the area of all skylight elements in the roof/ceiling assembly (See Section 502.2.1.2.1), Btu/h · ft² · °F [W/(m² · K)].

A_s = The area (including frame) of all skylights in the roof/ceiling assembly, square feet (m²). (see Section 502.2.1.2.1).

Notes: (1) When more than one type of roof/ceiling and/or skylight is used, the U and A terms for those items shall be expanded into their subelements as in Equation 5-6:

$(U_{R1} \times A_{R1}) + (U_{R2} \times A_{R2}) + ...$ etc. **(Equation 5-6)**

(2) Access doors or hatches in a roof/ceiling assembly shall be included as a subelement of the roof/ceiling assembly.

(3) When the roof/ceiling assembly contains cold-formed steel truss framing, the U_R value to be used in Equation 5-5 shall be determined by Equation 5-7, 5-8, or 5-9. These equations apply to cold-formed steel truss roof framing spaced at 24 inches (609 mm) on-center and where the penetrations of the truss members through the cavity insulation do not exceed three penetrations for each 4-foot (1220 mm) length of the truss.

For constructions without foam between the drywall and bottom chord of the steel truss use Equation 5-7:

$$U_R = \frac{1}{(0.864 \times R_{ins}) + 0.330}$$ **(Equation 5-7)**

where:

R_{ins} = The R-value of the cavity insulation, h · ft² · °F/Btu.

RESIDENTIAL — COMPONENT PERFORMANCE APPROACH

TABLE 502.2.1.1.2(1)
REQUIRED U_w FOR WALL WITH A HEAT CAPACITY EQUAL
TO OR EXCEEDING 6 Btu/ft² · °F WITH INSULATION
PLACED ON THE EXTERIOR OF THE WALL MASS

HEATING DEGREE DAYS	U_w REQUIRED FOR WALLS WITH A HEAT CAPACITY LESS THAN 6 Btu/ft² · °F AS DETERMINED BY USING EQUATION 5-1 AND FIGURE 502.2(1)										
	0.24	0.22	0.20	0.18	0.16	0.14	0.12	0.10	0.08	0.06	0.04
0 - 2,000	0.33	0.31	0.28	0.26	0.23	0.21	0.18	0.16	0.13	0.11	0.08
2,001 - 4,000	0.32	0.30	0.27	0.25	0.22	0.20	0.17	0.15	0.13	0.10	0.08
4,001 - 5,500	0.30	0.28	0.25	0.23	0.21	0.18	0.16	0.14	0.11	0.09	0.07
5,501 - 6,500	0.28	0.26	0.23	0.21	0.19	0.17	0.15	0.12	0.10	0.08	0.06
6,501 - 8,000	0.26	0.24	0.22	0.19	0.17	0.15	0.13	0.11	0.09	0.07	0.05
> 8,000	0.24	0.22	0.20	0.18	0.16	0.14	0.12	0.10	0.08	0.06	0.04

For SI: °C = [(°F)-32]/1.8, 1 Btu/ft² · °F = 0.176 kJ/(m² · °K).

TABLE 502.2.1.1.2(2)
REQUIRED U_w FOR WALL WITH A HEAT CAPACITY EQUAL
TO OR EXCEEDING 6 Btu/ft² · °F WITH INSULATION
PLACED ON THE INTERIOR OF THE WALL MASS

HEATING DEGREE DAYS	U_w REQUIRED FOR WALLS WITH A HEAT CAPACITY LESS THAN 6 Btu/ft² · °F AS DETERMINED BY USING EQUATION 5-1 AND FIGURE 502.2(1)										
	0.24	0.22	0.20	0.18	0.16	0.14	0.12	0.10	0.08	0.06	0.04
0 - 2,000	0.29	0.27	0.25	0.22	0.20	0.17	0.15	0.12	0.09	0.07	0.04
2,001 - 4,000	0.28	0.26	0.24	0.21	0.19	0.16	0.14	0.12	0.09	0.07	0.04
4,001 - 5,500	0.27	0.25	0.23	0.21	0.19	0.16	0.14	0.11	0.09	0.07	0.04
5,501 - 6,500	0.26	0.24	0.22	0.20	0.17	0.15	0.13	0.11	0.09	0.06	0.04
6,501 - 8,000	0.25	0.23	0.21	0.19	0.17	0.14	0.12	0.10	0.08	0.06	0.04
> 8,000	0.24	0.22	0.20	0.18	0.16	0.14	0.12	0.10	0.08	0.06	0.04

For SI: °C = [(°F)-32]/1.8, 1 Btu/ft² · °F = 0.176 kJ/(m² · °K).

TABLE 502.2.1.1.2(3)
REQUIRED U_w FOR WALL WITH A HEAT CAPACITY EQUAL
TO OR EXCEEDING 6 Btu/ft² · °F WITH INTEGRAL
INSULATION (INSULATION AND MASS MIXED, SUCH AS A LOG WALL)

HEATING DEGREE DAYS	U_w REQUIRED FOR WALLS WITH A HEAT CAPACITY LESS THAN 6 Btu/ft² · °F AS DETERMINED BY USING EQUATION 5-1 AND FIGURE 502.2(1)										
	0.24	0.22	0.20	0.18	0.16	0.14	0.12	0.10	0.08	0.06	0.04
0 - 2,000	0.33	0.31	0.28	0.25	0.23	0.20	0.17	0.15	0.12	0.09	0.07
2,001 - 4,000	0.32	0.30	0.27	0.24	0.22	0.19	0.17	0.14	0.11	0.09	0.06
4,001 - 5,500	0.30	0.28	0.26	0.23	0.21	0.18	0.16	0.13	0.11	0.08	0.06
5,501 - 6,500	0.28	0.26	0.24	0.21	0.19	0.17	0.14	0.12	0.10	0.08	0.05
6,501 - 8,000	0.26	0.24	0.22	0.20	0.18	0.15	0.13	0.11	0.09	0.07	0.05
> 8,000	0.24	0.22	0.20	0.18	0.16	0.14	0.12	0.10	0.08	0.06	0.04

For SI: °C = [(°F)-32]/1.8, 1 Btu/ft² · °F = 0.176 kJ/(m² · °K).

RESIDENTIAL — COMPONENT PERFORMANCE APPROACH

For constructions with R-3 foam between the drywall and bottom chord of the steel truss use Equation 5-8:

$$U_R = \frac{1}{(0.864 \times R_{ins}) + 4.994}$$ (Equation 5-8)

For constructions with R-5 foam between the drywall and bottom chord of the steel truss use Equation 5-9:

$$U_R = \frac{1}{(0.864 \times R_{ins}) + 7.082}$$ (Equation 5-9)

Exception: Overall system tested U_R values for roof/ceiling assemblies from approved laboratories, when such data are acceptable to the code official.

(4) When the roof/ceiling assembly contains conventional C-shaped cold-formed joist/rafter steel framing, the U_R value to be used in equation 5-5 shall be determined Equation 5-10 as follows:

$$U_R = \frac{1}{R_s + (R_{ins} \times F_{cor})}$$ (Equation 5-10)

where:

R_S = The total thermal resistance of the elements of roof/ceiling construction, in a series along the path of heat transfer, excluding the cavity insulation and the steel framing, h · ft² · °F/Btu.

R_{ins} = The R-value of the cavity insulation, h · ft² · °F/Btu.

F_{cor} = The correction factor listed in Table 502.2.1.2, dimensionless.

Exception: Overall system tested U_R values for roof/ceiling assemblies from approved laboratories, when such data are acceptable to the code official.

502.2.1.2.1 Skylights. Skylight shafts, 12 inches (305 mm) in depth and greater, shall be insulated to no less than R-13 in climates 0 - 4,000 HDD and R-19 in climates greater than 4,000 HDD. The skylight shaft thermal performance shall not be included in the roof thermal transmission coefficient calculation.

502.2.1.3 Floors over unheated spaces. The combined thermal transmittance factor (U_o) of the gross area of floors over unheated spaces shall not exceed the value given in Table 502.2. For floors over outdoor air (i.e., overhangs), U_o-factors shall not exceed the value for roofs given in Table 502.2. Equation 5-11 shall be used to determine acceptable combinations to meet this requirement.

$$U_o = \frac{(U_{f1} \times A_{f1}) + (U_{f2} \times A_{f2}) + \mathrm{K} + (U_{fn} \times A_{fn})}{A_o}$$

(Equation 5-11)

where:

U_o = The average thermal transmittance of the gross floor area, Btu/h · ft² · °F [W/(m² · K)].

A_o = The gross area of the different floor assemblies, square feet (m²).

U_{fn} = The combined thermal transmittance of the various paths of heat transfer through the nth floor assembly, Btu/h · ft² · °F [W/(m² · K)].

A_{fn} = The area associated with the nth floor assembly, square feet (m²).

Note: Access doors or hatches in a floor assembly shall be included as a subelement of the floor assembly.

Exceptions: When the floor assembly contains C-shaped, cold-formed steel framing, the value of U_{fn} used in Equation 5-11 shall be recalculated using a se-

TABLE 502.2.1.2
CORRECTION FACTORS (F_{cor}) FOR ROOF/CEILING ASSEMBLIES

MEMBER SIZE[a]	SPACING OF FRAMING MEMBERS[b] (INCHES)	CAVITY INSULATION R-VALUE			
		R-19	R-30	R-38	R-49
2 × 4	16 o.c.	0.90	0.94	0.95	0.96
2 × 6		0.70	0.81	0.85	0.88
2 × 8		0.35	0.65	0.72	0.78
2 × 10		0.35	0.27	0.62	0.70
2 × 12		0.35	0.27	0.51	0.62
2 × 4	24 o.c.	0.95	0.96	0.97	0.97
2 × 6		0.78	0.86	0.88	0.91
2 × 8		0.44	0.72	0.78	0.83
2 × 10		0.44	0.35	0.69	0.76
2 × 12		0.44	0.35	0.61	0.69

For SI: 1 inch = 25.4 mm.
a. Applies to steel framing members up to a maximum thickness of 0.064 inches (16 gage).
b. Linear interpolation for determining correction factors which are intermediate between those given in the table is permitted.

RESIDENTIAL — COMPONENT PERFORMANCE APPROACH

ries of path procedure to correct for parallel path thermal bridging. The U_{fn} for purposes of Equation 5-11 for C-shaped, cold-formed steel-framing construction shall be determined using Equation 5-12 as follows:

$$U_{fn} = \frac{1}{R_{fn} + (R_{ins} \times F_{cor})} \quad \textbf{(Equation 5-12)}$$

where:

R_{fn} = The total thermal resistance of the elements of floor construction, in series along the path of heat transfer, excluding the cavity insulation and the steel joist, h · ft² · °F/Btu.

R_{ins} = The R-value of the cavity insulation, h · ft² · °F/Btu.

F_{cor} = The correction factor listed in Table 502.2.1.3, dimensionless.

Exception: Overall system tested U_{fn} values for steel-framed floors from approved laboratories, when such data are acceptable to the code official.

502.2.1.4 Slab-on-grade floors. The thermal resistance of the insulation around the perimeter of the floor shall not be less than the value given in Table 502.2. Where insulation is not required in accordance with Footnote d to Table 502.2, building envelope compliance shall be demonstrated by using Section 502.2.2 or Chapter 4 with the actual slab insulation R-value in Table 502.2; or using Section 502.2.4.

Insulation shall be of an approved type, and placed on the outside of the foundation or on the inside of a foundation wall. In climates below 6,000 annual Fahrenheit HDD, the insulation shall extend downward from the elevation of the top of the slab for a minimum distance of 24 inches (610 mm) or downward to at least the bottom of the slab and then horizontally to the interior or exterior for a minimum total distance of 24 inches (610 mm). In all climates equal to or greater than 6,000 HDD, the insulation shall extend downward from the elevation of the top of the slab for a minimum of 48 inches (1219 mm) or downward to at least the bottom of the slab and then horizontally to the interior or exterior for a minimum total distance of 48 inches (1219 mm). In all climates, horizontal insulation extending outside of the foundation shall be covered by pavement or by soil a minimum of 10 inches (254 mm) thick. The top edge of the insulation installed between the exterior wall and the edge of the interior slab shall be permitted to be cut at a 45-degree (0.8 rad) angle away from the exterior wall.

502.2.1.5 Crawl space walls. If the floor above a crawl space does not meet the requirements of Section 502.2.1.3 and the crawl space does not have ventilation openings that communicate directly with the outside air, then the exterior walls of the crawl space shall have a thermal transmittance value not exceeding the value given in Table 502.2. Where the inside ground surface is 12 inches (305 mm) or greater below the outside finish ground level, insulation shall extend from the top of the wall to at least the inside ground surface [see Appendix Detail 502.2.1.5(1) and the DOE *Foundation Design Handbook*]. Where the inside ground surface is less than 12 inches (305 mm) below the outside finish ground level, insulation shall extend from the top of the crawl space wall to the top of the footing [see Appendix Detail 502.2.1.5(2) and the DOE *Foundation Design Handbook*].

TABLE 502.2.1.3
CORRECTION FACTORS (F_{cor}) FOR STEEL FLOOR ASSEMBLIES

MEMBER SIZE[a]	SPACING OF FRAMING MEMBERS[b] (INCHES)	CAVITY INSULATION R-VALUE		
		R-19	R-30	R-38
2 × 6	16 o.c.	0.70	Not Applicable	Not Applicable
2 × 8		0.35	Not Applicable	Not Applicable
2 × 10		0.35	0.27	Not Applicable
2 × 12		0.35	0.27	0.24
2 × 6	24 o.c.	0.78	Not Applicable	Not Applicable
2 × 8		0.44	Not Applicable	Not Applicable
2 × 10		0.44	0.35	Not Applicable
2 × 12		0.44	0.35	0.32

For SI: 1 inch = 25.4 mm.
a. Applies to steel framing members up to a maximum thickness of 0.064 inches (16 gage).
b. Linear interpolation is permitted for determining correction factors which are between those given in the table.

502.2.1.6 Basement walls. The exterior walls of conditioned basements shall have a transmittance value not exceeding the value given in Table 502.2 from the top of the basement wall to a depth of 10 feet (3048 mm) below the outside finish ground level, or to the level of the basement floor, whichever is less.

502.2.2 Compliance by total building envelope performance. The building envelope design of a proposed building shall be permitted to deviate from the U_o-factors, U-factors, or R-values specified in Table 502.2, provided the total thermal transmission heat gain or loss for the proposed building envelope does not exceed the total heat gain or loss resulting from the proposed building's conformance to the values specified in Table 502.2. For basement and crawl space walls that are part of the building envelope, the U-factor of the proposed foundation shall be adjusted by the R-value of the adjacent soil where the corresponding U-factor in Table 502.2 is similarly adjusted. Heat gain or loss calculations for slab edge and basement or crawl space wall foundations shall be determined using methods consistent with the ASHRAE *Fundamentals Handbook*.

502.2.3 Compliance by acceptable practice on an individual component basis. Each component of the building envelope shall meet the provisions of Table 502.2 as provided in Sections 502.2.3.1 through 502.2.3.6. The various walls, roof and floor assemblies described in Section 502.2.3 are typical and are not intended to be all inclusive. Other assemblies shall be permitted, provided documentation is submitted indicating the thermal transmittance value of the opaque section. Documentation shall be in accordance with accepted engineering practice.

502.2.3.1 Walls. The U_o of the exterior wall shall be determined in accordance with Equation 5-13.

$$U_o \frac{(U_f \times A_f) + U_w \times (100 - A_f)}{100}$$

(Equation 5-13)

where:

U_o = The overall thermal transmittance of the gross exterior wall area.

U_f = The average thermal transmittance of the glazing area.

$$A_f = \frac{\text{Glazing Area}}{\text{Gross Exterior Wall Area} \times 100}$$

(Equation 5-14)

U_w = The average thermal transmittance of the opaque exterior wall area.

The U-factor for the opaque portion of the exterior wall (U_w) shall meet the provisions of Table 502.2 as determined by Equation 5-13, and be selected from Table 502.2.3.1(1), 502.2.3.1(2) or 502.2.3.1(3) listed in the Appendix. The glazing U-factor (U_f) and the percentage of glazing area (A_f) shall consist of all glazed surfaces in the building envelope measured using the rough opening and including the sash, curbing and other framing elements that enclose conditioned spaces. The value of U_f shall be determined in accordance with Section 102.5.2. Opaque doors in the building envelope shall have a maximum U-factor of 0.35. One door shall be exempt from this requirement.

Exceptions:

1. When the exterior wall(s) is comprised of steel stud framing members, the procedure contained in Section 502.2.1.1.1 shall be used to adjust the U-factor of the opaque sections of such walls prior to selection of the appropriate acceptable practice(s) from Appendix Table 502.2.3.1(1).
2. When the thermal mass of the exterior building walls is considered, the procedure contained in Section 502.2.1.1.2 shall be used to adjust the U-factor of the opaque sections of such walls prior to the selection of the appropriate acceptable practice(s) from Table 502.2.3.1(2) or 502.2.3.1(3) listed in the Appendix.

502.2.3.2 Roof/ceiling. The roof/ceiling assembly shall be selected from Appendix Table 502.2.3.2 for a thermal transmittance value not exceeding the value specified for roofs/ceilings in Table 502.2.

Exception: When the roof/ceiling is comprised of assemblies containing truss type or C-shaped, cold-formed steel-framing members, the procedure outlined in Section 502.2.1.2 shall be used to adjust the roof/ceiling U-factor before selecting a roof/ceiling assembly from Appendix Table 502.2.3.2.

502.2.3.3 Floors over unheated spaces. The floor section over an unheated space shall be selected from Appendix Table 502.2.3.3 for the overall thermal transmittance factor (U_o) not exceeding the value specified for floors over unheated spaces in Table 502.2. For floors over outdoor air (i.e., overhangs), U_o-factors for heating shall meet the same requirement as shown for roofs/ceilings in Table 502.2.

Exception: When the floor is comprised of C-shaped, cold-formed steel-framing members, the procedure outlined in Section 502.2.1.3 shall be used to adjust the floor U-factor before selecting a floor assembly from Appendix Table 502.2.3.3.

502.2.3.4 Slab-on-grade floors. Slab-on-grade floors shall meet the provisions of Table 502.2 as determined by Section 502.2.1.4.

502.2.3.5 Crawl space walls. Where the floor above a crawl space does not meet the requirements of Section 502.2.3.3 and the crawl space does not have ventilation openings that communicate directly with the outside air, then the exterior walls of the crawl space shall have a thermal transmittance value not exceeding the value given in Table 502.2. The U-factor of the exterior crawl space wall shall be determined by selecting the U-factor for the appropriate crawl space wall section for Appendix Table 502.2.3.5. Where the inside ground surface is 12 inches (305 mm) or greater below the outside finish ground level, insulation shall extend from the top of the wall to at least the inside ground surface [see Appendix Detail 502.2.1.5(1) and the DOE *Foundation Design Handbook*]. Where the inside ground surface is less than 12 inches (305 mm) below the outside finish ground level, insulation shall extend from the top of the crawl space wall to the top of the footing [see Appendix Detail 502.2.1.5(2) and the DOE *Foundation Design Handbook*].

502.2.3.6 Basement walls. The exterior walls of conditioned basements shall have a thermal transmittance value not exceeding the value given in Table 502.2 from the top of the basement wall to a depth of 10 feet (3048 mm) below grade, or to the level of the basement floor, whichever is less. The U-factor of the wall shall be determined by selecting the U-factor for the wall section from Appendix Table 502.2.3.6.

502.2.4 Compliance by prescriptive specification on an individual component basis. For buildings with a window area less than or equal to 8 percent, 12 percent, 15 percent, 18 percent, 20 percent or 25 percent (detached one- and two-family dwellings) or 20 percent, 25 percent or 30 percent (Group R-2, R-4 or townhouse residential buildings) of the gross exterior wall area, the thermal resistance of insulation applied to the opaque building envelope components shall be greater than or equal to the minimum R-values, and the area-weighted average thermal transmittance (U-factor) of all fenestration assemblies (other than opaque doors which are governed by Section 502.2.4.6) shall be less than or equal to the maximum U-factors shown in Table 502.2.4(1), 502.2.4(2), 502.2.4(3), 502.2.4(4), 502.2.4(5), 502.2.4(6), 502.2.4(7), 502.2.4(8), or 502.2.4(9), as applicable. Sections 502.2.4.1 through 502.2.4.19 shall apply to the use of these tables.

502.2.4.1 Walls. The sum of the thermal resistance of cavity insulation plus insulating sheathing (if used) shall meet or exceed the "Exterior wall R-value."

502.2.4.2 Wood construction only. The tables shall only be used for wood construction.

502.2.4.3 Window area. The actual window area of a proposed design shall be computed using the rough opening area of all skylights, above-grade windows and, where the basement is conditioned space, any basement windows.

502.2.4.4 Window area, exempt. One percent of the total window area computed under Section 502.2.4.3 shall be exempt from the "Glazing U-factor" requirement.

TABLE 502.2.4(1)
PRESCRIPTIVE BUILDING ENVELOPE REQUIREMENTS, DETACHED ONE- AND TWO-FAMILY DWELLINGS
WINDOW AREA 8 PERCENT OF GROSS EXTERIOR WALL AREA

ZONE	HEATING DEGREE DAYS	MAXIMUM Glazing U-factor	MINIMUM Ceiling R-value	Exterior wall R-value	Floor R-value	Basement wall R-value	Slab perimeter R-value and depth	Crawl space wall R-value
1	0 - 499	Any	R-13	R-11	R-11	R-0	R-0	R-0
2	500 - 999	Any	R-19	R-11	R-11	R-0	R-0	R-4
3	1,000 - 1,499	Any	R-19	R-11	R-11	R-0	R-0	R-5
4	1,500 - 1,999	Any	R-19	R-11	R-11	R-5	R-0	R-5
5	2,000 - 2,499	0.90	R-19	R-11	R-11	R-5	R-0	R-6
6	2,500 - 2,999	0.70	R-26	R-11	R-11	R-5	R-0	R-6
7	3,000 - 3,499	0.70	R-26	R-11	R-13	R-5	R-0	R-6
8	3,500 - 3,999	0.65	R-30	R-11	R-13	R-6	R-2, 2 ft.	R-7
9	4,000 - 4,499	0.59	R-30	R-11	R-15	R-8	R-2, 2 ft.	R-9
10	4,500 - 4,999	0.55	R-30	R-13	R-15	R-8	R-2, 2 ft.	R-12
11	5,000 - 5,499	0.52	R-30	R-13	R-19	R-9	R-7, 2 ft.	R-16
12	5,500 - 5,999	0.45	R-38	R-13	R-19	R-9	R-7, 2 ft.	R-16
13	6,000 - 6,499	0.45	R-38	R-16	R-19	R-10	R-7, 4 ft.	R-16
14	6,500 - 6,999	0.43	R-38	R-16	R-19	R-10	R-7, 4 ft.	R-16
15	7,000 - 8,499	0.42	R-38	R-16	R-19	R-11	R-8, 4 ft.	R-16
16	8,500 - 8,999	0.42	R-38	R-16	R-19	R-16	R-8, 4 ft.	R-16
17	9,000 - 12,999	0.42	R-38	R-16	R-19	R-16	R-11, 4 ft.	R-16

For SI: 1 foot = 304.8 mm.

RESIDENTIAL — COMPONENT PERFORMANCE APPROACH

TABLE 502.2.4(2)
PRESCRIPTIVE BUILDING ENVELOPE REQUIREMENTS, DETACHED ONE- AND TWO-FAMILY DWELLINGS
WINDOW AREA 12 PERCENT OF GROSS EXTERIOR WALL AREA

ZONE	HEATING DEGREE DAYS	MAXIMUM Glazing U-factor	Ceiling R-value	Exterior wall R-value	Floor R-value	Basement wall R-value	Slab perimeter R-value and depth	Crawl space wall R-value
1	0 - 499	Any	R-13	R-11	R-11	R-0	R-0	R-0
2	500 - 999	Any	R-19	R-11	R-11	R-0	R-0	R-4
3	1,000 - 1,499	0.75	R-19	R-11	R-11	R-0	R-0	R-5
4	1,500 - 1,999	0.75	R-19	R-11	R-11	R-4	R-0	R-5
5	2,000 - 2,499	0.65	R-19	R-13	R-11	R-5	R-0	R-5
6	2,500 - 2,999	0.60	R-26	R-13	R-13	R-5	R-0	R-5
7	3,000 - 3,499	0.60	R-30	R-13	R-15	R-6	R-0	R-6
8	3,500 - 3,999	0.60	R-30	R-13	R-19	R-8	R-4, 2 ft.	R-10
9	4,000 - 4,499	0.55	R-38	R-13	R-19	R-9	R-4, 2 ft.	R-12
10	4,500 - 4,999	0.50	R-38	R-14	R-19	R-9	R-5, 2 ft.	R-16
11	5,000 - 5,499	0.45	R-38	R-16	R-19	R-9	R-6, 2 ft.	R-16
12	5,500 - 5,999	0.45	R-38	R-17	R-19	R-9	R-6, 2 ft.	R-16
13	6,000 - 6,499	0.40	R-38	R-18	R-19	R-10	R-6, 4 ft.	R-16
14	6,500 - 6,999	0.40	R-49	R-21	R-19	R-10	R-7, 4 ft.	R-17
15	7,000 - 8,499	0.40	R-49	R-21	R-19	R-10	R-9, 4 ft.	R-17
16	8,500 - 8,999	0.40	R-49	R-21	R-19	R-16	R-9, 4 ft.	R-17
17	9,000 - 12,999	0.40	R-49	R-21	R-19	R-16	R-11, 4 ft.	R-17

For SI: 1 foot = 304.8 mm.

TABLE 502.2.4(3)
PRESCRIPTIVE BUILDING ENVELOPE REQUIREMENTS, DETACHED ONE- AND TWO-FAMILY DWELLINGS
WINDOW AREA 15 PERCENT OF GROSS EXTERIOR WALL AREA

ZONE	HEATING DEGREE DAYS	MAXIMUM Glazing U-factor	Ceiling R-value	Exterior wall R-value	Floor R-value	Basement wall R-value	Slab perimeter R-value and depth	Crawl space wall R-value
1	0 - 499	Any	R-13	R-11	R-11	R-0	R-0	R-0
2	500 - 999	0.90	R-19	R-11	R-11	R-0	R-0	R-4
3	1,000 - 1,499	0.75	R-19	R-11	R-11	R-0	R-0	R-5
4	1,500 - 1,999	0.75	R-26	R-13	R-11	R-5	R-0	R-5
5	2,000 - 2,499	0.65	R-30	R-13	R-11	R-5	R-0	R-6
6	2,500 - 2,999	0.60	R-30	R-13	R-19	R-6	R-4, 2 ft.	R-7
7	3,000 - 3,499	0.55	R-30	R-13	R-19	R-7	R-4, 2 ft.	R-8
8	3,500 - 3,999	0.50	R-30	R-13	R-19	R-8	R-5, 2 ft.	R-10
9	4,000 - 4,499	0.45	R-38	R-13	R-19	R-8	R-5, 2 ft.	R-11
10	4,500 - 4,999	0.45	R-38	R-16	R-19	R-9	R-6, 2 ft.	R-17
11	5,000 - 5,499	0.45	R-38	R-18	R-19	R-9	R-6, 2 ft.	R-17
12	5,500 - 5,999	0.40	R-38	R-18	R-21	R-10	R-9, 2 ft.	R-19
13	6,000 - 6,499	0.35	R-38	R-18	R-21	R-10	R-9, 4 ft.	R-20
14	6,500 - 6,999	0.35	R-49	R-21	R-21	R-11	R-11, 4 ft.	R-20
15	7,000 - 8,499	0.35	R-49	R-21	R-21	R-11	R-13, 4 ft.	R-20
16	8,500 - 8,999	0.35	R-49	R-21	R-21	R-18	R-14, 4 ft.	R-20
17	9,000 - 12,999	0.35	R-49	R-21	R-21	R-19	R-18, 4 ft.	R-20

For SI: 1 foot = 304.8 mm.

RESIDENTIAL — COMPONENT PERFORMANCE APPROACH

TABLE 502.2.4(4)
PRESCRIPTIVE BUILDING ENVELOPE REQUIREMENTS, DETACHED ONE- AND TWO-FAMILY DWELLINGS
WINDOW AREA 18 PERCENT OF GROSS EXTERIOR WALL AREA

ZONE	HEATING DEGREE DAYS	MAXIMUM Glazing U-factor	MINIMUM					
			Ceiling R-value	Exterior wall R-value	Floor R-value	Basement wall R-value	Slab perimeter R-value and depth	Crawl space wall R-value
1	0 - 499	0.80	R-19	R-11	R-11	R-0	R-0	R-0
2	500 - 999	0.75	R-19	R-11	R-11	R-0	R-0	R-4
3	1,000 - 1,499	0.70	R-26	R-13	R-11	R-0	R-0	R-5
4	1,500 - 1,999	0.65	R-30	R-13	R-11	R-5	R-0	R-5
5	2,000 - 2,499	0.55	R-30	R-13	R-11	R-5	R-0	R-6
6	2,500 - 2,999	0.52	R-30	R-13	R-19	R-6	R-0	R-7
7	3,000 - 3,499	0.50	R-38	R-13	R-19	R-7	R-0	R-8
8	3,500 - 3,999	0.46	R-38	R-13	R-19	R-8	R-6, 2 ft.	R-11
9	4,000 - 4,499	0.40	R-38	R-13	R-19	R-9	R-6, 2 ft.	R-13
10	4,500 - 4,999	0.37	R-38	R-15	R-19	R-9	R-6, 2 ft.	R-16
11	5,000 - 5,499	0.37	R-38	R-16	R-19	R-9	R-7, 2 ft.	R-17
12	5,500 - 5,999	0.37	R-38	R-19	R-19	R-10	R-8, 2 ft.	R-17
13	6,000 - 6,499	0.34	R-49	R-22	R-19	R-10	R-8, 4 ft.	R-17
14	6,500 - 6,999	0.33	R-49	R-22	R-25	R-11	R-14, 4 ft.	R-19
15	7,000 - 8,499	0.33	R-49	R-25	R-30	R-15	Note a	R-25
16	8,500 - 8,999	0.33	R-49	R-25	R-30	R-19	Note a	R-25
17	9,000 - 12,999	0.33	R-49	R-25	R-30	R-19	Note a	R-25

For SI: 1 foot = 304.8 mm.
a. See Section 502.2.4.13.

TABLE 502.2.4(5)
PRESCRIPTIVE BUILDING ENVELOPE REQUIREMENTS, DETACHED ONE- AND TWO-FAMILY DWELLINGS
WINDOW AREA 20 PERCENT OF GROSS EXTERIOR WALL AREA

ZONE	HEATING DEGREE DAYS	MAXIMUM Glazing U-factor	MINIMUM					
			Ceiling R-value	Exterior wall R-value	Floor R-value	Basement wall R-value	Slab perimeter R-value and depth	Crawl space wall R-value
1	0 - 499	0.80	R-19	R-11	R-11	R-0	R-0	R-0
2	500 - 999	0.75	R-30	R-13	R-11	R-0	R-0	R-4
3	1,000 - 1,499	0.70	R-30	R-13	R-11	R-0	R-0	R-5
4	1,500 - 1,999	0.60	R-30	R-13	R-11	R-5	R-0	R-5
5	2,000 - 2,499	0.52	R-38	R-13	R-11	R-5	R-0	R-6
6	2,500 - 2,999	0.50	R-38	R-13	R-19	R-6	R-0	R-7
7	3,000 - 3,499	0.46	R-38	R-13	R-19	R-7	R-0	R-9
8	3,500 - 3,999	0.42	R-38	R-13	R-19	R-8	R-6, 2 ft.	R-10
9	4,000 - 4,499	0.37	R-38	R-13	R-19	R-9	R-6, 2 ft.	R-13
10	4,500 - 4,999	0.37	R-38	R-16	R-19	R-9	R-6, 2 ft.	R-16
11	5,000 - 5,499	0.36	R-38	R-19	R-19	R-9	R-6, 2 ft.	R-16
12	5,500 - 5,999	0.33	R-49	R-20	R-19	R-10	R-7, 2 ft.	R-17
13	6,000 - 6,499	0.31	R-49	R-24	R-19	R-10	R-7, 4 ft.	R-17
14	6,500 - 6,999	0.30	R-49	R-26	R-21	R-11	R-10, 4 ft.	R-17
15	7,000 - 8,499	0.30	R-49	R-26	R-21	R-11	R-12, 4 ft.	R-19
16	8,500 - 8,999	0.30	R-49	R-26	R-21	R-19	R-12, 4 ft.	R-19
17	9,000 - 12,999	0.30	R-49	R-26	R-21	R-19	R-16, 4 ft.	R-19

For SI: 1 foot = 304.8 mm.

RESIDENTIAL — COMPONENT PERFORMANCE APPROACH

TABLE 502.2.4(6)
PRESCRIPTIVE BUILDING ENVELOPE REQUIREMENTS, DETACHED ONE- AND TWO-FAMILY DWELLINGS
WINDOW AREA 25 PERCENT OF GROSS EXTERIOR WALL AREA

ZONE	HEATING DEGREE DAYS	MAXIMUM Glazing U-factor	MINIMUM Ceiling R-value	Exterior wall R-value	Floor R-value	Basement wall R-value	Slab perimeter R-value and depth	Crawl space wall R-value
1	0 - 499	0.70	R-30	R-11	R-11	R-0	R-0	R-0
2	500 - 999	0.65	R-30	R-13	R-11	R-0	R-0	R-4
3	1,000 - 1,499	0.55	R-30	R-13	R-11	R-0	R-0	R-5
4	1,500 - 1,999	0.52	R-30	R-13	R-13	R-6	R-0	R-6
5	2,000 - 2,499	0.50	R-38	R-13	R-19	R-8	R-0	R-10
6	2,500 - 2,999	0.46	R-38	R-16	R-19	R-6	R-0	R-7
7	3,000 - 3,499	0.45	R-38	R-19	R-19	R-7	R-0	R-9
8	3,500 - 3,999	0.41	R-38	R-19	R-19	R-8	R-6, 2 ft.	R-10
9	4,000 - 4,499	0.37	R-38	R-19	R-19	R-9	R-6, 2 ft.	R-13
10	4,500 - 4,999	0.33	R-38	R-19	R-19	R-9	R-6, 2 ft.	R-17
11	5,000 - 5,499	0.29	R-38	R-19	R-19	R-9	R-6, 2 ft.	R-17
12	5,500 - 5,999	0.27	R-38	R-19	R-21	R-10	Note a	R-22
13	6,000 - 6,499	0.25	R-49	R-19	R-21	R-10	R-9, 4 ft.	R-20
14	6,500 - 6,999	0.25	R-49	R-19	R-30	R-14	Note a	Note a
15	7,000 - 8,499	0.25	R-49	R-19	R-30	R-15	Note a	Note a
16	8,500 - 8,999	0.25	R-49	R-19	R-30	R-28	Note a	Note a
17	9,000 - 12,999	0.25	R-49	R-19	R-30	R-28	Note a	Note a

For SI: 1 foot = 304.8 mm.
a. See Section 502.2.4.13.

TABLE 502.2.4(7)
PRESCRIPTIVE BUILDING ENVELOPE REQUIREMENTS, GROUP R-2, R-4 OR TOWNHOUSE RESIDENTIAL BUILDINGS
WINDOW AREA 20 PERCENT OF GROSS EXTERIOR WALL AREA

ZONE	HEATING DEGREE DAYS	MAXIMUM Glazing U-factor	MINIMUM Ceiling R-value	Exterior wall R-value	Floor R-value	Basement wall R-value	Slab perimeter R-value and depth	Crawl space wall R-value
1	0 - 499	Any	R-13	R-11	R-11	R-0	R-0	R-0
2	500 - 999	Any	R-19	R-11	R-11	R-0	R-0	R-5
3	1,000 - 1,499	Any	R-19	R-11	R-11	R-0	R-0	R-5
4	1,500 - 1,999	0.85	R-19	R-11	R-11	R-5	R-0	R-5
5	2,000 - 2,499	0.70	R-19	R-11	R-11	R-5	R-0	R-5
6	2,500 - 2,999	0.55	R-30	R-13	R-11	R-5	R-0	R-5
7	3,000 - 3,499	0.55	R-30	R-13	R-11	R-5	R-0	R-5
8	3,500 - 3,999	0.55	R-30	R-13	R-11	R-5	R-0	R-5
9	4,000 - 4,499	0.55	R-38	R-13	R-11	R-5	R-0	R-5
10	4,500 - 4,999	0.50	R-26	R-11	R-13	R-6	R-0	R-7
11	5,000 - 5,499	0.50	R-26	R-13	R-11	R-5	R-0	R-6
12	5,500 - 5,999	0.50	R-30	R-13	R-11	R-5	R-0	R-6
13	6,000 - 6,499	0.50	R-26	R-13	R-19	R-9	R-5, 4 ft.	R-14
14	6,500 - 6,999	0.45	R-30	R-13	R-19	R-10	R-7, 4 ft.	R-16
15	7,000 - 8,499	0.35	R-38	R-16	R-19	R-11	R-9, 4 ft.	R-18
16	8,500 - 8,999	0.35	R-38	R-16	R-19	R-17	R-10, 4 ft.	R-18
17	9,000 - 12,999	Note a	Note a	Note a	Note a	Note a	Note a	Note a

For SI: 1 foot = 304.8 mm.
a. See Section 502.2.4.13.

RESIDENTIAL — COMPONENT PERFORMANCE APPROACH

TABLE 502.2.4(8)
PRESCRIPTIVE BUILDING ENVELOPE REQUIREMENTS, GROUP R-2, R-4 OR TOWNHOUSE RESIDENTIAL BUILDINGS
WINDOW AREA 25 PERCENT OF GROSS EXTERIOR WALL AREA

ZONE	HEATING DEGREE DAYS	MAXIMUM Glazing U-factor	MINIMUM					
			Ceiling R-value	Exterior wall R-value	Floor R-value	Basement wall R-value	Slab perimeter R-value and depth	Crawl space wall R-value
1	0 - 499	Any	R-13	R-11	R-11	R-0	R-0	R-0
2	500 - 999	Any	R-19	R-11	R-11	R-0	R-0	R-5
3	1,000 - 1,499	Any	R-19	R-11	R-11	R-0	R-0	R-5
4	1,500 - 1,999	0.85	R-19	R-11	R-11	R-5	R-0	R-5
5	2,000 - 2,499	0.70	R-19	R-11	R-11	R-5	R-0	R-5
6	2,500 - 2,999	0.55	R-30	R-13	R-11	R-5	R-0	R-5
7	3,000 - 3,499	0.55	R-30	R-13	R-11	R-5	R-0	R-5
8	3,500 - 3,999	0.55	R-30	R-13	R-11	R-5	R-0	R-5
9	4,000 - 4,499	0.54	R-30	R-13	R-11	R-5	R-0	R-5
10	4,500 - 4,999	0.53	R-30	R-13	R-11	R-5	R-0	R-6
11	5,000 - 5,499	0.52	R-30	R-13	R-11	R-5	R-0	R-6
12	5,500 - 5,999	0.51	R-30	R-13	R-11	R-6	R-0	R-6
13	6,000 - 6,499	0.51	R-30	R-13	R-19	R-10	R-7, 4 ft.	R-16
14	6,500 - 6,999	0.45	R-30	R-13	R-19	R-10	R-7, 4 ft.	R-16
15	7,000 - 8,499	0.35	R-38	R-16	R-19	R-11	R-9, 4 ft.	R-18
16	8,500 - 8,999	0.35	R-38	R-16	R-19	R-17	R-10, 4 ft.	R-18
17	9,000 - 12,999	Note a	Note a	Note a	Note a	Note a	Note a	Note a

For SI: 1 foot = 304.8 mm.
a. See Section 502.2.4.13

TABLE 502.2.4(9)
PRESCRIPTIVE BUILDING ENVELOPE REQUIREMENTS, GROUP R-2, R-4 OR TOWNHOUSE RESIDENTIAL BUILDINGS
WINDOW AREA 30 PERCENT OF GROSS EXTERIOR WALL AREA

ZONE	HEATING DEGREE DAYS	MAXIMUM Glazing U-factor	MINIMUM					
			Ceiling R-value	Exterior wall R-value	Floor R-value	Basement wall R-value	Slab perimeter R-value and depth	Crawl space wall R-value
1	0 - 499	0.90	R-13	R-11	R-11	R-0	R-0	R-0
2	500 - 999	0.75	R-19	R-11	R-11	R-0	R-0	R-3
3	1,000 - 1,499	0.70	R-19	R-11	R-11	R-0	R-0	R-4
4	1,500 - 1,999	0.65	R-26	R-11	R-11	R-5	R-0	R-5
5	2,000 - 2,499	0.57	R-38	R-13	R-11	R-5	R-0	R-6
6	2,500 - 2,999	0.47	R-38	R-13	R-19	R-7	R-0	R-8
7	3,000 - 3,499	0.47	R-38	R-13	R-19	R-7	R-0	R-9
8	3,500 - 3,999	0.46	R-38	R-13	R-19	R-8	R-4, 2 ft.	R-9
9	4,000 - 4,499	0.46	R-38	R-13	R-19	R-9	R-6, 2 ft.	R-13
10	4,500 - 4,999	0.45	R-38	R-13	R-19	R-9	R-6, 2 ft.	R-15
11	5,000 - 5,499	0.45	R-38	R-13	R-19	R-10	R-8, 2 ft.	R-18
12	5,500 - 5,999	0.44	R-38	R-13	R-19	R-10	R-8, 2 ft.	R-18
13	6,000 - 6,499	0.44	R-38	R-19	R-19	R-10	R-8, 4 ft.	R-18
14	6,500 - 6,999	0.38	R-38	R-19	R-19	R-10	R-8, 4 ft.	R-18
15	7,000 - 8,499	0.32	R-49	R-21	R-30	R-18	Note a	Note a
16	8,500 - 8,999	0.32	R-49	R-21	R-30	Note a	Note a	Note a
17	9,000 - 12,999	Note a	Note a	Note a	Note a	Note a	Note a	Note a

For SI: 1 foot = 304.8 mm.
a. See Section 502.2.4.13

RESIDENTIAL — COMPONENT PERFORMANCE APPROACH

502.2.4.5 Truss/rafter construction. "Ceiling R-value" assumes standard truss or rafter construction. Where raised-heel trusses or other construction techniques are employed to obtain the full height of ceiling insulation over the exterior wall top plate, R-30 shall be permitted to be used where R-38 is required in the table, and R-38 shall be permitted to be used where R-49 is required.

502.2.4.6 Doors. Opaque doors in the building envelope shall have a maximum U-factor of 0.35. One door shall be exempt from this requirement.

502.2.4.7 Ceilings. "Ceiling R-value" shall be required for flat or "cathedral" (inclined) ceilings.

502.2.4.8 Floors. "Floor R-value" shall apply to floors over unconditioned spaces. A floor over outside air shall meet the requirement for "Ceiling R-value."

502.2.4.9 Basement walls. Basement wall insulation shall be installed in accordance with Section 502.2.1.6.

502.2.4.10 Unheated slabs. Slab perimeter insulation shall be installed in accordance with Section 502.2.1.4.

502.2.4.11 Heated slabs. R-2 shall be added to the "Slab perimeter R-value" where the slab is heated.

502.2.4.12 Crawl space walls. "Crawl space wall R-value" shall apply to unventilated crawl spaces only. Crawl space insulation shall be installed in accordance with Section 502.2.1.5.

502.2.4.13 Tables not applicable. The particular climate range indicated by Note a in Tables 502.2.4(4), 502.2.4(6), 502.2.4(7), 502.2.4(8) and 502.2.4(9) shall not be used with the indicated envelope component(s) to demonstrate compliance under Section 502.2.4.

502.2.4.14 Climates greater than 13,000 HDD. These tables shall not be used for climates greater than or equal to 13,000 HDD.

502.2.4.15 Fenestration solar heat gain coefficient. In locations with HDD less than 3,500, fenestration products shall also meet the requirements of Section 502.1.5.

502.2.4.16 Steel-framed wall construction. Where steel framing is used in wall construction, the wall assembly shall meet the equivalent wall cavity and sheathing R-values in Table 502.2.4.16(1) or 502.2.4.16(2), based on the "on-center" (o.c.) dimension of the steel studs and the required R-value for wood-framed walls determined in accordance with Section 502.2.4, and utilizing any combination of cavity and sheathing insulation set off by commas in Table 502.2.4.16(1) or 502.2.4.16(2).

502.2.4.17 High-mass wall construction. Exterior walls constructed of high-mass materials having heat capacity greater than or equal to 6 Btu/ ft² · °F [1.06 kJ/(m² · K)] of exterior wall area shall meet the equivalent insulation R-values in Table 502.2.4.17(1) or 502.2.4.17(2), based on the placement of the insulation, the HDD of the building location, and the required R-value for wood-framed walls determined in accordance with Section 502.2.4.

TABLE 502.2.4.16(1)
16-INCH O.C. STEEL-FRAMED WALL EQUIVALENT R-VALUES

WOOD-FRAMED WALL R-VALUE [a]	EQUIVALENT STEEL-FRAMED WALL CAVITY AND SHEATHING R-VALUE
R-11	R-0+R-9, R-11+R-4, R-15+R-3, R-21+R-2
R-13	R-11+R-5, R-15+R-4, R-21+R-3
R-14	R-11+R-6, R-13+R-5, R-19+R-4
R-15	R-11+R-6, R-15+R-5, R-19+R-4
R-16	R-11+R-8, R-15+R-7, R-21+R-6
R-17	R-11+R-9, R-13+R-8, R-19+R-7
R-18	R-11+R-9, R-15+R-8, R-21+R-7
R-19	R-11+R-10, R-13+R-9, R-19+R-8, R-25+R-7
R-20	R-11+R-10, R-13+R-9, R-19+R-8
R-21	R-13+R-10, R-19+R-9, R-25+R-8
R-22	R-13+R-10, R-19+R-9
R-24	R-19+R-10, R-25+R-9
R-25	R-19+R-10
R-26	R-19+R-11, R-21+R-10

For SI: 1 inch = 25.4 mm.

a. As required by Section 502.2.4 and the tabular entry for "Exterior wall R-value" shown in Tables 502.2.4(1) through 502.2.4(9), as applicable.

TABLE 502.2.4.16(2)
24-INCH O.C. STEEL FRAMED WALL EQUIVALENT R-VALUES

WOOD-FRAMED WALL R-VALUE [a]	EQUIVALENT STEEL-FRAMED WALL CAVITY AND SHEATHING R-VALUE
R-11	R-0+R-9, R-11+R-3, R-15+R-2, R-25+R-0
R-13	R-11+R-4, R-15+R-3, R-19+R-2
R-14	R-11+R-5, R-13+R-4, R-15+R-3, R-21+R-2
R-15	R-11+R-5, R-13+R-4, R-19+R-3, R-21+R-2
R-16	R-11+R-7, R-13+R-6, R-19+R-5, R-25+R-4
R-17	R-11+R-8, R-13+R-7, R-15+R-6, R-21+R-5
R-18	R-11+R-8, R-13+R-7, R-19+R-6, R-25+R-5
R-19	R-11+R-9, R-13+R-8, R-15+R-7, R-21+R-6
R-20	R-11+R-9, R-13+R-8, R-19+R-7, R-21+R-6
R-21	R-11+R-9, R-15+R-8, R-21+R-7
R-22	R-11+R-10, R-13+R-9, R-19+R-8, R-21+R-7
R-24	R-11+R-10, R-15+R-9, R-19+R-8
R-25	R-13+R-10, R-19+R-9, R-21+R-8
R-26	R-15+R-10, R-19+R-9, R-25+R-8

For SI: 1 inch = 25.4 mm.

a. As required by Section 502.2.4 and the tabular entry for "Exterior wall R-value" shown in Tables 502.2.4(1) through 502.2.4(9), as applicable.

TABLE 502.2.4.17(1)
HIGH-MASS WALL EQUIVALENT R-VALUES
INSULATION PLACED ON THE EXTERIOR OF THE WALL
OR WITH INTEGRAL INSULATION

WOOD-FRAMED WALL R-VALUE[a]	EQUIVALENT HIGH-MASS WALL R-VALUE					
	HDD 0 - 1,999	HDD 2,000 - 3,999	HDD 4,000 - 5,499	HDD 5,500 - 6,499	HDD 6,500 - 8,499	HDD ≥ 8,500
R-11	R-6	R-6	R-7	R-8	R-9	R-10
R-13	R-6	R-6	R-8	R-9	R-10	R-11
R-14	R-6	R-7	R-8	R-9	R-10	R-11
R-15	R-7	R-7	R-8	R-9	R-10	R-12
R-16	R-7	R-7	R-8	R-9	R-11	R-12
R-17	R-7	R-7	R-9	R-10	R-11	R-13
R-18	R-7	R-7	R-9	R-10	R-11	R-13
R-19	R-8	R-9	R-10	R-11	R-13	R-15
R-20	R-8	R-9	R-10	R-11	R-13	R-16
R-21	R-8	R-9	R-10	R-12	R-14	R-16
R-22	R-8	R-9	R-10	R-12	R-14	R-17
R-23	R-9	R-9	R-11	R-12	R-14	R-17
R-24	R-9	R-9	R-11	R-12	R-14	R-17
R-25	R-9	R-10	R-11	R-13	R-15	R-18
R-26	R-9	R-10	R-11	R-13	R-15	R-18

a. As required by Section 502.2.4 and the tabular entry for "Exterior wall R-value" shown in Tables 502.2.4(1) through 502.2.4(9), as applicable.

TABLE 502.2.4.17(2)
HIGH-MASS WALL EQUIVALENT R-VALUES
INSULATION PLACED ON THE INTERIOR OF THE WALL

WOOD-FRAMED WALL R-VALUE[a]	EQUIVALENT HIGH-MASS WALL R-VALUE					
	HDD 0 - 1,999	HDD 2,000 - 3,999	HDD 4,000 - 5,499	HDD 5,500 - 6,499	HDD 6,500 - 8,499	HDD ≥ 8,500
R-11	R-10	R-10	R-11	R-11	R-12	R-12
R-13	R-11	R-11	R-12	R-12	R-14	R-14
R-14	R-12	R-12	R-12	R-13	R-15	R-15
R-15	R-13	R-13	R-13	R-14	R-15	R-15
R-16	R-13	R-13	R-13	R-15	R-15	R-15
R-17	R-14	R-14	R-14	R-15	R-16	R-16
R-18	R-15	R-15	R-15	R-19	R-16	R-16
R-19	R-16	R-16	R-16	R-20	R-19	R-19
R-20	R-16	R-16	R-16	R-21	R-20	R-20
R-21	R-17	R-17	R-17	R-21	R-21	R-21
R-22	R-17	R-17	R-17	R-22	R-21	R-21
R-23	R-18	R-18	R-18	R-22	R-22	R-22
R-24	R-19	R-19	R-19	R-22	R-22	R-22
R-25	R-20	R-20	R-20	R-22	R-22	R-22
R-26	R-21	R-21	R-21	R-23	R-23	R-23

a. As required by Section 502.2.4 and the tabular entry for "Exterior wall R-value" shown in Tables 502.2.4(1) through 502.2.4(9), as applicable.

502.2.4.18 Steel-framed roof/ceiling construction. When truss-type, cold-formed steel framing is used in roof/ceiling construction, the roof/ceiling assembly shall meet the equivalent insulation R-values in Table 502.2.4.18(1).

When C-shaped, cold-formed steel framing is used in roof/ceiling construction, the steel roof/ceiling assembly shall meet the equivalent wood framed U_R-factors in Table 502.2.4.18(2).

TABLE 502.2.4.18(1)
TRUSS TYPE COLD-FORMED STEEL ROOF/CEILING EQUIVALENT R-VALUES[a]

WOOD-FRAMED ROOF/CEILING R-VALUE[b]	TRUSS TYPE COLD-FORMED STEEL CAVITY AND CONTINUOUS INSULATION R-VALUE, 24 INCHES ON CENTER[c]
R-13	R-19, R-13 + R-3
R-19	R-26, R-19 + R-3
R-26	R-38, R-26 + R-3
R-30	R-38, R-30 + R-3
R-38	R-49, R-38 + R-5
R-49	Not applicable

For SI: 1 inch = 25.4 mm, 1 foot = 304.8 mm.

a. This table applies to cold-formed, steel-truss roof framing spaced at 24 inches on center and where the penetrations of the truss members through the cavity insulation do not exceed three penetrations of the truss members through the cavity insulation for each 4-foot length of the truss.
b. As required by Section 502.2.4 and the tabular entry for "Ceiling R-value" shown in Tables 502.2.4(1) through 502.2.4(9).
c. The cavity R-value requirement is listed first, followed by the continuous insulation R-value requirement.

502.2.4.19 Steel-framed floor construction. When C-shaped, cold-formed steel framing is used in floor construction, the steel floor assembly shall meet the equivalent wood framed U_f-factors in Table 502.2.4.19.

502.2.5 Prescriptive path for additions and window replacements. As an alternative to demonstrating compliance with Section 402 or 502.2, additions with a conditioned floor area less than 500 square feet (46.5 m^2) to existing single-family residential buildings and structures shall meet the prescriptive envelope component criteria in Table 502.2.5 for the designated heating degree days (HDD) applicable to the location. The U-factor of each individual fenestration product (windows, doors and skylights) shall be used to calculate an area-weighted average fenestration product U-factor for the addition, which shall not exceed the applicable listed values in Table 502.2.5. For additions, other than sunroom additions, the total area of fenestration products shall not exceed 40 percent of the gross wall and roof area of the addition. The R-values for opaque thermal envelope components shall be equal to or greater than the applicable listed values in Table 502.2.5. Replacement fenestration products (where some or all of an existing fenestration unit is replaced with an entire new replacement unit, including the frame, sash and glazing) shall meet the prescriptive fenestration U-factor criteria in Table 502.2.5 for the designated HDD applicable to the location.

Conditioned sunroom additions shall maintain thermal isolation; shall not be used as kitchens or sleeping rooms; and shall be served by a separate heating or cooling system, or be thermostatically controlled as a separate zone of the existing system.

Fenestration products used in additions and as replacement windows in accordance with this section shall also meet the requirements of Section 502.1.5 in locations with HDD less than 3,500.

Exception: Replacement skylights shall have a maximum U-factor of 0.60 when installed in any location above 1,999 HDD.

SECTION 503
BUILDING MECHANICAL SYSTEMS AND EQUIPMENT

503.1 General. This section covers mechanical systems and equipment used to provide heating, ventilating and air-conditioning functions. This section assumes that residential buildings and dwelling units therein will be designed with individual HVAC systems. Where equipment not shown in Table 503.2 is specified, it shall meet the provisions of Sections 803.2.2 and 803.3.2.

503.2 Mechanical equipment efficiency. Equipment shown in Table 503.2 shall meet the specified minimum performance. Data furnished by the equipment supplier, or certified under a nationally recognized certification procedure, shall be used to satisfy these requirements. All such equipment shall be installed in accordance with the manufacturer's instructions.

503.3 HVAC systems. HVAC systems shall meet the criteria set forth in Sections 503.3.1 through 503.3.3.

[M] 503.3.1 Load calculations. Heating and cooling system design loads for the purpose of sizing systems and equipment shall be determined in accordance with the procedures described in the ASHRAE *Fundamentals Handbook*. Heating and cooling loads shall be adjusted to account for load reductions that are achieved when energy recovery systems are utilized in the HVAC system in accordance with the ASHRAE *HVAC Systems and Equipment Handbook*. Alternatively, design loads shall be determined by an approved equivalent computation procedure, using the design parameters specified in Chapter 3.

503.3.2 Temperature and humidity controls. Temperature and humidity controls shall be provided in accordance with Sections 503.3.2.1 through 503.3.2.4.

503.3.2.1 System controls. Each dwelling unit shall be considered a zone and be provided with thermostatic controls responding to temperature within the dwelling unit. Each heating and cooling system shall include at least one temperature control device.

RESIDENTIAL — COMPONENT PERFORMANCE APPROACH

TABLE 502.2.4.18(2)
C-SHAPED COLD-FORMED STEEL ROOF/CEILING EQUIVALENT U_R-FACTORS[a]

FRAMING[b] Steel / Wood Equivalent	SPACING	R-13[c]	R-19[c]	R-26[c]	R-30[c]	R-38[c]	R-49[c]
Wood Equivalent	16 inches o.c.	0.0773	0.0537	0.0405	0.0355	0.0285	0.0223
2 × 4		0.1328	0.0530	0.0387	0.0336	0.0265	0.0206
2 × 6		0.1328	0.0667	0.0456	0.0386	0.0295	0.0223
2 × 8		0.1328	0.1208	0.0585	0.0475	0.0345	0.0251
2 × 10		0.1328	0.1208	0.1094	0.1037	0.0398	0.0277
2 × 12		0.1328	0.1208	0.1094	0.1037	0.0471	0.0311
Wood Equivalent	24 inches o.c.	0.0742	0.0519	0.0390	0.0342	0.0274	0.0215
2 × 4		0.1129	0.0510	0.0376	0.0327	0.0260	0.0202
2 × 6		0.1129	0.0610	0.0428	0.0366	0.0284	0.0216
2 × 8		0.1129	0.0994	0.0517	0.0429	0.0320	0.0237
2 × 10		0.1129	0.0994	0.0873	0.0816	0.0357	0.0257
2 × 12		0.1129	0.0994	0.0873	0.0816	0.0403	0.0280

For SI: 1 inch = 25.4 mm.
a. Linear interpolation is permitted for determining U-factors which are between those given in the table.
b. Applies to steel framing up to a maximum thickness of 0.064 inches (16 gage.)
c. As required by Section 502.2.4 and the tabular entry for "Ceiling R-value" shown in Tables 502.2.4(1) through 502.2.4(9), as applicable.

TABLE 502.2.4.19
C-SHAPED COLD-FORMED STEEL FLOOR EQUIVALENT U_f-FACTORS[a]

FRAMING[b] Steel / Wood Equivalent	SPACING	R-11[c]	R-13[c]	R-15[c]	R-19[c]	R-21[c]	R-25[c]	R-30
Wood Equivalent	16 inches o.c.	0.0725	0.0652	0.0595	0.0477	0.0452	0.0382	0.0327
2 × 6		0.1058	0.1031	0.1005	0.0583	0.0523	NA	NA
2 × 8		0.1058	0.1031	0.1005	0.0957	0.0935	0.0548	NA
2 × 10		0.1058	0.1031	0.1005	0.0957	0.0935	0.0894	0.0838
2 × 12		0.1058	0.1031	0.1005	0.0957	0.0935	0.0894	0.0838
Wood Equivalent	24 inches o.c.	0.0708	0.0633	0.0574	0.0464	0.0436	0.0370	0.0317
2 × 6		0.0941	0.0907	0.0875	0.0538	0.0486	NA	NA
2 × 8		0.0941	0.0907	0.0875	0.0818	0.0792	0.0488	NA
2 × 10		0.0941	0.0907	0.0875	0.0818	0.0792	0.0745	0.0697
2 × 12		0.0941	0.0907	0.0875	0.0818	0.0792	0.0745	0.0697

For SI: 1 inch = 25.4 mm.
NA = Not applicable.
a. Linear interpolation is permitted for determining U-factors which are between those given in the table.
b. Applies to steel framing up to a maximum thickness of 0.064 inches (16 gage.)
c. As required by Section 502.2.4 and the tabular entry for "Floor R-value" shown in Tables 502.2.4(1) through 502.2.4(9), as applicable.

RESIDENTIAL — COMPONENT PERFORMANCE APPROACH

TABLE 502.2.5
PRESCRIPTIVE ENVELOPE COMPONENT CRITERIA
ADDITIONS TO AND REPLACEMENT WINDOWS FOR EXISTING
DETACHED ONE- AND TWO-FAMILY DWELLINGS

HEATING DEGREE DAYS	MAXIMUM Fenestration U-factor[e]	MINIMUM Ceiling R-value[a,e]	Wall R-value[e]	Floor R-value	Basement wall R-value[b]	Slab perimeter R-value and depth[c]	Crawl space wall R-value[d]
0 - 1,999	0.75	R-26	R-13	R-11	R-5	R-0	R-5
2,000 - 3,999	0.50	R-30	R-13	R-19	R-8	R-5, 2 ft.	R-10
4,000 - 5,999	0.40	R-38	R-18	R-21	R-10	R-9, 2 ft.	R-19
6,000 - 8,499	0.35	R-49	R-21	R-21	R-11	R-13, 4 ft.	R-20
8,500 - 12,999	0.35	R-49	R-21	R-21	R-19	R-18, 4 ft.	R-20

For SI: 1 foot = 304.8 mm.
a. "Ceiling R-value" shall be required for flat or inclined (cathedral) ceilings. Floors over outside air shall meet "Ceiling R-value" requirements.
b. Basement wall insulation shall be installed in accordance with Section 502.2.1.6.
c. Slab perimeter insulation shall be installed in accordance with Section 502.2.1.4. An additional R-2 shall be added to "Slab perimeter R-value" in the table if the slab is heated.
d. "Crawl space wall R-value" shall apply to unventilated crawl spaces only. Crawl space insulation shall be installed in accordance with Section 502.2.1.5.
e. Sunroom additions shall be required to have a maximum fenestration U-factor of 0.50 in locations with 2,000 - 12,999 HDD. In locations with 0-5,999 HDD, the minimum ceiling R-value shall be R-19 and the minimum wall R-value shall be R-13. In locations with 6,000 - 12,999 HDD, the minimum ceiling R-value shall be R-24 and the minimum wall R-value shall be R-13.

TABLE 503.2
MINIMUM EQUIPMENT PERFORMANCE

EQUIPMENT CATEGORY	SUBCATEGORY[e]	REFERENCED STANDARD	MINIMUM PERFORMANCE
Air-cooled heat pumps, Heating mode < 65,000 Btu/h cooling capacity	Split systems	ARI 210/240	6.8 HSPF[a,b]
	Single package		6.6 HSPF[a,b]
Gas-fired or oil-fired furnace < 225,000 Btu/h	—	DOE 10 CFR Part 430, Subpart B, Appendix N	AFUE 78%[b] E_t 80%[c]
Gas-fired or oil-fired steam and hot-water boilers < 300,000 Btu/h	—	DOE 10 CFR Part 430, Subpart B, Appendix N	AFUE 80%[b,d]
Air-cooled air conditioners and heat pumps, Cooling mode <65,000 Btu/h cooling capacity	Split systems	ARI 210/240	10.0 SEER[b]
	Single package		9.7 SEER[b]

For SI: 1 British thermal unit per hour = 0.2931 W.
a. For multicapacity equipment, the minimum performance shall apply to each capacity step provided. Multicapacity refers to manufacturer-published ratings for more than one capacity mode allowed by the product's controls.
b. This is used to be consistent with the National Appliance Energy Conservation Act (NAECA) of 1987 (Public Law 100-12).
c. These requirements apply to combination units not covered by NAECA (three-phase power or cooling capacity 65,000 Btu/h).
d. Except for gas-fired steam boilers for which the minimum AFUE shall be 75 percent.
e. Seasonal rating.

503.3.2.2 Thermostatic control capabilities. Where used to control comfort heating, thermostatic controls shall be capable of being set locally or remotely by adjustment or selection of sensors down to 55°F (13°C) or lower.

Where used to control comfort cooling, thermostatic controls shall be capable of being set locally or remotely by adjustment or selection of sensors up to 85°F (29°C) or higher.

Where used to control both comfort heating and cooling, thermostatic controls shall be capable of providing a temperature range or deadband of at least 5°F (Δ3°C) within which the supply of heating and cooling energy is shut off or reduced to a minimum.

Exceptions:

1. Special occupancy or special usage conditions approved by the code official.
2. Thermostats that require manual changeover between heating and cooling modes.

503.3.2.3 Heat pump auxiliary heat. Heat pumps having supplementary electric resistance heaters shall have controls that prevent heater operation when the heating load is capable of being met by the heat pump. Supplemental heater operation is not allowed except during outdoor coil defrost cycles not exceeding 15 minutes.

RESIDENTIAL — COMPONENT PERFORMANCE APPROACH

503.3.2.4 Humidistat. Humidistats used for comfort purposes shall be capable of being set to prevent the use of fossil fuel or electricity to reduce relative humidity below 60 percent or increase relative humidity above 30 percent.

503.3.3 Distribution system, construction and insulation. Distribution systems shall be constructed and insulated in accordance with Sections 503.3.3.1 through 503.3.3.7.

503.3.3.1 Piping insulation. All HVAC system piping shall be thermally insulated in accordance with Table 503.3.3.1.

Exceptions:

1. Factory-installed piping within HVAC equipment tested and rated in accordance with Section 503.2.
2. Piping that conveys fluids which have a design operating temperature range between 55°F and 105°F (13°C and 41°C).
3. Piping that conveys fluids which have not been heated or cooled through the use of fossil fuels or electricity.

503.3.3.2 Other insulation thicknesses. Insulation thicknesses in Table 503.3.3.1 are based on insulation having thermal resistivity in the range of 4.0 to 4.6 h · ft² · °F/Btu/inch (0.704 to 0.810 m² · K/W per 25 mm) of thickness on a flat surface at a mean temperature of 75°F (24°C).

Minimum insulation thickness shall be increased for materials having values less than 4.0, or shall be permitted to be reduced for materials having thermal resistivity values greater than 4.6 in accordance with Equation 5-15.

$$\frac{4.6 \times \text{Table 503.3.3.1 Thickness}}{\text{Actual Resistivity}} = \text{New Minumum Thickness}$$

(Equation 5-15)

For materials with thermal resistivity values less than 4.0, the minimum insulation thickness shall be permitted to be increased in accordance with Equation 5-16.

$$\frac{4.0 \times \text{Table 503.3.3.1 Thickness}}{\text{Actual Resistivity}} = \text{New Minumum Thickness}$$

(Equation 5-16)

503.3.3.3 Duct and plenum insulation. All supply and return-air ducts and plenums installed as part of an HVAC air-distribution system shall be thermally insulated in accordance with Table 503.3.3.3, or where such ducts or plenums operate at static pressures greater than 2 inches w.g. (500 Pa), in accordance with Section 503.3.3.4.1.

Exceptions:

1. Factory-installed plenums, casings or ductwork furnished as a part of the HVAC equipment tested and rated in accordance with Section 503.2.
2. Ducts within the conditioned space that they serve.

[M] 503.3.3.4 Duct construction. Ductwork shall be constructed and erected in accordance with the *International Mechanical Code*.

TABLE 503.3.3.1
MINIMUM PIPE INSULATION
(thickness in inches)

PIPING SYSTEM TYPES	FLUID TEMPERATURE RANGE, °F	PIPE SIZES[a]					
		Runouts up to 2"[b]	1" and less	1.25" to 2"	2.5" to 4"	5" to 6"	8" and larger
HEATING SYSTEMS							
Steam and hot water							
High pressure/temperature	306-450	1 1/2	2 1/2	2 1/2	3	3 1/2	3 1/2
Medium pressure/temperature	251-305	1 1/2	2	2 1/2	2 1/2	3	3
Low pressure/temperature	201-250	1	1 1/2	1 1/2	2	2	2
Low temperature	106-200	1/2	1	1	1 1/2	1 1/2	1 1/2
Steam condensate (for feed water)	Any	1	1	1 1/2	2	2	2
COOLING SYSTEMS							
Chilled water, refrigerant and brine	40-55	1/2	1/2	3/4	1	1	1
	Below 40	1	1	1 1/2	1 1/2	1 1/2	1 1/2

For SI: 1 inch = 25.4 mm, 1 foot = 304.8 mm, °C = [(°F)-32]/1.8.
a. For piping exposed to outdoor air, increase insulation thickness by 0.5 inch.
b. Runouts not exceeding 12 feet in length to individual terminal units.

TABLE 503.3.3.3
MINIMUM DUCT INSULATION[a]

ANNUAL HEATING DEGREE DAYS	INSULATION R-VALUE (h · ft² · °F.)/Btu[d]			
	Ducts in unconditioned attics or outside building		Ducts in unconditioned basements, crawl spaces, garages, and other unconditioned spaces[c]	
	Supply	Return	Supply	Return[b]
< 1,500	8	4	4	0
1,500 to 3,500	8	4	6	2
3,501 to 7,500	8	4	8	2
> 7,500	11	6	11	2

For SI: °C= [(°F)-32]/1.8, 1 (h·ft²·°F)/Btu= 0.176 (m²·K)/W, 1 foot= 304.5 mm.

a. Insulation R-values shown are for the insulation as installed and do not include film resistance. The required minimum R-values do not consider water vapor transmission and condensation. Where control of condensation is required, additional insulation, vapor retarders or both shall be provided to limit vapor transmission and condensation. For ducts that are designed to convey both heated and cooled air, duct insulation shall be as required by the most restrictive condition. Where exterior walls are used as plenums, wall insulation shall be as required by the most restrictive condition of this section.
b. Insulation on return ducts in basements is not required.
c. Unconditioned spaces include ventilated crawl spaces and framed cavities in those floors, wall and ceiling assemblies which separate conditioned space from unconditioned space or outside air, and are uninsulated on the side facing away from the condition space.
d. Insulation resistance measured on a horizontal plane in accordance with ASTM C 518, at a mean temperature of 75°F.

503.3.3.4.1 High-and medium-pressure duct systems. All ducts and plenums operating at static pressures greater than 2 inches w.g. (500 Pa) shall be insulated and sealed in accordance with Section 803.2.8. Ducts operating at static pressures in excess of 3 inches w.g. (750 Pa) shall be leak tested in accordance with Section 803.3.6. Pressure classifications specific to the duct system shall be clearly indicated on the construction documents in accordance with the *International Mechanical Code*.

503.3.3.4.2 Low-pressure duct systems. All longitudinal and transverse joints, seams and connections of supply and return ducts operating at static pressures less than or equal to 2 inches w.g. (500 Pa) shall be securely fastened and sealed with welds, gaskets, mastics (adhesives), mastic-plus-embedded-fabric systems or tapes installed in accordance with the manufacturer's installation instructions. Pressure classifications specific to the duct system shall be clearly indicated on the construction documents in accordance with the *International Mechanical Code*.

> **Exception:** Continuously welded and locking-type longitudinal joints and seams on ducts operating at static pressures less than 2 inches w.g. (500 Pa) pressure classification.

503.3.3.4.3 Sealing required. All joints, longitudinal and transverse seams, and connections in ductwork, shall be securely fastened and sealed with welds, gaskets, mastics (adhesives), mastic-plus-embedded-fabric systems or tapes. Tapes and mastics used to seal ductwork shall be listed and labeled in accordance with UL 181A or UL 181B. Duct connections to flanges of air distribution system equipment shall be sealed and mechanically fastened. Unlisted duct tape is not permitted as a sealant on any metal ducts.

503.3.3.5 Mechanical ventilation. Each mechanical ventilation system (supply or exhaust, or both) shall be equipped with a readily accessible switch or other means for shutoff, or volume reduction and shutoff, when ventilation is not required. Automatic or gravity dampers that close when the system is not operating shall be provided for outdoor air intakes and exhausts.

503.3.3.6 Transport energy. The air transport factor for each all-air system shall be not less than 5.5 when calculated in accordance with Equation 5-17. Energy for transfer of air through heat-recovery devices shall not be included in determining the air transport factor.

$$\text{Air Transport Factor} = \frac{\text{Space Sensible Heat Removal}^a}{\text{Supply + Return Fans(s) Power Input}^a}$$

(Equation 5-17)

a. Expressed in consistent units, either Btu/h or Watts.

For purposes of these calculations, space sensible heat removal is equivalent to the maximum coincident design sensible cooling load of all spaces served for which the system provides cooling. Fan power input is the rate of energy delivered to the fan prime mover.

Air and water, all-water and unitary systems employing chilled, hot, dual-temperature or condenser water-transport systems to space terminals shall not require greater transport energy (including central and terminal fan power and pump power) than an equivalent all-air system providing the same space sensible heat removal and having an air transport factor of not less than 5.5.

503.3.3.7 Balancing. The HVAC system design shall provide means for balancing air and water systems. Balancing mechanisms shall include, but not be limited to, dampers, temperature and pressure test connections, and balancing valves.

SECTION 504
SERVICE WATER HEATING

504.1 Scope. The purpose of this section is to provide criteria for design and equipment selection that will produce energy savings when applied to service water heating. Water supplies to ice-making machines and refrigerators shall be taken from a cold-water line of the water distribution system.

504.2 Water heaters, storage tanks and boilers. Water heaters, storage tanks and boilers shall meet the performance criteria set forth in Sections 504.2.1 and 504.2.2.

504.2.1 Performance efficiency. Water heaters and hot water storage tanks shall meet the minimum performance of water-heating equipment specified in Table 504.2.1. Where multiple criteria are listed, all criteria shall be met.

Exception: Storage water heaters and hot water storage tanks having more than 140 gallons (530 L) of storage capacity need not meet the standby loss (SL) or heat loss (HL) requirements of Table 504.2.1 if the tank surface area is thermally insulated to R-12.5 and if a standing pilot light is not used.

TABLE 504.2.1
MINIMUM PERFORMANCE OF WATER-HEATING EQUIPMENT

CATEGORY	TYPE	FUEL	INPUT RATING	V_T[a] (gallons)	INPUT TO V_T RATIO (Btuh/gal)	TEST METHOD	ENERGY FACTOR[b]	THERMAL EFFICIENCY E_t (percent)	STANDBY LOSS (percent/hour)[a]
NAECA-covered water-heating equipment[c]	All	Electric	≤ 12kW	All[e]	—	Note f	≥ 0.93-0.00132V*	—	—
	Storage	Gas	≤ 75,000 Btu/h	All[e]	—	Note f	≥ 0.62-0.0019V*	—	—
	Instantaneous	Gas	≤ 200,000 Btu/h[e]	All	—	Note f	≥ 0.62-0.0019V*	—	—
	Storage	Oil	≤ 105,000 Btu/h	All	—	Note f	≥ 0.59-0.0019V*	—	—
	Instantaneous	Oil	≤ 210,000 Btu/h	All	—	Note f	≥ 0.59-0.0019V*	—	—
	Pool heater	Gas/oil	All	All	—	Note g	—	≥ 78%	—
Other water-heating equipment[d]	Storage	Electric	All	all	—	Note h	—	—	≤ 0.30+27/V_T*
	Storage/ instantaneous	Gas/oil	≤ 155,000 Btu/h > 155,000 Btu/h	All	< 4,000	Note h	—	≥ 78%	≤ 1.3+114/V_T*
				All	< 4,000	Note h	—	≥ 78%	≤ 1.3+95/V_T*
				< 10 ≥ 10	≥ 4,000 ≥ 4,000	Note h	—	≥ 80% ≥ 77%	— ≤ 2.3+67/V_T*
Unfired storage tanks	—	—	—	All	—	—	—	—	≤ 6.5 Btuh/ft^2 [i]*

For SI: 1 British thermal unit per square foot = 3.155 W/m^2, 1 British thermal unit per hour = 0.2931 W, 1 gallon = 3.785 L, °C = [(°F)-32]/1.8.

a. V_T is the storage volume in gallons as measured during the standby loss test. For the purpose of estimating the standby loss requirement using the rated volume shown on the rating plate, V_T should be no less than 0.95V for gas and oil water heaters and no less than 0.90V for electric water heaters.
b. V is rated storage volume in gallons as specified by the manufacturer.
c. Consistent with National Appliance Energy Conservation Act (NAECA) of 1987.
d. All except those water heaters covered by NAECA.
e. DOE CFR 10; Part 430, Subpart B, Appendix E applies to electric and gas storage water heaters with rated volumes 20 gallons and gas instantaneous water heaters with input ratings of 50,000 to 200,000 Btu/h.
f. DOE CFR 10; Part 430, Subpart B, Appendix E.
g. ANSI Z21.56.
h. ANSI Z21.10.3. When testing an electric storage water heater for standby loss using the test procedure of Section 2.9 of ANSI Z21.10.3, the electrical supply voltage shall be maintained within ± 1 percent of the center of the voltage range specified on the water heater nameplate. Also, when needed for calculations, the thermal efficiency (E_t) shall be 98 percent. When testing an oil water heater using the test procedures of Sections 2.8 and 2.9 of ANSI Z21.10.3, the following modifications will be made: A vertical length of the flue pipe shall be connected to the flue gas outlet of sufficient height to establish the minimum draft specified in the manufacturer's installation instructions. All measurements of oil consumption will be taken by instruments with an accuracy of ± 1 percent or better. The burner shall be adjusted to achieve an hourly Btu input rate within ± 2 percent of the manufacturer's specified input rate with the CO_2 reading as specified by the manufacturer with smoke no greater than 1 and the fuel pump pressure within ± 1 percent of the manufacturer's specification.
i. Heat loss of tank surface area (Btu/h · ft^2) based on 80°F water-air temperature difference.
*Minimum efficiencies marked with an asterisk are established by preemptive federal law and are printed for the convenience of the user.

RESIDENTIAL — COMPONENT PERFORMANCE APPROACH

504.2.2 Combination service water-heating/space-heating boilers. Service water-heating equipment shall not be dependent on year-round operation of space-heating boilers; that is, boilers that have as another function winter space heating.

Exceptions:

1. Systems with service/space-heating boilers having a standby loss (Btu/h) (W) less than that calculated in equation 5-18:

$$SL \leq \frac{(13.3 \cdot pmd) + 400}{n}$$

(Equation 5-18)

as determined by the fixture count method where:

pmd = Probable maximum demand in gallons/hour as determined in accordance with the ASHRAE *HVAC Applications Handbook*.

n = Fraction of year when outdoor daily mean temperature exceeds 64.9°F (18°C).

The standby loss is to be determined for a test period of 24-hour duration while maintaining a boiler water temperature of 90°F (32°C) above an ambient of 60 to 90°F (16 to 32°C) and a 5-foot (1524 mm) stack on appliance.

2. For systems where the use of a single heating unit will lead to energy savings, such unit shall be utilized.

504.3 Swimming pools. Swimming pools shall be provided with energy-conserving measures in accordance with Sections 504.3.1 through 504.3.3.

504.3.1 On-off switch. All pool heaters shall be equipped with an ON-OFF switch mounted for easy access to allow shutting off the operation of the heater without adjusting the thermostat setting and to allow restarting without relighting the pilot light.

504.3.2 Pool covers. Heated swimming pools shall be equipped with a pool cover.

Exception: Outdoor pools deriving more than 20 percent of the energy for heating from renewable sources (computed over an operating season) are exempt from this requirement.

504.3.3 Time clocks. Time clocks shall be installed so that the pump can be set to run in the off-peak electric demand period and can be set for the minimum time necessary to maintain the water in a clear and sanitary condition in keeping with applicable health standards.

504.4 Hot water system controls. Automatic-circulating hot water system pumps or heat trace shall be arranged to be conveniently turned off, automatically or manually, when the hot water system is not in operation.

504.5 Pipe insulation. For automatic-circulating hot water systems, piping heat loss shall be limited to a maximum of 17.5 Btu/h per linear foot (16.8 W/m) of pipe in accordance with Table 504.5, which is based on design external temperature no lower than 65°F (18°C). For external design temperatures lower than 65°F (18°C) insulation thickness must be calculated in accordance with Section 503.3.3.2.

Exception: Piping insulation is not required when the heat loss of the piping, without insulation, does not increase the annual energy requirements of the building.

TABLE 504.5
MINIMUM PIPE INSULATION
(thickness in inches)

SERVICE WATER-HEATING TEMPERATURES (°F)	PIPE SIZES[a]			
	Noncirculating runouts	Circulating mains and runouts		
	Up to 1"	Up to 1.25"	1.5" to 2"	Over 2"
170 -180	1/2	1	1 1/2	2
140 -169	1/2	1/2	1	1 1/2
100 -139	1/2	1/2	1/2	1

For SI: 1 inch = 25.4 mm, °C = [(°F)-32]/1.8,
1 Btuh/inch · ft² · °F = 0.144 W/(m · K).

a. Nominal iron pipe size and insulation thickness. Conductivity, $k = 0.27$.

504.6 Conservation of hot water. Hot water shall be conserved in accordance with Section 504.6.1.

504.6.1 Showers. Shower heads shall have a maximum flow rate of 2.5 gallons per minute (gpm) (0.158 L/s) at a pressure of 80 pounds per square inch (psi) (551 kPa) when tested in accordance with ASME A112.18.1.

504.7 Heat traps. Water heaters with vertical pipe risers shall have a heat trap on both the inlet and outlet of the water heater unless the water heater has an integral heat trap or is part of a circulating system.

SECTION 505
ELECTRICAL POWER AND LIGHTING

505.1 Electrical energy consumption. In residential buildings having individual dwelling units, provisions shall be made to determine the electrical energy consumed by each tenant by separately metering individual dwelling units.

505.2 Lighting power budget. The lighting system shall meet the applicable provisions of Section 805.

Exception: Detached one- and two- family dwellings and townhouses and the dwelling portion of Group R-2 and R-4 residential buildings.

CHAPTER 6

SIMPLIFIED PRESCRIPTIVE REQUIREMENTS FOR DETACHED ONE- AND TWO-FAMILY DWELLINGS AND GROUP R-2, R-4 OR TOWNHOUSE RESIDENTIAL BUILDINGS

SECTION 601
GENERAL

601.1 Scope. This chapter sets forth energy-efficiency-related requirements for the design and construction of detached one- and two-family dwellings and Group R-2, R-4 or townhouse residential buildings.

Exception: Portions of the building envelope that do not enclose conditioned space.

601.2 Compliance. Compliance shall be demonstrated in accordance with Section 601.2.1 or 601.2.2.

601.2.1 Residential buildings, detached one- and two-family dwellings. Compliance for detached one- and two-family dwellings shall be demonstrated by either:

1. Meeting the requirements of this chapter for buildings with a glazing area that does not exceed 15 percent of the gross area of exterior walls; or

2. Meeting the requirements of Chapter 4, or Chapter 5 for detached one- and two-family dwellings.

601.2.2 Residential buildings, Groups R-2, R-4 or townhouses. Compliance for Group R-2, R-4 or townhouse residential buildings shall be demonstrated by either:

1. Meeting the requirements of this chapter for buildings with a glazing area that does not exceed 25 percent of the gross area of exterior walls; or

2. Meeting the requirements of Chapter 4, or Chapter 5 for Group R-2, R-4 or townhouse residential buildings.

601.3 Materials and equipment. Materials and equipment shall be identified in a manner that will allow a determination of their compliance with the applicable provisions of this chapter. Materials and equipment used to conform to the applicable provisions of this chapter shall be installed in accordance with the manufacturer's installation instructions.

601.3.1 Insulation. The thermal resistance (R-value) shall be indicated on all insulation and the insulation installed such that the R-value can be verified during inspection, or a certification of the installed R-value shall be provided at the job site by the insulation installer. Where blown-in or sprayed insulation is applied in walls, the installer shall provide a certification of the installed density and R-value. Where blown-in or sprayed insulation is applied in the roof/ceiling assembly, the installer shall provide a certification of the initial installed thickness, settled thickness, coverage area, and number of bags of insulating material installed. Markers shall be provided for every 300 square feet (28 m^2) of area, attached to the trusses, rafters or joists, and indicate in 1-inch-high (25 mm) numbers the installed thickness of the insulation.

601.3.2 Fenestration. The U-factor of fenestration shall be determined in accordance with NFRC 100 by an accredited, independent laboratory, and labeled and certified by the manufacturer. The solar heat gain coefficient (SHGC) of fenestration shall be determined in accordance with NFRC 200 by an accredited, independent laboratory, and labeled and certified by the manufacturer.

601.3.2.1 Default fenestration performance. Where a manufacturer has not determined a fenestration product's U-factor in accordance with NFRC 100, compliance shall be determined by assigning such products a default U-factor from Tables 102.5.2(1) and 102.5.2(2). When a manufacturer has not determined a fenestration product's SHGC in accordance with NFRC 200, compliance shall be determined by assigning such products a default SHGC from Table 102.5.2(3).

601.3.2.2 Air leakage. The air leakage of prefabricated fenestration shall be determined in accordance with AAMA/WDMA 101/I.S.2 or NFRC 400 by an accredited, independent laboratory, and labeled and certified by the manufacturer and shall not exceed the values in Table 502.1.4.1. Alternatively, the manufacturer shall certify that the fenestration is installed in accordance with Section 502.1.4.

601.3.3 Maintenance. Where mechanical or plumbing system components require preventive maintenance for efficient operation, regular maintenance requirements shall be clearly stated and affixed to the component, or the source for such information shall be shown on a label attached to the component.

SECTION 602
BUILDING ENVELOPE

602.1 Thermal performance criteria. The minimum required insulation R-value or the area-weighted average maximum required fenestration U-factor (other than opaque doors which are governed by Section 602.1.3) for each element in the building thermal envelope (fenestration, roof/ceiling, opaque wall, floor, slab edge, crawl space wall and basement wall) shall be in accordance with the criteria in Table 602.1.

The building envelope requirements of Chapter 4 or 5 shall be used to determine compliance with detached one- and two-family dwellings with greater than 15-percent glazing area; Group R-2, R-4 or townhouse residential buildings with greater than 25-percent glazing area; and any residential building in climates with heating degree days (HDD) equal to or greater than 13,000.

SIMPLIFIED PRESCRIPTIVE REQUIREMENTS

TABLE 602.1
SIMPLIFIED PRESCRIPTIVE BUILDING ENVELOPE THERMAL COMPONENT CRITERIA
MINIMUM REQUIRED THERMAL PERFORMANCE (*U*-FACTOR AND *R*-VALUE)

CLIMATE ZONE	HEATING DEGREE DAYS	MAXIMUM Glazing *U*-factor	MINIMUM					
			Ceiling *R*-value	Wall *R*-value	Floor *R*-value	Basement wall *R*-value	Slab perimeter *R*-value and depth	Crawl space wall *R*-value
1	0 - 499	Any	R-13	R-11	R-11	R-0	R-0	R-0
2	500 - 999	0.90	R-19	R-11	R-11	R-0	R-0	R-4
3	1,000 - 1,499	0.75	R-19	R-11	R-11	R-0	R-0	R-5
4	1,500 - 1,999	0.75	R-26	R-13	R-11	R-5	R-0	R-5
5	2,000 - 2,499	0.65	R-30	R-13	R-11	R-5	R-0	R-6
6	2,500 - 2,999	0.60	R-30	R-13	R-19	R-6	R-4, 2 ft.	R-7
7	3,000 - 3,499	0.55	R-30	R-13	R-19	R-7	R-4, 2 ft.	R-8
8	3,500 - 3,999	0.50	R-30	R-13	R-19	R-8	R-5, 2 ft.	R-10
9	4,000 - 4,499	0.45	R-38	R-13	R-19	R-8	R-5, 2 ft.	R-11
10	4,500 - 4,999	0.45	R-38	R-16	R-19	R-9	R-6, 2 ft.	R-17
11	5,000 - 5,499	0.45	R-38	R-18	R-19	R-9	R-6, 2 ft.	R-17
12	5,500 - 5,999	0.40	R-38	R-18	R-21	R-10	R-9, 4 ft.	R-19
13	6,000 - 6,499	0.35	R-38	R-18	R-21	R-10	R-9, 4 ft.	R-20
14	6,500 - 6,999	0.35	R-49	R-21	R-21	R-11	R-11, 4 ft.	R-20
15	7,000 - 8,499	0.35	R-49	R-21	R-21	R-11	R-13, 4 ft.	R-20
16	8,500 - 8,999	0.35	R-49	R-21	R-21	R-18	R-14, 4 ft.	R-20
17	9,000 - 12,999	0.35	R-49	R-21	R-21	R-19	R-18, 4 ft.	R-20

For SI: 1 foot = 304.8 mm.

602.1.1 Exterior walls. The sum of the *R*-values of the insulation materials installed in framing cavities and insulating sheathing (where used) shall meet or exceed the minimum required "Wall *R*-value" in Table 602.1. Framing, drywall, structural sheathing or exterior siding materials shall not be considered as contributing, in any way, to the thermal performance of exterior walls. Insulation separated from the conditioned space by a vented space shall not be counted towards the required *R*-value.

602.1.1.1 Mass walls. Mass walls shall be permitted to meet the criteria in Table 602.1.1.1(1) based on the insulation position and the climate zone where the building is located. Other mass walls shall meet the frame wall criteria for the building type and the climate zone where the building is located, based on the sum of interior and exterior insulation. Walls with "exterior insulation" position have the entire effective mass layer interior to an insulation layer. Walls with "integral insulation" position have either insulation and mass materials well mixed as in wood (logs); or substantially equal amounts of mass material on the interior and exterior of insulation as in concrete masonry units with insulated cores or masonry cavity walls. Walls with interior insulation position have the mass material located exterior to the insulating material(s). Walls not meeting the above descriptions for exterior or integral positions shall meet the requirements for "other mass walls" in Table 602.1.1.1(1). The *R*-value of the mass assembly for typical masonry construction shall be taken from Table 602.1.1.1(2). The mass assembly *R*-value for a solid concrete wall with a thickness of 4 inches (102 mm) or greater is R-1.1. *R*-values for other assemblies are permitted to be based on the hot box tests referenced in ASTM C 236 or ASTM C 976, two-dimensional calculations or isothermal plane calculations.

602.1.1.2 Steel-frame walls. The minimum required *R*-values for steel-frame walls shall be in accordance with Table 602.1.1.2.

TABLE 602.1.1.2
STEEL-FRAME WALL MINIMUM
PERFORMANCE REQUIREMENTS (*R-VALUE*)

HDD	EQUIVALENT STEEL-FRAME WALL CAVITY AND SHEATHING R-VALUE[a]
0 - 1,999	R-11 + R-5, R-15 + R-4, R-21 + R-3
2,000 - 3,999	R-11 + R-5, R-15 + R-4, R-21 + R-3
4,000 - 5,999	R-11 + R-9, R-15 + R-8, R-21 + R-7
6,000 - 8,499	R-13 + R-10, R-19 + R-9, R-25 + R-8
8,500 - 12,999	R-13 + R-10, R-19 + R-9, R-25 + R-8

a. The cavity insulation *R*-value requirement is listed first, followed by the sheathing *R*-value requirement.

SIMPLIFIED PRESCRIPTIVE REQUIREMENTS

TABLE 602.1.1.1(1)
MASS WALL PRESCRIPTIVE BUILDING ENVELOPE REQUIREMENTS

MASS WALL ASSEMBLY R-VALUE[a]			
Building Location		Exterior or Integral Insulation	Other Mass Walls
Zone	Heating Degree Days	Residential Buildings	Residential Buildings
1	0 - 499	R-3.8	R-9.7
2	500 - 999	R-4.8	R-9.7
3	1,000 - 1,499	R-4.8	R-9.7
4	1,500 - 1,999	R-8.1	R-10.8
5	2,000 - 2,499	R-8.9	R-10.8
6	2,500 - 2,999	R-8.9	R-10.8
7	3,000 - 3,499	R-8.9	R-10.8
8	3,500 - 3,999	R-8.9	R-10.8
9	4,000 - 4,499	R-8.9	R-10.9
10	4,500 - 4,999	R-10.4	R-12.3
11	5,000 - 5,499	R-11.9	R-15.2
12	5,500 - 5,999	R-11.9	R-15.2
13	6,000 - 6,499	R-11.9	R-15.2
14	6,500 - 6,999	R-15.5	R-18.4
15	9,000 - 8,499	R-15.5	R-18.4
16	8,500 - 8,999	R-18.4	R-18.4
17	9,000 - 12,999	R-18.4	R-18.4

a. The sum of the value in Table 602.1.1.1(2) and additional insulation layers.

TABLE 602.1.1.1(2)
MASS ASSEMBLY R-VALUES

ASSEMBLY TYPE	UNGROUTED CELLS, NOT INSULATED	UNGROUTED CELLS INSULATED		
		No grout	Vertical cells grouted at 10' o.c. or greater	Vertical cells grouted at less than 10' o.c.
6" Lightweight concrete block	2.3	5.0	4.5	3.8
6" Medium-weight concrete block	2.1	4.2	3.8	3.2
6" Normal-weight concrete block	1.9	3.3	3.1	2.7
8" Lightweight concrete block	2.6	6.7	5.9	4.8
8" Medium-weight concrete block	2.3	5.3	4.8	4.0
8" Normal-weight concrete block	2.1	4.2	3.8	3.3
12" Lightweight concrete block	2.9	9.1	7.9	6.3
12" Medium-weight concrete block	2.6	7.1	6.4	5.2
12" Normal-weight concrete block	2.3	5.6	5.1	4.3
Brick cavity wall	3.7	6.7	6.2	5.4
Hollow clay brick	2.0	2.7	2.6	2.4

For SI: 1 inch = 25.4 mm, 1 foot = 304.8 mm.

SIMPLIFIED PRESCRIPTIVE REQUIREMENTS

602.1.2 Ceilings. The required "Ceiling R-value" in Table 602.1 assumes standard truss or rafter construction, and shall apply to all roof/ceiling portions of the building thermal envelope, including cathedral ceilings. Where the construction technique allows the required R-value of ceiling insulation to be obtained over the exterior wall top plate, R-30 shall be permitted to be used where R-38 is required in the table, and R-38 shall be permitted to be used where R-49 is required.

602.1.2.1 Steel-framed ceiling. The maximum required U_R-factor for cold-formed steel truss roof/ceiling assemblies shall be in accordance with Table 602.1.2.1(1) and compliance shall be determined by using the U_R-factors in Table 602.1.2.1(2). This table applies to cold-formed steel truss roof framing spaced at 24 inches (609 mm) on center and where the penetrations of the truss members through the cavity insulation do not exceed three penetrations for each 4-foot (1220 mm) length of the truss. The maximum required U_R-factor for C-shaped cold-formed steel roof/ceiling assemblies shall be in accordance with Table 602.1.2.1(3) and compliance shall be determined by using the U_R-factors in Table 602.1.2.1(4).

TABLE 602.1.2.1(2)
COLD-FORMED STEEL ROOF/CEILING TRUSS U_R-FACTORS

CAVITY INSULATION R-VALUE	CONTINUOUS INSULATION BETWEEN DRYWALL AND BOTTOM CHORD		
	R-0	R-3	R-5
R-13	0.0865	0.0616	0.0546
R-19	0.0597	0.0467	0.0426
R-26	0.0439	0.0364	0.0338
R-30	0.0382	0.0324	0.0303
R-38	0.0302	0.0265	0.0251
R-49	0.0235	0.0212	0.0203

TABLE 602.1.2.1(1)
MAXIMUM COLD-FORMED STEEL ROOF/CEILING TRUSS U_R-FACTORS

HEATING DEGREE DAYS	U_R-FACTOR
0 - 499	0.0742
500 - 1,499	0.0504
1,500 - 1,999	0.0372
2,000 - 3,999	0.0323
4,000 - 6,499	0.0257
6,500 - 12,999	0.0200

TABLE 602.1.2.1(3)
MAXIMUM C-SHAPED, COLD-FORMED STEEL ROOF/CEILING U_R-FACTORS

HEATING DEGREE DAYS	U_R-FACTOR	
	16 inches o.c.	24 inches o.c.
0 - 499	0.0773	0.0742
500 - 1,499	0.0537	0.0519
1,500 - 1,999	0.0405	0.0390
2,000 - 3,999	0.0355	0.0342
4,000 - 6,499	0.0285	0.0274
6,500 - 12,999	0.0223	0.0215

For SI: 1 inch = 25.4 mm.

TABLE 602.1.2.1(4)
C-SHAPED, COLD-FORMED STEEL ROOF/CEILING U_R-FACTORS[a]

FRAMING[b]	SPACING	R-13	R-19	R-26	R-30	R-38	R-49
2 × 4	16 inches o.c.	0.1328	0.0530	0.0387	0.0336	0.0265	0.0206
2 × 6		0.1328	0.0667	0.0456	0.0386	0.0295	0.0223
2 × 8		0.1328	0.1208	0.0585	0.0475	0.0345	0.0251
2 × 10		0.1328	0.1208	0.1094	0.1037	0.0398	0.0277
2 × 12		0.1328	0.1208	0.1094	0.1037	0.0471	0.0311
2 × 4	24 inches o.c.	0.1129	0.0510	0.0376	0.0327	0.0260	0.0202
2 × 6		0.1129	0.0610	0.0428	0.0366	0.0284	0.0216
2 × 8		0.1129	0.0994	0.0517	0.0429	0.0320	0.0237
2 × 10		0.1129	0.0994	0.0873	0.0816	0.0357	0.0257
2 × 12		0.1129	0.0994	0.0873	0.0816	0.0403	0.0280

For SI: 1 inch = 25.4 mm.
a. Linear interpolation is permitted for determining U-factors which are between those given in the table.
b. Applies to steel framing up to a maximum thickness of 0.064 inches (16 gage.)

602.1.3 Opaque doors. Opaque doors in the building envelope shall have a maximum U-factor of 0.35. One opaque door shall be exempt from this U-factor requirement.

602.1.4 Floor. The required R-value in Table 602.1 shall apply to all floors.

Exception: Any individual floor assembly with more than 25 percent of its conditioned floor area exposed directly to outside air shall meet the R-value requirement in Table 602.1 for "Ceiling R-value."

602.1.4.1 Steel-framed floors. The maximum required U_f-factor for C-shaped, cold-formed, steel-framed floors shall be in accordance with Table 602.1.4.1(1) and compliance shall be determined by using the U_f-factors in Table 602.1.4.1(2).

TABLE 602.1.4.1(1)
MAXIMUM C-SHAPED, COLD-FORMED STEEL FLOOR
U_f-FACTORS

HEATING DEGREE DAYS	U_f-FACTOR	
	16 inches o.c.	24 inches o.c.
0 - 2,499	0.0725	0.0708
2,500 - 5,499	0.0477	0.0464
5,500 - 12,999	0.0452	0.0436

For SI: 1 inch = 25.4 mm.

602.1.5 Basement walls. Where the basement is considered a conditioned space, the basement walls shall be insulated in accordance with Table 602.1. Where the basement is not considered a conditioned space, either the basement wall or the ceiling(s) separating the basement from conditioned space shall be insulated in accordance with Table 602.1. Where basement walls are required to be insulated, the required R-value shall be applied from the top of the basement wall to a depth of 10 feet (3048 mm) below grade or to the top of the basement floor, whichever is less.

602.1.6 Slab-on-grade floors. For slabs with a top edge 12 inches (305 mm) or less below finished grade, the required "Slab perimeter R-value and depth" in Table 602.1 shall be applied to the outside of the foundation or the inside of the foundation wall. The insulation shall extend downward from the top of the slab or downward from the top of the slab to the bottom of the slab and then horizontally to the interior or exterior, until the distance listed in Table 602.1 is reached.

Where installed between the exterior wall and the edge of the interior slab, the top edge of the insulation shall be permitted to be cut at a 45-degree (0.79 rad) angle away from the exterior wall. Insulation extending horizontally outside of the foundation shall be protected by pavement or by a minimum of 10 inches (254 mm) of soil.

In locations of 500 HDD or greater, R-2 shall be added to the "Slab perimeter R-value" in Table 602.1 where uninsulated hot water pipes, air distribution ducts or electric heating cables are installed within or under the slab.

Exception: Slab perimeter insulation is not required for unheated slabs in areas of very heavy termite infestation probability as shown in Figure 502.2(7). Where this exception is used, building envelope compliance shall be demonstrated by using Section 502.2.2 or Chapter 4 with the actual "Slab perimeter R-value and depth" in Table 602.1, or by using Section 502.2.4.

602.1.7 Crawl space walls. Where the floor above the crawl space is uninsulated, insulation shall be installed on crawl space walls when the crawl space is not vented to outside air. The required "Crawl space wall R-value" in Table 602.1 shall be applied inside of the crawl space wall, downward from the sill plate to the exterior finished grade level and then vertically or horizontally or both for 24 inches (610 mm). The exposed earth in all crawl space foundations shall be covered with a continuous vapor retarder having a maximum permeance rating of 1.0 perm ($5.7 \cdot 10^{-11}$ kg/Pa $\cdot$ s $\cdot$ m^2), when tested in accordance with ASTM E 96.

602.1.8 Masonry veneer. For exterior foundation insulation, the horizontal portion of the foundation which supports a masonry veneer is not required to be insulated.

602.1.9 Protection. Exposed insulating materials applied to the exterior of foundation walls shall have a rigid, opaque and weather-resistant protective covering. The protective

TABLE 602.1.4.1(2)
C-SHAPED, COLD-FORMED STEEL FLOOR U_f-FACTORS[a]

FRAMING[b]	SPACING	R-11	R-13	R-15	R-19	R-21	R-25	R-30
2 × 6	16 inches o.c.	0.1058	0.1031	0.1005	0.0583	0.0523	NA	NA
2 × 8		0.1058	0.1031	0.1005	0.0957	0.0935	0.0548	NA
2 × 10		0.1058	0.1031	0.1005	0.0957	0.0935	0.0894	0.0838
2 × 12		0.1058	0.1031	0.1005	0.0957	0.0935	0.0894	0.0838
2 × 6	24 inches o.c.	0.0941	0.0907	0.0875	0.0538	0.0486	NA	NA
2 × 8		0.0941	0.0907	0.0875	0.0818	0.0792	0.0488	NA
2 × 10		0.0941	0.0907	0.0875	0.0818	0.0792	0.0745	0.0697
2 × 12		0.0941	0.0907	0.0875	0.0818	0.0792	0.0745	0.0697

For SI: 1 inch = 25.4 mm.
NA = Not applicable
a. Linear interpolation is permitted for determining U-factors which are between those given in the table.
b. Applies to steel framing up to a maximum thickness of 0.064 inches (16 gage).

covering shall extend 6 inches (152 mm) below finished grade level.

602.1.10 Caulking, sealants and gasketing. All joints, seams, penetrations (site-built windows, doors and skylights), openings between window and door assemblies and their respective jambs and framing, and other sources of air leakage (infiltration and exfiltration) through the building envelope shall be caulked, gasketed, weatherstripped, wrapped or otherwise sealed to limit uncontrolled air movement.

602.1.11 Moisture control. Provisions for moisture control shall be in accordance with Section 502.1.1.

602.1.12 Recessed lighting fixtures. Where provided, recessed lighting fixtures shall be installed in accordance with Section 502.1.3.

602.2 Maximum solar heat gain coefficient for fenestration products. In locations with heating degree days (HDD) less than 3,500, the area-weighted-average solar heat gain coefficient (SHGC) for glazed fenestration installed in the building envelope shall not exceed 0.40.

602.3 Fenestration exemption. Up to 1 percent of the total glazing area shall be exempt from the "Glazing U-factor" requirement in Table 602.1.

602.4 Replacement fenestration. Where some or all of an existing fenestration unit is replaced with an entirely new replacement fenestration product, including frame, sash and glazed portion, the replacement fenestration product shall have a U-factor that does not exceed the "Fenestration U-factor" requirement in Table 502.2.5 applicable to the climate zone (HDD) where the building is located. The replacement fenestration product(s) must also satisfy the air leakage requirements and SHGC of Sections 601.3.2.2 and 602.2, respectively.

Exception: Replacement skylights shall have a maximum U-factor of 0.60 when installed in any location above 1,999 HDD.

SECTION 603
MECHANICAL SYSTEMS

603.1 Heating and air-conditioning equipment and appliances. Heating and air-conditioning equipment and appliances shall comply with the applicable requirements of Section 503.

SECTION 604
SERVICE WATER HEATING

604.1 Water-heating equipment and appliances. Water-heating equipment and appliances shall comply with the applicable requirements of Section 504.

SECTION 605
ELECTRICAL POWER AND LIGHTING

605.1 Electrical energy consumption. In residential buildings having individual dwelling units, provisions shall be made to determine the electrical energy consumed by each tenant by separately metering individual dwelling units.

CHAPTER 7
BUILDING DESIGN FOR ALL COMMERCIAL BUILDINGS

SECTION 701
GENERAL

701.1 Scope. Commercial buildings shall meet the requirements of ASHRAE/IESNA 90.1.

Exception: Commercial buildings that comply with Chapter 8.

CHAPTER 8

DESIGN BY ACCEPTABLE PRACTICE FOR COMMERCIAL BUILDINGS

SECTION 801
GENERAL

801.1 Scope. The requirements contained in this chapter are applicable to commercial buildings, or portions of commercial buildings. Buildings constructed in accordance with this chapter are deemed to comply with this code.

801.2 Application. The requirements in Sections 802, 803, 804 and 805 shall each be satisfied on an individual basis. Where one or more of these sections is not satisfied, compliance for that section(s) shall be demonstrated in accordance with the applicable provisions of ASHRAE/IESNA 90.1.

> **Exception:** Buildings conforming to Section 806, provided Sections 802.1.2, 802.3, 803.2.1 or 803.3.1 as applicable, 803.2.2 or 803.3.2 as applicable, 803.2.3 or 803.3.3 as applicable, 803.2.8 or 803.3.6 as applicable, 803.2.9 or 803.3.7 as applicable, 804, 805.2, 805.3, 805.4, 805.6 and 805.7 are each satisfied.

SECTION 802
BUILDING ENVELOPE REQUIREMENTS

802.1 General. Walls, roof assemblies, floors, glazing and slabs on grade which are part of the building envelope for buildings where the window and glazed door area is not greater than 50 percent of the gross area of above-grade walls shall meet the requirements of Sections 802.2.1 through 802.2.9, as applicable. Buildings with more glazing shall meet the applicable provisions of ASHRAE/IESNA 90.1.

802.1.1 Classification of walls. Walls associated with the building envelope shall be classified in accordance with Section 802.1.1.1, 802.1.1.2 or 802.1.1.3.

802.1.1.1 Above-grade walls. Above-grade walls are those walls covered by Section 802.2.1 on the exterior of the building and completely above grade or the above-grade portion of a basement or first-story wall that is more than 15 percent above grade.

802.1.1.2 Below-grade walls. Below-grade walls covered by Section 802.2.8 are basement or first-story walls associated with the exterior of the building that are at least 85 percent below grade.

802.1.1.3 Interior walls. Interior walls covered by Section 802.2.9 are those walls not on the exterior of the building and that separate conditioned and unconditioned space.

802.1.2 Moisture control. All framed walls, floors and ceilings not ventilated to allow moisture to escape shall be provided with an approved vapor retarder having a permeance rating of 1 perm (5.7×10^{-11} kg/Pa · s · m^2) or less, when tested in accordance with the dessicant method using Procedure A of ASTM E 96. The vapor retarder shall be installed on the warm-in-winter side of the insulation.

> **Exceptions:**
> 1. Buildings located in Climate Zones 1 through 7 as indicated in Table 302.1.
> 2. In construction where moisture or its freezing will not damage the materials.
> 3. Where other approved means to avoid condensation in unventilated framed wall, floor, roof and ceiling cavities are provided.

802.2 Criteria. The building envelope components shall meet each of the applicable requirements in Tables 802.2(1), 802.2(2), 802.2(3) and 802.2(4) based on the percentage of wall that is glazed. The percentage of wall that is glazed shall be determined by dividing the aggregate area of rough openings for glazing (windows and glazed doors) in all above-grade walls associated with the building envelope by the total gross area of all above-grade exterior walls that are a part of the building envelope. In buildings with multiple types of building envelope construction, each building envelope construction type shall be evaluated separately. Where Table 802.2(1), 802.2(2), 802.2(3) or 802.2(4) does not list a particular construction type, the applicable provisions of ASHRAE/IESNA 90.1 shall be used in lieu of Section 802.

802.2.1 Above-grade walls. The minimum thermal resistance (R-value) of the insulating material(s) installed in the wall cavity between the framing members and continuously on the walls shall be as specified in Table 802.2(1), 802.2(2), 802.2(3) or 802.2(4), based on framing type and construction materials used in the wall assembly. Where both cavity and continuous insulation values are provided in Table 802.2(1), 802.2(2), 802.2(3) or 802.2(4), both requirements shall be met. Concrete masonry units (CMU) at least 8 inches (203 mm) nominal in thickness with essentially equal amounts of mass on either side of the insulation layer are considered as having integral insulation; however, the thermal resistance of that insulation shall not be considered when determining compliance with Table 802.2(1), 802.2(2), 802.2(3) or 802.2(4). "Other masonry walls" shall include walls weighing at least 35 pounds per square foot (170 kg/m^2) of wall surface area and do not include CMUs less than 8 inches (203 mm) nominal in thickness.

802.2.2 Nonglazed doors. Nonglazed doors shall meet the applicable requirements for windows and glazed doors and be considered as part of the gross area of above-grade walls that are part of the building envelope.

ACCEPTABLE PRACTICE FOR COMMERCIAL BUILDINGS

TABLE 802.2(1)
BUILDING ENVELOPE REQUIREMENTS[a through e]
WINDOW AND GLAZED DOOR AREA 10 PERCENT OR LESS OF ABOVE-GRADE WALL AREA

ELEMENT	CONDITION/VALUE		
Skylights (U-factor)			
Slab or below-grade wall (R-value)			
Windows and glass doors	SHGC		U-factor
PF < 0.25			
0.25 ≤ PF < 0.50			
PF ≥ 0.50			
Roof assemblies (R-value)	Insulation between framing		Continuous insulation
All-wood joist/truss			
Metal joist/truss			
Concrete slab or deck			
Metal purlin with thermal block			
Metal purlin without thermal block			
Floors over outdoor air or unconditioned space (R-value)	Insulation between framing		Continuous insulation
All-wood joist/truss			
Metal joist/truss			
Concrete slab or deck			
Above-grade walls (R-value)	No framing	Metal framing	Wood framing
Framed R-value cavity	NA		
R-value continuous	NA		
CMU, ≥ 8 in, with integral insulation R-value cavity	NA		
R-value continuous			
Other masonry walls R-value cavity	NA		
R-value continuous			

For SI: 1 inch = 25.4 mm.

a. Values shall be determined from Tables 802.2(5) through 802.2(37) using the climate zone(s) specified in Table 302.1. (Note: The tables begin on page 78.)
b. "NA" indicates the condition is not applicable.
c. An R-value of zero indicates no insulation is required.
d. "Any" indicates any available product will comply.
e. "X" indicates no complying option exists for this condition.

ACCEPTABLE PRACTICE FOR COMMERCIAL BUILDING

TABLE 802.2(2)
BUILDING ENVELOPE REQUIREMENTS[a through e]
WINDOW AND GLAZED DOOR AREA GREATER THAN 10 PERCENT
BUT NOT GREATER THAN 25 PERCENT OF ABOVE-GRADE WALL AREA

ELEMENT	CONDITION/VALUE		
Skylights (U-factor)			
Slab or below-grade wall (R-value)			
Windows and glass doors	SHGC		U-factor
PF < 0.25			
0.25 ≤ PF < 0.50			
PF ≥ 0.50			
Roof assemblies (R-value)	Insulation between framing		Continuous insulation
All-wood joist/truss			
Metal joist/truss			
Concrete slab or deck			
Metal purlin with thermal block			
Metal purlin without thermal block			
Floors over outdoor air or unconditioned space (R-value)	Insulation between framing		Continuous insulation
All-wood joist/truss			
Metal joist/truss			
Concrete slab or deck			
Above-grade walls (R-value)	No framing	Metal framing	Wood framing
Framed R-value cavity	NA		
R-value continuous	NA		
CMU, ≥ 8 in, with integral insulation R-value cavity	NA		
R-value continuous			
Other masonry walls R-value cavity	NA		
R-value continuous			

For SI: 1 inch = 25.4 mm.

a. Values shall be determined from Tables 802.2(5) through 802.2(37) using the climate zone(s) specified in Table 302.1. (Note: The tables begin on page 78.)
b. "NA" indicates the condition is not applicable.
c. An R-value of zero indicates no insulation is required.
d. "Any" indicates any available product will comply.
e. "X" indicates no complying option exists for this condition.

ACCEPTABLE PRACTICE FOR COMMERCIAL BUILDINGS

TABLE 802.2(3)
BUILDING ENVELOPE REQUIREMENTS[a through e]
WINDOW AND GLAZED DOOR AREA GREATER THAN 25 PERCENT
BUT NOT GREATER THAN 40 PERCENT OF ABOVE-GRADE WALL AREA

ELEMENT	CONDITION/VALUE			
Skylights (U-factor)				
Slab or below-grade wall (R-value)				
Windows and glass doors	SHGC		U-factor	
PF < 0.25				
0.25 ≤ PF < 0.50				
PF ≥ 0.50				
Roof assemblies (R-value)	Insulation between framing		Continuous insulation	
All-wood joist/truss				
Metal joist/truss				
Concrete slab or deck				
Metal purlin with thermal block				
Metal purlin without thermal block				
Floors over outdoor air or unconditioned space (R-value)	Insulation between framing		Continuous insulation	
All-wood joist/truss				
Metal joist/truss				
Concrete slab or deck				
Above-grade walls (R-value)	No framing	Metal framing		Wood framing
Framed R-value cavity	NA			
R-value continuous	NA			
CMU, ≥ 8 in, with integral insulation R-value cavity	NA			
R-value continuous				
Other masonry walls R-value cavity	NA			
R-value continuous				

For SI: 1 inch = 25.4 mm.

a. Values shall be determined from Tables 802.2(5) through 802.2(37) using the climate zone(s) specified in Table 302.1. (Note: The tables begin on page 78.)
b. "NA" indicates the condition is not applicable.
c. An R-value of zero indicates no insulation is required.
d. "Any" indicates any available product will comply.
e. "X" indicates no complying option exists for this condition.

**TABLE 802.2(4)
BUILDING ENVELOPE REQUIREMENTS[a through e]
WINDOW AND GLAZED DOOR AREA GREATER THAN 40 PERCENT
BUT NOT GREATER THAN THAN 50 PERCENT OF ABOVE-GRADE WALL AREA**

ELEMENT	CONDITION/VALUE		
Skylights (U-factor)			
Slab or below-grade wall (R-value)			
Windows and glass doors	SHGC		U-factor
PF < 0.25			
0.25 ≤ PF < 0.50			
PF ≥ 0.50			
Roof assemblies (R-value)	Insulation between framing		Continuous insulation
All-wood joist/truss			
Metal joist/truss			
Concrete slab or deck			
Metal purlin with thermal block			
Metal purlin without thermal block			
Floors over outdoor air or unconditioned space (R-value)	Insulation between framing		Continuous insulation
All-wood joist/truss			
Metal joist/truss			
Concrete slab or deck			
Above-grade walls (R-value)	No framing	Metal framing	Wood framing
Framed R-value cavity	NA		
R-value continuous	NA		
CMU, ≥ 8 in, with integral insulation R-value cavity	NA		
R-value continuous			
Other masonry walls R-value cavity	NA		
R-value continuous			

For SI: 1 inch = 25.4 mm.

a. Values shall be determined from Tables 802.2(5) through 802.2(37) using the climate zone(s) specified in Table 302.1. (Note: The tables begin on page 78.)
b. "NA" indicates the condition is not applicable.
c. An R-value of zero indicates no insulation is required.
d. "Any" indicates any available product will comply.
e. "X" indicates no complying option exists for this condition.

802.2.3 Windows and glass doors. The maximum solar heat gain coefficient (SHGC) and thermal transmittance (*U*-factor) of window assemblies and glass doors located in the building envelope shall be as specified in Table 802.2(1), 802.2(2), 802.2(3) or 802.2(4), based on the window projection factor.

The window projection factor shall be determined in accordance with Equation 8-1.

$PF = A/B$ **(Equation 8-1)**

where:

PF = Projection factor (decimal).

A = Distance measured horizontally from the furthest continuous extremity of any overhang, eave, or permanently attached shading device to the vertical surface of the glazing.

B = Distance measured vertically from the bottom of the glazing to the underside of the overhang, eave, or permanently attached shading device.

Where different windows or glass doors have different *PF* values, they shall each be evaluated separately, or an area-weighted *PF* value shall be calculated and used for all windows and glass doors.

802.2.4 Roof assembly. The minimum thermal resistance (*R*-value) of the insulating material installed either between the roof framing or continuously on the roof assembly shall be as specified in Table 802.2(1), 802.2(2), 802.2(3) or 802.2(4), based on construction materials used in the roof assembly.

802.2.5 Skylights. Skylights located in the building envelope shall be limited to 3 percent of the gross roof assembly area and shall have a maximum thermal transmittance (*U*-factor) of the skylight assembly as specified in Table 802.2(1), 802.2(2), 802.2(3) or 802.2(4).

802.2.6 Floors over outdoor air or unconditioned space. The minimum thermal resistance (*R*-value) of the insulating material installed either between the floor framing or continuously on the floor assembly shall be as specified in Table 802.2(1), 802.2(2), 802.2(3) or 802.2(4) based on construction materials used in the floor assembly.

802.2.7 Slabs on grade. The minimum thermal resistance (*R*-value) of the insulation around the perimeter of the slab floor shall be as specified in Table 802.2(1), 802.2(2), 802.2(3) or 802.2(4). The insulation shall be placed on the outside of the foundation or on the inside of a foundation wall. The insulation shall extend downward from the top of the slab for a minimum of 48 inches (1219 mm) or downward to at least the bottom of the slab and then horizontally to the interior or exterior for a minimum total distance of 48 inches (1219 mm).

802.2.8 Below-grade walls. The minimum thermal resistance (*R*-value) of the insulating material installed in, or continuously on, the below-grade walls shall be as specified in Table 802.2(1), 802.2(2), 802.2(3) or 802.2(4) and shall extend to a depth of 10 feet (3048 mm) below the outside finish ground level, or to the level of the floor, whichever is less.

802.2.9 Interior walls. The minimum thermal resistance (*R*-value) of the insulating material installed in the wall cavity or continuously on the interior walls shall be as specified in Table 802.2(1) for above-grade walls, regardless of glazing area, based on framing type and construction materials used in the wall assembly.

802.3 Air leakage. The requirements for air leakage shall be as specified in Sections 802.3.1 and 802.3.2.

802.3.1 Window and door assemblies. The air leakage of window and sliding or swinging door assemblies that are part of the building envelope shall be determined in accordance with AAMA/WDMA 101/I.S.2 or 101/I.S.2/NAFS-02, or NFRC 400 by an accredited, independent laboratory, and labeled and certified by the manufacturer and shall not exceed the values in Table 502.1.4.1.

> **Exception:** Site-constructed windows and doors that are weatherstripped or sealed in accordance with Section 802.3.3.

802.3.2 Curtain wall, storefront glazing and commercial entrance doors. Curtain wall, storefront glazing and commercial-glazed swinging entrance doors and revolving doors shall be tested for air leakage at 1.57 pounds per square inch (psi) (75 Pa) in accordance with ASTM E 283. For curtain walls and storefront glazing, the maximum air leakage rate shall be 0.3 cubic feet per minute per square foot (cfm/ft^2) (5.5 m^3/h · m^2) of fenestration area. For commercial glazed swinging entrance doors and revolving doors, the maximum air leakage rate shall be 1.00 cfm/ft^2 (18.3 m^3/h · m^2) of door area when tested in accordance with ASTM E 283.

802.3.3 Sealing of the building envelope. Openings and penetrations in the building envelope shall be sealed with caulking materials or closed with gasketing systems compatible with the construction materials and location. Joints and seams shall be sealed in the same manner or taped or covered with a moisture vapor-permeable wrapping material. Sealing materials spanning joints between construction materials shall allow for expansion and contraction of the construction materials.

802.3.4 Dampers integral to the building envelope. Stair, elevator shaft vents, and other dampers integral to the building envelope shall be equipped with motorized dampers with a maximum leakage rate of 3 cfm/ft^2 [5.1 L/s · m^2] at 1.0 inch water gauge (w.g.) (250 Pa) when tested in accordance with AMCA 500.

> **Exception:** Gravity (nonmotorized) dampers are permitted to be used in buildings less than three stories in height above grade.

802.3.5 Loading dock weatherseals. Cargo doors and loading dock doors shall be equipped with weatherseals to restrict infiltration when vehicles are parked in the doorway.

802.3.6 Vestibules. A door that separates conditioned space from the exterior shall be protected with an enclosed vestibule, with all doors opening into and out of the vestibule equipped with self-closing devices. Vestibules shall be designed so that in passing through the vestibule it is not necessary for the interior and exterior doors to open at the same time.

Exceptions:

1. Buildings in Climate Zones 1a through 4b as indicated in Table 302.1.
2. Doors not intended to be used as a building entrance door, such as doors to mechanical or electrical equipment rooms.
3. Doors opening directly from a guestroom or dwelling unit.
4. Doors that open directly from a space less than 3,000 square feet (298 m^2) in area.
5. Revolving doors.
6. Doors used primarily to facilitate vehicular movement or material handling and adjacent personnel doors.

802.3.7 Recessed lighting fixtures. When installed in the building envelope, recessed lighting fixtures shall meet one of the following requirements:

1. Type IC rated, manufactured with no penetrations between the inside of the recessed fixture and ceiling cavity and sealed or gasketed to prevent air leakage into the unconditioned space.
2. Type IC or non-IC rated, installed inside a sealed box constructed from a minimum 0.5-inch-thick (12.7 mm) gypsum wallboard or constructed from a preformed polymeric vapor barrier, or other air-tight assembly manufactured for this purpose, while maintaining required clearances of not less than 0.5 inch (12.7 mm) from combustible material and not less than 3 inches (76 mm) from insulation material.
3. Type IC rated, in accordance with ASTM E 283 admitting no more than 2.0 cubic feet per minute (cfm) (0.944 L/s) of air movement from the conditioned space to the ceiling cavity. The lighting fixture shall be tested at 1.57 psi (75 Pa) pressure difference and shall be labeled.

SECTION 803
BUILDING MECHANICAL SYSTEMS

803.1 General. This section covers the design and construction of mechanical systems and equipment serving the building heating, cooling or ventilating needs.

803.1.1 Compliance. Compliance with Section 803 shall be achieved by meeting either Section 803.2 or 803.3.

803.2 Simple HVAC systems and equipment. This section applies to buildings served by unitary or packaged HVAC equipment listed in Tables 803.2.2(1) through 803.2.2(5), each serving one zone and controlled by a single thermostat in the zone served. It also applies to two-pipe heating systems serving one or more zones, where no cooling system is installed.

This section does not apply to fan systems serving multiple zones, nonunitary or nonpackaged HVAC equipment and systems or hydronic or steam heating and hydronic cooling equipment and distribution systems that provide cooling or cooling and heating which are covered by Section 803.3.

[M] 803.2.1 Calculation of heating and cooling loads. Design loads shall be determined in accordance with the procedures described in the ASHRAE *Fundamentals Handbook*. Heating and cooling loads shall be adjusted to account for load reductions that are achieved when energy recovery systems are utilized in the HVAC system in accordance with the ASHRAE *HVAC Systems and Equipment Handbook*. Alternatively, design loads shall be determined by an approved equivalent computation procedure, using the design parameters specified in Chapter 3.

803.2.1.1 Equipment and system sizing. Heating and cooling equipment and systems capacity shall not exceed the loads calculated in accordance with Section 803.2.1. A single piece of equipment providing both heating and cooling must satisfy this provision for one function with the capacity for the other function as small as possible, within available equipment options.

803.2.2 HVAC equipment performance requirements. Equipment shall meet the minimum efficiency requirements of Tables 803.2.2(1), 803.2.2(2), 803.2.2(3), 803.2.2(4) and 803.2.2(5), when tested and rated in accordance with the applicable test procedure. The efficiency shall be verified through data furnished by the manufacturer or through certification under an approved certification program. Where multiple rating conditions or performance requirements are provided, the equipment shall satisfy all stated requirements.

803.2.3 Temperature and humidity controls. Requirements for temperature and humidity controls shall be as specified in Sections 803.2.3.1 and 803.2.3.2.

803.2.3.1 Temperature controls. Each heating and cooling system shall have at least one solid-state programmable thermostat. The thermostat shall have the capability to set back or shut down the system based on day of the week and time of day, and provide a readily accessible manual override that will return to the presetback or shutdown schedule without reprogramming.

Exceptions:

1. HVAC systems serving hotel/motel guestrooms.
2. Packaged terminal air conditioners, packaged terminal heat pumps and room air conditioner systems.

803.2.3.2 Heat pump supplementary heat. Heat pumps having supplementary electric-resistance heat shall have controls that, except during defrost, prevent supplemental heat operation when the heat pump can meet the heating load.

803.2.3.3 Humidity controls. When humidistats are installed, they shall have the capability to prevent the use of fossil fuel or electric power to achieve a humidity below

60 percent when the system controlled is cooling, and above 30 percent when the system controlled is heating.

Exceptions:

1. Systems serving spaces where specific humidity levels are required to satisfy process needs, such as computer rooms, museums, surgical suites and buildings with refrigerating systems, such as supermarkets, refrigerated warehouses and ice arenas.
2. Systems where humidity is removed as the result of the use of a desiccant system with energy recovery.
3. Reheat systems utilizing site-recovered (including condenser heat) or site-solar energy sources.

803.2.4 Hydronic system controls. Hydronic systems of at least 600,000 British thermal units per hour (Btu/h) (175 860 W) design capacity supplying heated water to comfort conditioning systems shall include controls that meet the requirements of Section 803.3.3.7.

803.2.5 Ventilation. Ventilation, either natural or mechanical, shall be provided in accordance with Chapter 4 of the *International Mechanical Code*. Where mechanical ventilation is provided, the system shall provide the capability to reduce the outdoor air supply to the minimum required by Chapter 4 of the *International Mechanical Code*.

803.2.6 Cooling with outdoor air. Each system with a cooling capacity greater than 65,000 Btu/h (19 kW) located in other than Climate Zones 1, 2, 3b, 5a or 6b as shown in Table 302.1 shall have an economizer that will automatically shut off the cooling system and allow all of the supply air to be provided directly from outdoors.

Economizers shall be capable of operating at 100-percent outside air, even if additional mechanical cooling is required to meet the cooling load of the building. Where a single room or space is supplied by multiple air systems, the aggregate capacity of those systems shall be used in applying this requirement.

Exceptions:

1. Where the cooling equipment is covered by the minimum efficiency requirements of Table 803.2.2(1) or 803.2.2(2) and meets the efficiency requirements of Table 803.2.6.
2. Systems with air or evaporatively cooled condensers and which serve spaces with open case refrigeration or that require filtration equipment in order to meet the minimum ventilation requirements of Chapter 4 of the *International Mechanical Code*.
3. Systems with a cooling capacity less than 135,000 Btu/h (40 kW) in Climate Zones 3c, 5b, 7, 13b, and 14.

TABLE 803.2.2 (1)
UNITARY AIR CONDITIONERS AND CONDENSING UNITS, ELECTRICALLY OPERATED,
MINIMUM EFFICIENCY REQUIREMENTS

EQUIPMENT TYPE	SIZE CATEGORY	SUBCATEGORY OR RATING CONDITION	MINIMUM EFFICIENCY [b]	TEST PROCEDURE [a]
Air conditioners, Air cooled	< 65,000 Btu/h [d]	Split system	10.0 SEER	ARI 210/240
		Single package	9.7 SEER	
	≥ 65,000 Btu/h and < 135,000 Btu/h	Split system and single package	10.3 EER [c]	ARI 340/360
	≥ 135,000 Btu/h and < 240,000 Btu/h	Split system and single package	9.7 EER [c]	
	≥ 240,000 Btu/h and < 760,000 Btu/h	Split system and single package	9.5 EER [c] 9.7 IPLV [c]	
	≥ 760,000 Btu/h	Split system and single package	9.2 EER [c] 9.4 IPLV [c]	
Air conditioners, Water and evaporatively cooled	< 65,000 Btu/h	Split system and single package	12.1 EER	ARI 210/240
	≥ 65,000 Btu/h and < 135,000 Btu/h	Split system and single package	11.5 EER [c]	ARI 340/360
	≥ 135,000 Btu/h and < 240,000 Btu/h	Split system and single package	11.0 EER [c]	
	≥ 240,000 Btu/h	Split system and single package	11.0 EER [c] 10.3 IPLV [c]	

For SI: 1 British thermal unit per hour = 0.2931 W.

a. Chapter 10 contains a complete specification of the referenced test procedure, including the referenced year version of the test procedure.
b. IPLVs are only applicable to equipment with capacity modulation.
c. Deduct 0.2 from the required EERs and IPLVs for units with a heating section other than electric resistance heat.
d. Single-phase air-cooled air conditioners < 65,000 Btu/h are regulated by the National Appliance Energy Conservation Act of 1987 (NAECA). SEER values are those set by NAECA.

TABLE 803.2.2(2)
UNITARY AND APPLIED HEAT PUMPS, ELECTRICALLY OPERATED,
MINIMUM EFFICIENCY REQUIREMENTS

EQUIPMENT TYPE	SIZE CATEGORY	SUBCATEGORY OR RATING CONDITION	MINIMUM EFFICIENCY[b]	TEST PROCEDURE[a]
Air cooled (Cooling mode)	< 65,000 Btu/h[d]	Split system	10.0 SEER	ARI 210/240
		Single package	9.7 SEER	
	≥ 65,000 Btu/h and < 135,000 Btu/h	Split system and single package	10.1 EER[c]	
	≥ 135,000 Btu/h and < 240,000 Btu/h	Split system and single package	9.3 EER[c]	ARI 340/360
	≥ 240,000 Btu/h	Split system and single package	9.0 EER[c] 9.2 IPLV[c]	
Water source (Cooling mode)	< 17,000 Btu/h	86°F entering water	11.2 EER	ARI/ASHRAE-13256-1
	≥ 17,000 Btu/h and < 135,000 Btu/h	86°F entering water	12.0 EER	ARI/ASHRAE-13256-1
Groundwater source (Cooling mode)	< 135,000 Btu/h	59°F entering water	16.2 EER	ARI/ASHRAE-13256-1
Ground source (Cooling mode)	< 135,000 Btu/h	77°F entering water	13.4 EER	ARI/ASHRAE-13256-1
Air cooled (Heating mode)	< 65,000 Btu/h[d] (Cooling capacity)	Split system	6.8 HSPF	ARI 210/240
		Single package	6.6 HSPF	
	≥ 65,000 Btu/h and < 135,000 Btu/h (Cooling capacity)	47°F db/43°F wb outdoor air	3.2 COP	
	≥ 135,000 Btu/h (Cooling capacity)	47°F db/43°F wb outdoor air	3.1 COP	ARI 340/360
Water source (Heating mode)	< 135,000 Btu/h (Cooling capacity)	68°F entering water	4.2 COP	ARI/ASHRAE-13256-1
Groundwater source (Heating mode)	< 135,000 Btu/h (Cooling capacity)	50°F entering water	3.6 COP	ARI/ASHRAE-13256-1
Ground source (Heating mode)	< 135,000 Btu/h (Cooling capacity)	32°F entering water	3.1 COP	ARI/ASHRAE-13256-1

For SI: °C = [(°F) - 32] / 1.8, 1British thermal unit per hour = 0.2931W.
db = dry-bulb temperature, °F
wb = wet-bulb temperature, °F

a. Chapter 10 contains a complete specification of the referenced test procedure, including the referenced year version of the test procedure.
b. IPLVs and Part load rating conditions are only applicable to equipment with capacity modulation.
c. Deduct 0.2 from the required EERs and IPLVs for units with a heating section other than electric resistance heat.
d. Single-phase air-cooled heat pumps < 65,000 Btu/h are regulated by the National Appliance Energy Conservation Act of 1987 (NAECA). SEER and HSPF values are those set by NAECA.

TABLE 803.2.2(3)
PACKAGED TERMINAL AIR CONDITIONERS AND PACKAGED TERMINAL HEAT PUMPS

EQUIPMENT TYPE	SIZE CATEGORY (INPUT)	SUBCATEGORY OR RATING CONDITION	MINIMUM EFFICIENCY[b]	TEST PROCEDURE[a]
PTAC (Cooling mode) New construction	All capacities	95°F db outdoor air	12.5 - (0.213 · Cap/1000) EER	ARI 310/380
PTAC (Cooling mode) Replacements[c]	All capacities	95°F db outdoor air	10.9 - (0.213 · Cap/1000) EER	
PTHP (Cooling mode) New construction	All capacities	95°F db outdoor air	12.3 - (0.213 · Cap/1000) EER	
PTHP (Cooling mode) Replacements[c]	All capacities	95°F db outdoor air	10.8 - (0.213 · Cap/1000) EER	
PTHP (Heating mode) New construction	All capacities	—	3.2 - (0.026 · Cap/1000) COP	
PTHP (Heating mode) Replacements[c]	All capacities	—	2.9 - (0.026 · Cap/1000) COP	

For SI: °C = [(°F) - 32] / 1.8, 1 British thermal unit per hour = 0.2931W.
db = dry-bulb temperature, °F
wb = wet-bulb temperature, °F

a. Chapter 10 contains a complete specification of the referenced test procedure, including the referenced year version of the test procedure.
b. Cap means the rated cooling capacity of the product in Btu/h. If the unit's capacity is less than 7,000 Btu/h, use 7,000 Btu/h in the calculation. If the unit's capacity is greater than 15,000 Btu/h, use 15,000 Btu/h in the calculation.
c. Replacement units must be factory labeled as follows: "MANUFACTURED FOR REPLACEMENT APPLICATIONS ONLY; NOT TO BE INSTALLED IN NEW CONSTRUCTION PROJECTS." Replacement efficiencies apply only to units with existing sleeves less than 16 inches (406 mm) high and less than 42 inches (1067 mm) wide.

TABLE 803.2.2(4)
WARM AIR FURNACES AND COMBINATION WARM AIR FURNACES/AIR-CONDITIONING UNITS, WARM AIR DUCT FURNACES AND UNIT HEATERS, MINIMUM EFFICIENCY REQUIREMENTS

EQUIPMENT TYPE	SIZE CATEGORY (INPUT)	SUBCATEGORY OR RATING CONDITION	MINIMUM EFFICIENCY[d,e]	TEST PROCEDURE[a]
Warm air furnaces, gas fired	< 225,000 Btu/h	—	78% AFUE or 80% E_t^c	DOE 10 CFR Part 430 or ANSI Z21.47
	≥ 225,000 Btu/h	Maximum capacity[c]	80% E_t^f	ANSI Z21.47
Warm air furnaces, oil fired	< 225,000 Btu/h	—	78% AFUE or 80% E_t^c	DOE 10 CFR Part 430 or UL 727
	≥ 225,000 Btu/h	Maximum capacity[b]	81% E_t^g	UL 727
Warm air duct furnaces, gas fired	All capacities	Maximum capacity[b]	80% E_c	ANSI Z83.9
Warm air unit heaters, gas fired	All capacities	Maximum capacity[b]	80% E_c	ANSI Z83.8
Warm air unit heaters, oil fired	All capacities	Maximum capacity[b]	80% E_c	UL 731

For SI: 1 British thermal unit per hour = 0.2931W.

a. Chapter 10 contains a complete specification of the referenced test procedure, including the referenced year version of the test procedure.
b. Minimum and maximum ratings as provided for and allowed by the unit's controls.
c. Combination units not covered by the National Appliance Energy Conservation Act of 1987 (NAECA) (3-phase power or cooling capacity greater than or equal to 65,000 Btu/h [19 kW]) shall comply with either rating.
d. E_t = Thermal efficiency. See test procedure for detailed discussion.
e. E_c = Combustion efficiency (100% less flue losses). See test procedure for detailed discussion.
f. E_c = Combustion efficiency. Units must also include an IID, have jacket losses not exceeding 0.75 percent of the input rating, and have either power venting or a flue damper. A vent damper is an acceptable alternative to a flue damper for those furnaces where combustion air is drawn from the conditioned space.
g. E_t = Thermal efficiency. Units must also include an IID, have jacket losses not exceeding 0.75 percent of the input rating, and have either power venting or a flue damper. A vent damper is an acceptable alternative to a flue damper for those furnaces where combustion air is drawn from the conditioned space.

ACCEPTABLE PRACTICE FOR COMMERCIAL BUILDINGS

TABLE 803.2.2(5)
BOILERS, GAS- AND OIL-FIRED, MINIMUM EFFICIENCY REQUIREMENTS

EQUIPMENT TYPE[f]	SIZE CATEGORY (INPUT)	SUBCATEGORY OR RATING CONDITION	MINIMUM EFFICIENCY[c, d, e]	TEST PROCEDURE[a]
Boilers, Gas fired	< 300,000 Btu/h	Hot water	80% AFUE	DOE 10 CFR Part 430
		Steam	75% AFUE	
	≥ 300,000 Btu/h and ≤ 2,500,000 Btu/h	Minimum capacity[b]	75% E_t	H.I. HBS 86
	> 2,500,000 Btu/h[f]	Hot water	80% E_c	
		Steam	80% E_c	
Boilers, Oil fired	< 300,000 Btu/h	—	80% AFUE	DOE 10 CFR Part 430
	≥ 300,000 Btu/h and ≤ 2,500,000 Btu/h	Minimum capacity[b]	78% E_t	H.I. HBS 86
	> 2,500,000 Btu/h[f]	Hot water	83% E_c	
		Steam	83% E_c	
Boilers, Oil fired (Residual)	≥ 300,000 Btu/h and ≤ 2,500,000 Btu/h	Minimum capacity[b]	78% E_t	H.I. HBS 86
	> 2,500,000 Btu/h[f]	Hot water	83% E_c	
		Steam	83% E_c	

For SI: 1 British thermal unit per hour = 0.2931W.

a. Chapter 10 contains a complete specification of the referenced test procedure, including the referenced year version of the test procedure.
b. Minimum ratings as provided for and allowed by the unit's controls.
c. E_c = Combustion efficiency (100 percent less flue losses). See reference document for detailed information.
d. E_t = Thermal efficiency. See reference document for detailed information.
e. Alternative test procedures used at the manufacturer's option are ASME PTC-4.1 for units greater than 5,000,000 Btu/h input, or ANSI Z21.13 for units greater than or equal to 300,000 Btu/h and less than or equal to 2,500,000 Btu/h input.
f. These requirements apply to boilers with rated input of 8,000,000 Btu/h or less that are not packaged boilers, and to all packaged boilers. Minimum efficiency requirements for boilers cover all capacities of packaged boilers.

TABLE 803.2.6
MINIMUM EQUIPMENT EFFICIENCY ECONOMIZER EXCEPTION

TOTAL COOLING CAPACITY OF EQUIPMENT	BUILDING LOCATION		
	Zones 6a, 9a, 10a, 11a, 12a, 12b, 13a, 13b, 14a, 14b, 15-19	Zones 3a, 3b, 4a, 7a, 8, 9b, 10b, 11b	Zones 4b, 5a, 5b, 6b, 7b
90,000 Btu/h to 134,999 Btu/h	Not Applicable	11.4 EER	10.4 EER
135,000 Btu/h to 759,999 Btu/h	Not Applicable	10.9 EER	9.9 EER
760,000 Btu/h or more	Not Applicable	10.5 EER	9.6 EER

For SI: °C = [(°F)-32]/1.8, 1 British thermal unit per hour = 0.2931 W.

803.2.7 Shutoff dampers. Outdoor air supply and exhaust ducts shall be provided with automatic means to reduce and shut off airflow.

Exceptions:

1. Systems serving areas designed for continuous operation.
2. Individual systems with a maximum 3,000 cfm (1416 L/s) airflow rate.
3. Systems with readily accessible manual dampers.
4. Where restricted by health and life safety codes.

803.2.8 Duct and plenum insulation and sealing. All supply and return air ducts and plenums shall be insulated with a minimum of R-5 insulation when located in unconditioned spaces and with a minimum of R-8 insulation when located outside the building. When located within a building envelope assembly, the duct or plenum shall be separated from the building exterior or unconditioned or exempt spaces by a minimum of R-8 insulation.

Exceptions:

1. When located within equipment.
2. When the design temperature difference between the interior and exterior of the duct or plenum does not exceed 15°F (8°C).

All joints, longitudinal and transverse seams and connections in ductwork, shall be securely fastened and sealed with welds, gaskets, mastics (adhesives), mastic-plus-embedded-fabric systems, or tapes. Tapes and mastics used to seal ductwork shall be listed and labeled in accordance with UL 181A or UL 181B. Duct connections to flanges of air distribution system equipment shall be sealed and mechanically fastened. Unlisted duct tape is not permitted as a sealant on any metal ducts.

803.2.8.1 Duct construction. Ductwork shall be constructed and erected in accordance with the *International Mechanical Code*.

803.2.8.1.1 High- and medium-pressure duct systems. All ducts and plenums operating at a static pressures greater than 2 inches w.g. (500 Pa) shall be insulated and sealed in accordance with Section 803.2.8. Ducts operating at a static pressures in excess of 3 inches w.g. (750 Pa) shall be leak tested in accordance with Section 803.3.6. Pressure classifications specific to the duct system shall be clearly indicated on the construction documents in accordance with the *International Mechanical Code*.

803.2.8.1.2 Low-pressure duct systems. All longitudinal and transverse joints, seams and connections of supply and return ducts operating at a static pressure less than or equal to 2 inches w.g. (500 Pa) shall be securely fastened and sealed with welds, gaskets, mastics (adhesives), mastic-plus-embedded-fabric systems or tapes installed in accordance with the manufacturer's installation instructions. Pressure classifications specific to the duct system shall be clearly indicated on the construction documents in accordance with the *International Mechanical Code*.

Exception: Continuously welded and locking-type longitudinal joints and seams on ducts operating at static pressures less than 2 inches w.g. (500 Pa) pressure classification.

803.2.9 Piping insulation. All piping serving as part of a heating or cooling system shall be thermally insulated in accordance with Section 803.3.7.

803.3 Complex HVAC systems and equipment. This section applies to buildings served by HVAC equipment and systems not covered in Section 803.2.

803.3.1 Calculation of heating and cooling loads. Design loads shall be determined in accordance with Section 803.2.1.

803.3.1.1 Equipment and system sizing. Heating and cooling equipment and system capacity shall not exceed the loads calculated in accordance with Section 803.2.1.

Exceptions:

1. Required standby equipment and systems provided with controls and devices that allow such systems or equipment to operate automatically only when the primary equipment is not operating.
2. Multiple units of the same equipment type with combined capacities exceeding the design load and provided with controls that have the capability to sequence the operation of each unit based on load.

803.3.2 HVAC equipment performance requirements. Equipment shall meet the minimum efficiency requirements of Tables 803.3.2(1) through 803.3.2(6) and Table 803.2.2(5), when tested and rated in accordance with the applicable test procedure. The efficiency shall be verified through certification under an approved certification program or, if no certification program exists, the equipment efficiency ratings shall be supported by data furnished by the manufacturer. Where multiple rating conditions or performance requirements are provided, the equipment shall satisfy all stated requirements. Where components, such as indoor or outdoor coils, from different manufacturers are used, calculations and supporting data shall be furnished by the designer that demonstrate that the combined efficiency of the specified components meets the requirements herein.

Where unitary or prepackaged equipment is used in a complex HVAC system and is not covered by Section 803.3.2, the equipment shall meet the applicable requirements of Section 803.2.2.

Exception: Equipment listed in Table 803.3.2(2) not designed for operation at ARI Standard test conditions of 44°F (7°C) leaving chilled water temperature and 85°F (29°C) entering condenser water temperature shall have a minimum full load COP and IPLV rating as shown in Tables 803.3.2(3) through 803.3.2(5) as applicable. The table values are only applicable over the following full load design ranges:

Leaving Chilled
Water Temperature: 40 to 48°F (4 to 9°C)

Entering Condenser
Water Temperature: 75 to 85°F (24 to 29°C)

Condensing Water
Temperature Rise: 5 to 15°F (Δ3 to Δ8°C)

Chillers designed to operate outside of these ranges are not covered by this code.

TABLE 803.3.2(1)
CONDENSING UNITS, ELECTRICALLY OPERATED, MINIMUM EFFICIENCY REQUIREMENTS

EQUIPMENT TYPE	SIZE CATEGORY	MINIMUM EFFICIENCY[b]	TEST PROCEDURE[a]
Condensing units, Air cooled	≥ 135,000 Btu/h	10.1 EER 11.2 IPLV	ARI 365
Condensing units, Water or evaporatively cooled	≥ 135,000 Btu/h	13.1 EER 13.1 IPLV	ARI 365

For SI: 1 British thermal unit per hour = 0.2931W.
a. Chapter 10 contains a complete specification of the referenced test procedure, including the referenced year version of the test procedure.
b. IPLVs are only applicable to equipment with capacity modulation.

TABLE 803.3.2(2)
WATER CHILLING PACKAGES, MINIMUM EFFICIENCY REQUIREMENTS

EQUIPMENT TYPE	SIZE CATEGORY	MINIMUM EFFICIENCY[b]	TEST PROCEDURE[a]
Air cooled, with condenser, Electrically operated	< 150 tons	2.80 COP 2.80 IPLV	ARI 550/590
	≥ 150 tons	2.50 COP 2.50 IPLV	
Air cooled, without condenser, Electrically operated	All capacities	3.10 COP 3.10 IPLV	ARI 550/590
Water cooled, Electrically operated, Positive displacement (reciprocating)	All capacities	4.20 COP 4.65 IPLV	ARI 550/590
Water cooled, Electrically operated, Positive displacement (rotary screw and scroll)	< 150 tons	4.45 COP 4.50 IPLV	ARI 550/590
	≥ 150 tons and < 300 tons	4.90 COP 4.95 IPLV	
	≥ 300 tons	5.50 COP 5.60 IPLV	
Water cooled, Electrically operated, centrifugal	< 150 tons	5.00 COP 5.00 IPLV	ARI 550/590
	≥ 150 tons and < 300 tons	5.55 COP 5.55 IPLV	
	≥ 300 tons	6.10 COP 6.10 IPLV	
Air cooled, absorption single effect	All capacities	0.60 COP	ARI 560
Water cooled, absorption single effect	All capacities	0.70 COP	
Absorption double effect, indirect-fired	All capacities	1.00 COP 1.05 IPLV	
Absorption double effect, direct-fired	All capacities	1.00 COP 1.00 IPLV	

For SI: 1 ton = 3.517 kW, °C = [(°F) − 32] / 1.8.
a. Chapter 10 contains a complete specification of the referenced test procedure, including the referenced year version of the test procedure.
b. The chiller equipment requirements do not apply for chillers used in low temperature applications where the design leaving fluid temperature is less than or equal to 40°F.

TABLE 803.3.2(3)
COPS AND IPLVS FOR NONSTANDARD CENTRIFUGAL CHILLERS < 150 TONS

CENTRIFUGAL CHILLERS < 150 TONS $COP_{std} = 5.4$							
Leaving chilled water temperature (°F)	Entering condenser water temperature (°F)	Lift[a] (°F)	Condenser flow rate					
			2 gpm/ton	2.5 gpm/ton	3 gpm/ton	4 gpm/ton	5 gpm/ton	6 gpm/ton
			Required COP and IPLV					
46	75	29	6.00	6.27	6.48	6.80	7.03	7.20
45	75	30	5.92	6.17	6.37	6.66	6.87	7.02
44	75	31	5.84	6.08	6.26	6.53	6.71	6.86
43	75	32	5.75	5.99	6.16	6.40	6.58	6.71
42	75	33	5.67	5.90	6.06	6.29	6.45	6.57
41	75	34	5.59	5.82	5.98	6.19	6.34	6.44
46	80	34	5.59	5.82	5.98	6.19	6.34	6.44
40	75	35	5.50	5.74	5.89	6.10	6.23	6.33
45	80	35	5.50	5.74	5.89	6.10	6.23	6.33
44	80	36	5.41	5.66	5.81	6.01	6.13	6.22
43	80	37	5.31	5.57	5.73	5.92	6.04	6.13
42	80	38	5.21	5.48	5.64	5.84	5.95	6.04
41	80	39	5.09	5.39	5.56	5.76	5.87	5.95
46	85	39	5.09	5.39	5.56	5.76	5.87	5.95
40	80	40	4.96	5.29	5.47	5.67	5.79	5.86
45	85	40	4.96	5.29	5.47	5.67	5.79	5.86
44	85	41	4.83	5.18	5.40	5.59	5.71	5.78
43	85	42	4.68	5.07	5.28	5.50	5.62	5.70
42	85	43	4.51	4.94	5.17	5.41	5.54	5.62
41	85	44	4.33	4.80	5.05	5.31	5.45	5.53
40	85	45	4.13	4.65	4.92	5.21	5.35	5.44
Condenser ΔT[b]			14.04	11.23	9.36	7.02	5.62	4.68

For SI: °C = [(°F) - 32] / 1.8, 1 gallon per minute = 3.785 L/min., 1 ton = 12,000 British thermal unit per hour = 3.517 kW.
a. Lift = Entering condenser water temperature (°F) – Leaving chilled water temperature (°F).
b. Condenser ΔT = Leaving condenser water temperature (°F) – Entering condenser water temperature (°F).

$K_{adj} = 6.1507 - 0.30244(X) + 0.0062692(X)^2 - 0.000045595(X)$

where: X = Condenser ΔT + Lift

$COP_{adj} = K_{adj} \times COP_{std}$

TABLE 803.3.2(4)
COPs AND IPLVS FOR NONSTANDARD CENTRIFUGAL CHILLERS ≥ 150 TONS, ≤ 300 TONS

Leaving chilled water temperature (°F)	Entering condenser water temperature (°F)	Lift[a] (°F)	\<CENTRIFUGAL CHILLERS ≥ 150 Tons, ≤ 300 Tons; COP_std = 5.55\> Condenser flow rate					
			2 gpm/ton	2.5 gpm/ton	3 gpm/ton	4 gpm/ton	5 gpm/ton	6 gpm/ton
			Required COP and IPLV					
46	75	29	6.17	6.44	6.66	6.99	7.23	7.40
45	75	30	6.08	6.34	6.54	6.84	7.06	7.22
44	75	31	6.00	6.24	6.43	6.71	6.90	7.05
43	75	32	5.91	6.15	6.33	6.58	6.76	6.89
42	75	33	5.83	6.07	6.23	6.47	6.63	6.75
41	75	34	5.74	5.98	6.14	6.36	6.51	6.62
46	80	34	5.74	5.98	6.14	6.36	6.51	6.62
40	75	35	5.65	5.90	6.05	6.26	6.40	6.51
45	80	35	5.65	5.90	6.05	6.26	6.40	6.51
44	80	36	5.56	5.81	5.97	6.17	6.30	6.40
43	80	37	5.46	5.73	5.89	6.08	6.21	6.30
42	80	38	5.35	5.64	5.80	6.00	6.12	6.20
41	80	39	5.23	5.54	5.71	5.91	6.03	6.11
46	85	39	5.23	5.54	5.71	5.91	6.03	6.11
40	80	40	5.10	5.44	5.62	5.83	5.95	6.03
45	85	40	5.10	5.44	5.62	5.83	5.95	6.03
44	85	41	4.96	5.33	5.55	5.74	5.86	5.94
43	85	42	4.81	5.21	5.42	5.66	5.78	5.86
42	85	43	4.63	5.08	5.31	5.56	5.69	5.77
41	85	44	4.45	4.93	5.19	5.46	5.60	5.69
40	85	45	4.24	4.77	5.06	5.35	5.50	5.59
Condenser ΔT[b]			14.04	11.23	9.36	7.02	5.62	4.68

For SI: °C = [(°F) - 32] / 1.8, 1 gallon per minute = 3.785 L/min., 1 ton = 12,000 British thermal unit per hour = 3.517 kW.

a. Lift = Entering condenser water temperature (°F) – Leaving chilled water temperature (°F).
b. Condenser ΔT = Leaving condenser water temperature (°F) – Entering condenser water temperature (°F).

$K_{adj} = 6.1507 - 0.30244(X) + 0.0062692(X)^2 - 0.000045595(X)$

where: X = Condenser ΔT + Lift

$COP_{adj} = K_{adj} \times COP_{std}$

ACCEPTABLE PRACTICE FOR COMMERCIAL BUILDINGS

TABLE 803.3.2(5)
COPS AND IPLVS FOR NONSTANDARD CENTRIFUGAL CHILLERS > 300 TONS

CENTRIFUGAL CHILLERS > 300 TONS $COP_{std} = 6.1$							
			Condenser Flow Rate				
			2 gpm/ton	2.5 gpm/ton	3 gpm/ton	4 gpm/ton	5 gpm/ton	6 gpm/ton
Leaving chilled water temperature (°F)	Entering Condenser Water Temperature (°F)	Lift[a] (°F)	Required COP and IPLV				
46	75	29	6.80	7.11	7.35	7.71	7.97	8.16
45	75	30	6.71	6.99	7.21	7.55	7.78	7.96
44	75	31	6.61	6.89	7.09	7.40	7.61	7.77
43	75	32	6.52	6.79	6.98	7.26	7.45	7.60
42	75	33	6.43	6.69	6.87	7.13	7.31	7.44
41	75	34	6.33	6.60	6.77	7.02	7.18	7.30
46	80	34	6.33	6.60	6.77	7.02	7.18	7.30
40	75	35	6.23	6.50	6.68	6.91	7.06	7.17
45	80	35	6.23	6.50	6.68	6.91	7.06	7.17
44	80	36	6.13	6.41	6.58	6.81	6.95	7.05
43	80	37	6.02	6.31	6.49	6.71	6.85	6.94
42	80	38	5.90	6.21	6.40	6.61	6.75	6.84
41	80	39	5.77	6.11	6.30	6.52	6.65	6.74
46	85	39	5.77	6.11	6.30	6.52	6.65	6.74
40	80	40	5.63	6.00	6.20	6.43	6.56	6.65
45	85	40	5.63	6.00	6.20	6.43	6.56	6.65
44	85	41	5.47	5.87	6.10	6.33	6.47	6.55
43	85	42	5.30	5.74	5.98	6.24	6.37	6.46
42	85	43	5.11	5.60	5.86	6.13	6.28	6.37
41	85	44	4.90	5.44	5.72	6.02	6.17	6.27
40	85	45	4.68	5.26	5.58	5.90	6.07	6.17
Condenser ΔT[b]			14.04	11.23	9.36	7.02	5.62	4.68

For SI: °C = [(°F) - 32] / 1.8, 1 gallon per minute = 3.785 L/min., 1 ton = 12,000 British thermal unit per hour = 3.517 kW.

a. Lift = Entering condenser water temperature (°F) – Leaving chilled water temperature (°F).
b. Condenser ΔT = Leaving condenser water temperature (°F) – Entering condenser water temperature (°F).

$K_{adj} = 6.1507 - 0.30244(X) + 0.0062692(X)^2 - 0.000045595(X)$

where: X = Condenser ΔT + Lift

$COP_{adj} = K_{adj} \times COP_{std}$

TABLE 803.3.2(6)
PERFORMANCE REQUIREMENTS FOR HEAT REJECTION EQUIPMENT

EQUIPMENT TYPE	TOTAL SYSTEM HEAT REJECTION CAPACITY AT RATED CONDITIONS	SUBCATEGORY OR RATING CONDITION	PERFORMANCE REQUIRED[a,b]	TEST PROCEDURE[c]
Propeller or axial fan cooling towers	All	95°F entering water 85°F leaving water 75°F wb outdoor air	≥ 38.2 gpm/hp	CTI ATC-105 and CTI STD-201
Centrifugal fan cooling towers	All	95°F entering water 85°F leaving water 75°F wb outdoor air	≥ 20.0 gpm/hp	CTI ATC-105 and CTI STD-201
Air-cooled condensers	All	125°F condensing temperature R-22 test fluid 190°F entering gas temperature 15°F subcooling 95°F entering db	≥ 176,000 Btu/h·hp (69 COP)	ARI 460

For SI: °C = [(°F) - 32] / 1.8, 1 British thermal unit per hour = 0.2931W, 1 gallon per minute per horsepower = 0.846 L/s · kW.
wb = wet-bulb temperature, °F

a. For purposes of this table, cooling tower performance is defined as the maximum flow rating of the tower units (gpm) divided by the fan nameplate rated motor power units (hp).
b. For purposes of this table, air-cooled condenser performance is defined as the heat rejected from the refrigerant units (Btu/h) divided by the fan nameplate rated motor power units (hp).
c. Chapter 10 contains a complete specification of the referenced test procedure, including the referenced year version of the test procedure.

803.3.3 HVAC system controls. Each heating and cooling system shall be provided with thermostatic controls as required in Sections 803.3.3.1 through 803.3.3.5.

803.3.3.1 Thermostatic controls. The supply of heating and cooling energy to each zone shall be controlled by individual thermostatic controls capable of responding to temperature within the zone. Where humidification or dehumidification or both is provided, at least one humidity control device shall be provided for each humidity control system

Exception: Independent perimeter systems that are designed to offset only building envelope heat losses or gains or both serving one or more perimeter zones also served by an interior system provided:

1. The perimeter system includes at least one thermostatic control zone for each building exposure having exterior walls facing only one orientation (within +/- 45 degrees) (0.8 rad) for more than 50 contiguous feet (15.2 m); and,

2. The perimeter system heating and cooling supply is controlled by a thermostat(s) located within the zone(s) served by the system.

803.3.3.1.1 Heat pump supplementary heat. Heat pumps having supplementary electric resistance heat shall have controls that, except during defrost, prevent supplementary heat operation when the heat pump can meet the heating load.

803.3.3.2 Set point overlap restriction. Where used to control both heating and cooling, zone thermostatic controls shall provide a temperature range or deadband of at least 5°F (Δ2.8°C) within which the supply of heating and cooling energy to the zone is capable of being shut off or reduced to a minimum.

Exception: Thermostats requiring manual changeover between heating and cooling modes.

803.3.3.3 Off-hour controls. Each zone shall be provided with thermostatic setback controls that are controlled by either an automatic time clock or programmable control system.

Exceptions:

1. Zones that will be operated continuously.
2. Zones with a full HVAC load demand not exceeding 6,800 Btu/h (2 kW) and having a readily accessible manual shutoff switch.

803.3.3.3.1 Thermostatic setback capabilities. Thermostatic setback controls shall have the capability to set back or temporarily operate the system to maintain zone temperatures down to 55°F (13°C) or up to 85°F (29°C).

803.3.3.3.2 Automatic setback and shutdown capabilities. Automatic time clock or programmable controls shall be capable of starting and stopping the system for seven different daily schedules per week and retaining their programming and time setting during a loss of power for at least 10 hours. Additionally, the controls shall have: a manual override that allows temporary operation of the system for up to 2 hours; a manually operated timer capable of being adjusted to operate the system for up to 2 hours; or an occupancy sensor.

803.3.3.4 Shutoff damper controls. Both outdoor air supply and exhaust ducts shall be equipped with gravity or motorized dampers that will automatically shut when the systems or spaces served are not in use.

Exception: Individual supply systems with a design airflow rate of 3,000 cfm (1416 L/s) or less.

803.3.3.5 Economizers. Economizers shall be provided on each system with a cooling capacity greater than 65,000 Btu/h (19 kW) in accordance with Section 803.2.6.

Exceptions:

1. Water economizers that are capable of cooling supply air by direct or indirect evaporation or both and providing up to 100 percent of the expected system cooling load at outside air temperatures of 50°F (10°C) dry bulb/45°F (7°C) wet bulb and below.
2. Systems with a cooling capacity less than 135,000 Btu/h (40 kW) in Climate Zones 3c, 5b, 7, 13b, and 14.

803.3.3.6 Variable air volume (VAV) fan control. Individual VAV fans with motors of 25 horsepower (18.8 kW) or greater shall be:

1. Driven by a mechanical or electrical variable speed drive; or
2. The fan motor shall have controls or devices that will result in fan motor demand of no more than 30 percent of their design wattage at 50 percent of design air flow when static pressure set point equals one-third of the total design static pressure, based on manufacturer's certified fan data.

803.3.3.7 Hydronic systems controls. The heating of fluids that have been previously mechanically cooled and the cooling of fluids that have been previously mechanically heated shall be limited in accordance with Sections 803.3.3.7.1 through 803.3.3.7.3. Hydronic heating systems comprised of multiple-packaged boilers and designed to deliver conditioned water or steam into a common distribution system shall include automatic controls capable of sequencing operation of the boilers. Hydronic heating systems comprised of a single boiler and greater than 500,000 Btu/h input design capacity shall include either a multistaged or modulating burner.

803.3.3.7.1 Three-pipe system. Hydronic systems that use a common return system for both hot water and chilled water are prohibited.

803.3.3.7.2 Two-pipe changeover system. Systems that use a common distribution system to supply both heated and chilled water shall be designed to allow a dead band between changeover from one mode to the other of at least 15°F (Δ8.3°C) outside air temperatures; be designed to and provided with controls that will allow operation in one mode for at least 4 hours before changing over to the other mode; and be provided with controls that allow heating and cooling supply temperatures at the changeover point to be no more than 30°F (Δ16.7°C) apart.

803.3.3.7.3 Hydronic (water loop) heat pump systems Hydronic heat pumps connected to a common heat pump water loop with central devices for heat rejection and heat addition shall have controls that are capable of providing a heat pump water supply temperature dead band of at least 20°F (Δ11.1°C) between initiation of heat rejection and heat addition by the central devices. For Climate Zones 5a to 19 as indicated in Table 302.1, if a closed-circuit cooling tower is used, either an automatic valve shall be installed to bypass all but a minimal flow of water around the tower, or lower leakage positive closure dampers shall be provided. If an open-circuit tower is used directly in the heat pump loop, an automatic valve shall be installed to bypass all heat pump water flow around the tower. If an open-circuit cooling tower is used in conjunction with a separate heat exchanger to isolate the cooling tower from the heat pump loop, then heat loss shall be controlled by shutting down the circulation pump on the cooling tower loop. Each hydronic heat pump on the hydronic system having a total pump system power exceeding 10 horsepower (hp) (7.5 kW) shall have a two-position valve.

Exception: Where a system loop temperature optimization controller is installed and can determine the most efficient operating temperature based on real time conditions of demand and capacity, dead bands of less than 20°F (11.1°C) shall be permitted.

803.3.3.7.4 Part load controls. Hydronic systems greater than or equal to 300,000 Btu/h (87,930 W) in design capacity supplying heated or chilled water to comfort conditioning systems shall include controls that have the capability to:

1. Automatically reset the supply water temperatures using zone return water temperature, building return water temperature, or outside air temperature as an indicator of building heating or cooling demand. The temperature shall be capable of being reset by at least 25 percent of the design supply-to-return water temperature difference; or
2. Reduce system pump flow by at least 50 percent of design flow rate utilizing adjustable speed drive(s) on pump(s), multiple staged pumps where at least one-half of the total pump horsepower is capable of being automatically turned off, control valves designed to modulate or step down, and close, as a function of load, or other approved means.

803.3.3.7.5 Pump isolation. Chilled water plants including more than one chiller shall have the capability to reduce flow automatically through the chiller plant when a chiller is shut down. Chillers piped in series for the purpose of increased temperature differential, shall be considered as one chiller.

Boiler plants including more than one boiler shall have the capability to reduce flow automatically through the boiler plant when a boiler is shut down.

803.3.3.8 Heat rejection equipment fan speed control. Each fan powered by a motor of 7.5 hp (5.6 kW) or larger shall have the capability to operate that fan at two-thirds of full speed or less, and shall have controls that automatically change the fan speed to control the leaving fluid temperature or condensing temperature/pressure of the heat rejection device.

> **Exception:** Factory-installed heat rejection devices within HVAC equipment tested and rated in accordance with Tables 803.3.2(1) through 803.3.2(6).

803.3.4 Requirements for complex mechanical systems serving multiple zones. Sections 803.3.4.1 through 803.3.4.3 shall apply to complex mechanical systems serving multiple zones. Supply air systems serving multiple zones shall be VAV systems which, during periods of occupancy, are designed and capable of being controlled to reduce primary air supply to each zone to one of the following before reheating, recooling or mixing takes place:

1. Thirty percent of the maximum supply air to each zone.
2. Three hundred cfm (142 L/s) or less where the maximum flow rate is less than 10 percent of the total fan system supply airflow rate.
3. The minimum ventilation requirements of Chapter 4 of the *International Mechanical Code*.

Exception: The following define when individual zones or when entire air distribution systems are exempted from the requirement for VAV control:

1. Zones where special pressurization relationships or cross-contamination requirements are such that VAV systems are impractical.
2. Zones or supply air systems where at least 75 percent of the energy for reheating or for providing warm air in mixing systems is provided from a site-recovered or site-solar energy source.
3. Zones where special humidity levels are required to satisfy process needs.
4. Zones with a peak supply air quantity of 300 cfm (142 L/s) or less and where the flow rate is less than 10 percent of the total fan system supply airflow rate.
5. Zones where the volume of air to be reheated, recooled or mixed is no greater than the volume of outside air required to meet the minimum ventilation requirements of Chapter 4 of the *International Mechanical Code*.
6. Zones or supply air systems with thermostatic and humidistatic controls capable of operating in sequence the supply of heating and cooling energy to the zone(s) and which are capable of preventing reheating, recooling, mixing or simultaneous supply of air that has been previously cooled, either mechanically or through the use of economizer systems, and air that has been previously mechanically heated.

803.3.4.1 Single duct variable air volume (VAV) systems, terminal devices. Single duct VAV systems shall use terminal devices capable of reducing the supply of primary supply air before reheating or recooling takes place.

803.3.4.2 Dual duct and mixing VAV systems, terminal devices. Systems that have one warm air duct and one cool air duct shall use terminal devices which are capable of reducing the flow from one duct to a minimum before mixing of air from the other duct takes place.

803.3.4.3 Single fan dual duct and mixing VAV systems, economizers. Individual dual duct or mixing heating and cooling systems with a single fan and with total capacities greater than 90,000 Btu/h [(26 375 W) 7.5 tons] shall not be equipped with air economizers.

803.3.5 Ventilation. Ventilation shall be in accordance with Section 803.2.5.

803.3.6 Duct and plenum insulation and sealing. All ducts and plenums shall be insulated and sealed in accordance with Section 803.2.8.

Ducts designed to operate at static pressures in excess of 3 inches w.g. (746 Pa) shall be leak-tested in accordance with the SMACNA *HVAC Air Duct Leakage Test Manual* with the rate of air leakage (CL) less than or equal to 6.0 as determined in accordance with Equation 8-2.

$$CL = F \times P^{0.65} \qquad \text{(Equation 8-2)}$$

where:

F = The measured leakage rate in cfm per 100 square feet of duct surface.

P = The static pressure of the test.

Documentation shall be furnished by the designer demonstrating that representative sections totaling at least 25 percent of the duct area have been tested and that all tested sections meet the requirements of this section.

803.3.7 Piping insulation. All piping serving as part of a heating or cooling system shall be thermally insulated in accordance with Table 803.3.7.

> **Exceptions:**
> 1. Factory-installed piping within HVAC equipment tested and rated in accordance with a test procedure referenced by this code.
> 2. Piping that conveys fluids that have a design operating temperature range between 55°F (13°C) and 105°F (41°C).
> 3. Piping that conveys fluids that have not been heated or cooled through the use of fossil fuels or electric power.
> 4. Runout piping not exceeding 4 feet (1219 mm) in length and 1 inch (25 mm) in diameter between the control valve and HVAC coil.

803.3.8 HVAC system completion. Prior to the issuance of a certificate of occupancy, the design professional shall provide evidence of system completion in accordance with Sections 803.3.8.1 through 803.3.8.3.

**TABLE 803.3.7
MINIMUM PIPE INSULATION[a]
(thickness in inches)**

FLUID	NOMINAL PIPE DIAMETER	
	≤ 1.5"	> 1.5"
Steam	$1^1/_2$	3
Hot water	1	2
Chilled water, brine or refrigerant	1	$1^1/_2$

For SI: 1 inch = 25.4 mm, British thermal unit per inch/h · ft² · °F = W per 25 mm/K · m².
a. Based on insulation having a conductivity (k) not exceeding 0.27 Btu per inch/h · ft² · °F.

803.3.8.1 Air system balancing. Each supply air outlet and zone terminal device shall be equipped with means for air balancing in accordance with the requirements of Chapter 6 of the *International Mechanical Code*. Discharge dampers are prohibited on constant volume fans and variable volume fans with motors 25 hp (18.6 kW) and larger.

803.3.8.2 Hydronic system balancing. Individual hydronic heating and cooling coils shall be equipped with means for balancing and pressure test connections.

803.3.8.3 Manuals. The construction documents shall require that an operating and maintenance manual be provided to the building owner by the mechanical contractor. The manual shall include, at least, the following:

1. Equipment capacity (input and output) and required maintenance actions.
2. Equipment operation and maintenance manuals.
3. HVAC system control maintenance and calibration information, including wiring diagrams, schematics, and control sequence descriptions. Desired or field-determined setpoints shall be permanently recorded on control drawings, at control devices or, for digital control systems, in programming comments.
4. A complete written narrative of how each system is intended to operate.

803.3.9 Heat recovery for service water heating. Condenser heat recovery shall be installed for heating or reheating of service hot water provided the facility operates 24 hours a day, the total installed heat capacity of water-cooled systems exceeds 6,000,000 Btu/hr of heat rejection, and the design service water heating load exceeds 1,000,000 Btu/h.

The required heat recovery system shall have the capacity to provide the smaller of:

1. Sixty percent of the peak heat rejection load at design conditions; or
2. The preheating required to raise the peak service hot water draw to 85°F (29°C).

Exceptions:
1. Facilities that employ condenser heat recovery for space heating or reheat purposes with a heat recovery design exceeding 30 percent of the peak water-cooled condenser load at design conditions.
2. Facilities that provide 60 percent of their service water heating from site solar or site recovered energy or from other sources

SECTION 804
SERVICE WATER HEATING

804.1 General. This section covers the minimum efficiency of, and controls for, service water-heating equipment and insulation of service hot water piping.

804.2 Service water-heating equipment performance efficiency. Water-heating equipment and hot water storage tanks shall meet the requirements of Table 804.2. The efficiency shall be verified through data furnished by the manufacturer or through certification under an approved certification program.

804.3 Temperature controls. Service water-heating equipment shall be provided with controls to allow a setpoint of 110°F (43°C) for equipment serving dwelling units and 90°F (32°C) for equipment serving other occupancies. The outlet temperature of lavatories in public facility rest rooms shall be limited to 110°F (43°C).

804.4 Heat traps. Water-heating equipment not supplied with integral heat traps and serving noncirculating systems shall be provided with heat traps on the supply and discharge piping associated with the equipment.

804.5 Pipe insulation. For automatic-circulating hot water systems, piping shall be insulated with 1 inch (25 mm) of insulation having a conductivity not exceeding 0.27 Btu per inch/h · ft² · °F (1.53 W per 25 mm/m² · K). The first 8 feet (2438 mm) of piping in noncirculating systems served by equipment without integral heat traps shall be insulated with 0.5 inch (12.7 mm) of material having a conductivity not exceeding 0.27 Btu per inch/h · ft² · °F (1.53 W per 25 mm/m² · K).

804.6 Hot water system controls. Automatic-circulating hot water system pumps or heat trace shall be arranged to be conveniently turned off automatically or manually when the hot water system is not in operation.

TABLE 804.2
MINIMUM PERFORMANCE OF WATER-HEATING EQUIPMENT

EQUIPMENT TYPE	SIZE CATEGORY (input)	SUBCATEGORY OR RATING CONDITION	PERFORMANCE REQUIRED[b]	TEST PROCEDURE
Water heaters, Electric	≤ 12 kW	Resistance	0.93 - 0.00132V, EF	DOE 10 CFR Part 430
	> 12 kW	Resistance	1.73V + 155 SL, Btu/h	ANSI Z21.10.3
	≤ 24 amps and ≤ 250 volts	Heat pump	0.93 - 0.00132V, EF	DOE 10 CFR Part 430
Storage water heaters, Gas	≤ 75,000 Btu/h	≥ 20 gal	0.62 - 0.0019V, EF	DOE 10 CFR Part 430
	> 75,000 Btu/h and ≤ 155,000 Btu/h	< 4,000 Btu/h/gal	80% E_t $\left(Q/800 + 110\sqrt{V}\right)$ SL, Btu/h	ANSI Z21.10.3
	> 155,000 Btu/h	< 4,000 Btu/h/gal	80% E_t $\left(Q/800 + 110\sqrt{V}\right)$ SL, Btu/h	
Instantaneous water heaters, Gas	> 50,000 Btu/h and < 200,000 Btu/h[d]	≥ 4,000 (Btu/h)/gal and < 2 gal	0.62 - 0.0019V EF	DOE 10 CFR Part 430
	≥ 200,000 Btu/h	≥ 4,000 Btu/h/gal and < 10 gal	80% E_t	ANSI Z21.10.3
	> 200,000 Btu/h	≥ 4,000 Btu/h/gal and ≥ 10 gal	80% E_t $\left(Q/800 + 110\sqrt{V}\right)$ SL, Btu/h	
Storage water heaters, Oil	≤ 105,000 Btu/h	≥ 20 gal	0.59 - 0.0019V, EF	DOE 10 CFR Part 430
	> 105,000 Btu/h	< 4,000 Btu/h/gal	78% E_t $\left(Q/800 + 110\sqrt{V}\right)$ SL, Btu/h	ANSI Z21.10.3
Instantaneous water heaters, Oil	≤ 210,000 Btu/h	≥ 4,000 Btu/h/gal and < 2 gal	0.59 - 0.0019V, EF	DOE 10 CFR Part 430
	> 210,000 Btu/h	≥ 4,000 Btu/h/gal and < 10 gal	80% E_t	ANSI Z21.10.3
	> 210,000 Btu/h	≥ 4,000 Btu/h/gal and ≥ 10 gal	78% E_t $\left(Q/800 + 110\sqrt{V}\right)$ SL, Btu/h	
Hot water supply boilers, Gas and Oil	≥ 300,000 Btu/h and <12,500,000 Btu/h	≥ 4,000 Btu/h/gal and < 10 gal	80% E_t	ANSI Z21.10.3
Hot water supply boilers, Gas and Oil		≥ 4,000 Btu/h/gal and ≥ 10 gal	80% E_t $\left(Q/800 + 110\sqrt{V}\right)$ SL, Btu/h	
Pool heaters, Gas and Oil	All	—	78% E_t	ASHRAE 146
Unfired storage tanks	All	—	≤ 6.5 Btu/h · ft^2	(none)

For SI: °C = [(°F) - 32] / 1.8, 1 British thermal unit per hour = 0.2931 W, 1 gallon = 3.785 L, 1 British thermal unit per hour per gallon = 0.078 W/L.

a. Energy factor (EF) and thermal efficiency (E_t) are minimum requirements. In the EF equation, V is the rated volume in gallons

b. Standby loss (SL) is the maximum Btu/h based on a nominal 70°F temperature difference between stored water and ambient requirements. In the SL equation, Q is the nameplate input rate in Btu/h. In the SL equation for electric water heaters, V is the rated volume in gallons. In the SL equation for oil and gas water heaters and boilers, V is the rated volume in gallons.

c. Instantaneous water heaters with input rates below 200,000 Btu/h must comply with these requirements if the water heater is designed to heat water to temperatures 180°F or higher.

SECTION 805
ELECTRICAL POWER AND LIGHTING SYSTEMS

805.1 General. This section covers lighting system controls, the connection of ballasts, the maximum lighting power for interior applications, and minimum acceptable lighting equipment for exterior applications.

Exception: Lighting within dwelling units.

805.2 Lighting controls. Lighting systems shall be provided with controls as required in Sections 805.2.1, 805.2.2 and 805.2.3.

805.2.1 Interior lighting controls. Each area enclosed by walls or floor-to-ceiling partitions shall have at least one manual control for the lighting serving that area. The required controls shall be located within the area served by the controls or be a remote switch that identifies the lights served and indicates their status.

Exceptions:

1. Areas designated as security or emergency areas that must be continuously lighted.
2. Lighting in stairways or corridors that are elements of the means of egress.

805.2.2 Additional controls. Each area that is required to have a manual control shall have additional controls that meet the requirements of Sections 805.2.2.1, 805.2.2.2 and 805.2.2.3.

Exceptions:

1. Areas that have only one luminaire.
2. Areas that are controlled by an occupant-sensing device.
3. Corridors, storerooms, restrooms or public lobbies.

805.2.2.1 Light reduction controls. Each area that is required to have a manual control shall also allow the occupant to reduce the connected lighting load in a reasonably uniform illumination pattern by at least 50 percent. Lighting reduction shall be achieved by one of the following or other approved method:

1. Controlling all lamps or luminaries;
2. Dual switching of alternate rows of luminaires, alternate luminaires or alternate lamps;
3. Switching the middle lamp luminaries independently of the outer lamps; or
4. Switching each luminaire or each lamp.

Exceptions:

1. Areas that have only one luminaire.
2. Areas that are controlled by an occupant-sensing device.
3. Corridors, storerooms, restrooms or public lobbies.
4. Guestrooms.
5. Spaces that use less than 0.6 Watts per square foot (6.5 W/m).

805.2.2.2 Automatic lighting shutoff. Buildings larger than 5,000 square feet (465 m^2) shall be equipped with an automatic control device to shut off lighting in those areas. This automatic control device shall function on either:

1. A scheduled basis, using time-of-day, with an independent program schedule that controls the interior lighting in areas that do not exceed 25,000 square feet (2323 m^2) and are not more than one floor; or
2. An unscheduled basis by occupant intervention.

805.2.2.2.1 Occupant override. Where an automatic time switch control device is installed to comply with Section 805.2.2.2, Item 1, it shall incorporate an override switching device that:

1. Is readily accessible.
2. Is located so that a person using the device can see the lights or the area controlled by that switch, or so that the area being lit is annunciated.
3. Is manually operated.
4. Allows the lighting to remain on for no more than 2 hours when an override is initiated.
5. Controls an area not exceeding 5,000 square feet (465 m^2).

Exceptions:

1. In malls and arcades, auditoriums, single-tenant retail spaces, industrial facilities and arenas, where captive-key override is utilized, override time may exceed 2 hours.
2. In malls and arcades, auditoriums, single-tenant retail spaces, industrial facilities and arenas, the area controlled may not exceed 20,000 square feet (1860 m^2).

805.2.2.2.2 Holiday scheduling. If an automatic time switch control device is installed in accordance with Section 805.2.2.2, Item 1, it shall incorporate an automatic holiday scheduling feature that turns off all loads for at least 24 hours, then resumes the normally scheduled operation.

Exception: Retail stores and associated malls, restaurants, grocery stores, churches and theaters.

805.2.2.3 Guestrooms. Guestrooms in hotels, motels, boarding houses or similar buildings shall have at least one master switch at the main entry door that controls all permanently wired lighting fixtures and switched receptacles, except those in the bathroom(s). Suites shall have a control meeting these requirements at the entry to each room or at the primary entry to the suite.

805.2.3 Exterior lighting controls. Automatic switching or photocell controls shall be provided for all exterior lighting not intended for 24-hour operation. Automatic time switches shall have a combination seven-day and seasonal daylight program schedule adjustment, and a minimum 4-hour power backup.

805.3 Tandem wiring. The following luminaires located within the same area shall be tandem wired:

1. Flourescent luminaires equipped with one, three or odd-numbered lamp configurations, that are recess-mounted within 10 feet (3048 mm) center-to-center of each other.
2. Flourescent luminaires equipped with one, three or any other odd-numbered lamp configuration, that are pendant- or surface-mounted within 1 foot (305 mm) edge-to-edge of each other.

Exceptions:

1. Where electronic high-frequency ballasts are used.
2. Luminaires on emergency circuits.
3. Luminaires with no available pair in the same area.

805.4 Exit signs. Internally illuminated exit signs shall not exceed 5 Watts per side.

805.5 Interior lighting power requirements. A building complies with this section if its total connected lighting power calculated under Section 805.5.1 is no greater than the interior lighting power calculated under Section 805.5.2.

805.5.1 Total connected interior lighting power. The total connected interior lighting power (Watts) shall be the sum of the watts of all interior lighting equipment as determined in accordance with Sections 805.5.1.1 through 805.5.1.4.

Exceptions: The connected power associated with the following lighting equipment is not included in calculating total connected lighting power.

1. Specialized medical, dental and research lighting.
2. Professional sports arena playing field lighting.
3. Display lighting for exhibits in galleries, museums and monuments.
4. Guestroom lighting in hotels, motels, boarding houses or similar buildings.
5. Emergency lighting automatically off during normal building operation.

805.5.1.1 Screw lamp holders. The wattage shall be the maximum labeled wattage of the luminaire.

805.5.1.2 Low-voltage lighting. The wattage shall be the specified wattage of the transformer supplying the system.

805.5.1.3 Other luminaires. The wattage of all other lighting equipment shall be the wattage of the lighting equipment verified through data furnished by the manufacturer or other approved sources.

805.5.1.4 Line-voltage lighting track and plug-in busway. The wattage shall be the greater of the wattage of the luminaires determined in accordance with Sections 805.5.1.1 through 805.5.1.3 or 30 W/linear foot (98W/lin m).

805.5.2 Interior lighting power. The interior lighting power shall be calculated using Section 805.5.2.1 or 805.5.2.2 as applicable.

805.5.2.1 Entire building method. Under this approach, the interior lighting power (Watts) is the value from Table 805.5.2 for the building type times the conditioned floor area of the entire building. The interior lighting power (Watts) shall not be increased by the allowances contained in the footnotes of Table 805.5.2 when using the entire building method.

805.5.2.2 Tenant area or portion of building method. The total interior lighting power (Watts) is the sum of all interior lighting powers for all areas in the building covered in this permit. The interior lighting power is the conditioned floor area for each area type listed in Table 805.5.2 times the value from Table 805.5.2 for that area. For the purposes of this method, an "area" shall be defined as all contiguous spaces that accommodate or are associated with a single area type as listed in Table 805.5.2. When this method is used to calculate the total interior lighting power for an entire building, each area type shall be treated as a separate area.

805.6 Exterior lighting. When the power for exterior lighting is supplied through the energy service to the building, all exterior lighting, other than low-voltage landscape lighting, shall have a source efficacy of at least 45 lumens per Watt.

Exception: Where approved because of historical, safety, signage or emergency considerations.

805.7 Electrical energy consumption. In buildings having individual dwelling units, provisions shall be made to determine the electrical energy consumed by each tenant by separately metering individual dwelling units.

SECTION 806
TOTAL BUILDING PERFORMANCE

806.1 General. The proposed design complies with this section where annual energy costs of the proposed design as determined in accordance with Section 806.3 do not exceed those of the standard design as determined in accordance with Section 806.4.

806.2 Analysis procedures. Sections 806.2.1 through 806.2.8 shall be applied in determining total building performance.

806.2.1 Energy analysis. Annual (8,760 hours) energy costs for the standard design and the proposed design shall each be determined using the same approved energy analysis simulation tool.

806.2.2 Climate data. The climate data used in the energy analysis shall cover a full calendar year (8,760 hours) and shall reflect approved coincident hourly data for temperature, solar radiation, humidity and wind speed for the building location.

806.2.3 Energy rates. The annual energy costs shall be estimated using energy rates published by the serving energy supplier and which would apply to the actual building or *DOE State-Average Energy Prices* published by DOE's Energy Information Administration and which would apply to the actual building.

ACCEPTABLE PRACTICE FOR COMMERCIAL BUILDINGS

**TABLE 805.5.2
INTERIOR LIGHTING POWER**

BUILDING OR AREA TYPE	ENTIRE BUILDING (W/ft^2)	TENANT AREA OR PORTION OF BUILDING (W/ft^2)
Auditorium	Not Applicable	1.8
Automotive facility	0.9	Not Applicable
Bank/financial institution [a]	Not Applicable	1.5
Classroom/lecture hall [b]	Not Applicable	1.4
Convention, conference or meeting center [a]	1.2	1.3
Corridor, restroom, support area	Not Applicable	0.9
Courthouse/town hall	1.2	Not Applicable
Dining [a]	Not Applicable	0.9
Dormitory	1.0	NA
Exercise center [a]	1.0	0.9
Exhibition hall	Not Applicable	1.3
Grocery store [c]	1.5	1.6
Gymnasium playing surface	Not Applicable	1.4
Hotel function [a]	1.0	1.3
Industrial work, < 20-foot ceiling height	Not Applicable	1.2
Industrial work, ≥ 20-foot ceiling height	Not Applicable	1.7
Kitchen	Not Applicable	1.2
Library [a]	1.3	1.7
Lobby—hotel [a]	Not Applicable	1.1
Lobby—other [a]	Not Applicable	1.3
Mall, arcade, or atrium	Not Applicable	0.6
Medical and clinical care [b, d]	1.2	1.2
Motel	1.0	Not Applicable
Multifamily	0.7	Not Applicable
Museum [b]	1.1	1.0
Office [b]	1.0	1.1
Parking garage	0.3	Not Applicable
Penitentiary	1.0	Not Applicable
Police/fire station	1.0	Not Applicable
Post office	1.1	Not Applicable
Religious worship [a]	1.3	2.4
Restaurant [a]	1.6	0.9
Retail sales, wholesale showroom [c]	1.5	1.7
School	1.2	Not Applicable
Storage, industrial and commercial	0.8	0.8
Theaters—motion picture	1.2	1.2
Theaters—performance [a]	1.6	2.6
Transportation	1.0	Not Applicable
Other	0.6	1.0

For SI: 1 foot = 304.8 mm, 1 Watts per square foot = W/0.0929 m^2.

a. Where lighting equipment is specified to be installed for decorative appearances in addition to lighting equipment specified for general lighting and is switched or dimmed on circuits different from the circuits for general lighting, the smaller of the actual wattage of the decorative lighting equipment or 1.0 W/ft^2 times the area of the space that the decorative lighting equipment is in shall be added to the interior lighting power determined in accordance with this line item.

b. Where lighting equipment is specified to be installed to meet requirements of visual display terminals as the primary viewing task, the smaller of the actual wattage of the lighting equipment or 0.35 W/ft^2 times the area of the space that the lighting equipment is in shall be added to the interior lighting power determined in accordance with this line item.

c. Where lighting equipment is specified to be installed to highlight specific merchandise in addition to lighting equipment specified for general lighting and is switched or dimmed on circuits different from the circuits for general lighting, the smaller of the actual wattage of the lighting equipment installed specifically for merchandise, or 1.6 W/ft^2 times the area of the specific display, or 3.9 W/ft^2 times the actual case or shelf area for displaying and selling fine merchandise such as jewelry, fine apparel and accessories, or china and silver, shall be added to the interior lighting power determined in accordance with this line item.

d. Where lighting equipment is specified to be installed, the smaller of the actual wattage of the lighting equipment, or 1.0 W/ft^2 times the area of the emergency, recovery, medical supply and pharmacy space shall be added to the interior lighting power determined in accordance with this line item.

806.2.4 Nondepletable energy. Nondepletable energy collected off site shall be treated and priced the same as purchased energy. Energy from nondepletable energy sources collected on site shall be omitted from the annual energy cost of the proposed design. The analysis and performance of any nondepletable energy system shall be determined in accordance with accepted engineering practice using approved methods.

806.2.5 Building operation. Building operation shall be simulated for a full calendar year (8,760 hours). Operating schedules shall include hourly profiles for daily operation and shall account for variations between weekdays, weekends, holidays, and any seasonal operation. Schedules shall model the time-dependent variations of occupancy, illumination, receptacle loads, thermostat settings, mechanical ventilation, HVAC equipment availability, service hot water usage, and any process loads.

806.2.6 Simulated loads. The following systems and loads shall be modeled in determining total building performance: heating systems, cooling systems, fan systems, lighting power, receptacle loads, and process loads that exceed 1.0 W/ft^2 (W/0.0929 m^2) of floor area of the room or space in which the process loads are located.

> **Exception:** Systems and loads serving required emergency power only.

806.2.7 Service water-heating systems. Service water-heating systems that are other than combined service hot water/space-heating systems shall be be omitted from the energy analysis provided all requirements in Section 804 have been met.

806.2.8 Exterior lighting. Exterior lighting systems shall be the same as in the standard and proposed designs.

806.3 Determining energy costs for the proposed design. Building systems and loads shall be simulated in the Proposed design in accordance with Sections 806.3.1 and 806.3.2.

806.3.1 HVAC and service water-heating equipment. All HVAC and service water-heating equipment shall be simulated in the proposed design using capacities, rated efficiencies and part-load performance data for the proposed equipment as provided by the equipment manufacturer.

806.3.2 Features not documented at time of permit. If any feature of the proposed design is not included in the building permit application, the energy performance of that feature shall be assumed to be that of the corresponding feature used in the calculations required in Section 806.4.

806.4 Determining energy costs for the standard design. Sections 806.4.1 through 806.4.7 shall be used in determining the annual energy costs of the Standard design.

806.4.1 Equipment efficiency. The space-heating, space-cooling, service water-heating, and ventilation systems and equipment shall meet, but not exceed, the minimum efficiency requirements of Sections 803 and 804.

806.4.2 HVAC system capacities. HVAC system capacities in the standard design shall be established such that no smaller number of unmet heating and cooling load hours and no larger heating and cooling capacity safety factors are provided than in the proposed design.

806.4.3 Envelope. The performance of elements of the thermal envelope of the standard design shall be determined in accordance with the requirements of Section 802.2 as applicable.

806.4.4 Identical characteristics. The heating/cooling system zoning, the orientation of each building feature, the number of floors and the gross envelope areas of the standard design shall be the same as those of the proposed design except as modified by Section 806.4.5 or 806.4.6.

> **Exception:** Permanent fixed or movable external shading devices for windows and glazed doors shall be excluded from the standard design.

806.4.5 Window area. The window area of the standard design shall be the same as the proposed design, or 35 percent of the above-grade wall area, whichever is less, and shall be distributed in a uniform pattern equally over each building facade.

806.4.6 Skylight area. The skylight area of the standard design shall be the same as the proposed design, or 3 percent of the gross area of the roof assembly, whichever is less.

806.4.7 Interior lighting. The lighting power for the standard design shall be the maximum allowed in accordance with Section 805.4. Where the occupancy of the building is not known, the lighting power density shall be 1.5 Watts per square foot (16.1 W/m^2).

806.5 Documentation. The energy analysis and supporting documentation shall be prepared by a registered design professional where required by the statutes of the jurisdiction in which the project is to be constructed. The information documenting compliance shall be submitted in accordance with Sections 806.5.1 through 806.5.4

806.5.1 Annual energy use and associated costs. The annual energy use and costs by energy source of the standard design and the proposed design shall be clearly indicated.

806.5.2 Energy-related features. A list of the energy-related features that are included in the proposed design and on which compliance with the provisions of the code are claimed shall be provided to the code official. This list shall include and prominently indicate all features that differ from those set forth in Section 806.4 and used in the energy analysis between the standard design and the proposed design.

806.5.3 Input and output report(s). Input and output report(s) from the energy analysis simulation program containing the complete input and output files, as applicable. The output file shall include energy use totals and energy use by energy source and end-use served, total hours that space conditioning loads are not met and any errors or warning messages generated by the simulation tool as applicable.

806.5.4 Written explanation(s). An explanation of any error or warning messages appearing in the simulation tool output shall be provided in a written, narrative format.

TABLE 802.2(5)
BUILDING ENVELOPE REQUIREMENTS[a through e] - CLIMATE ZONE 1a

WINDOW AND GLAZED DOOR AREA 10 PERCENT OR LESS OF ABOVE-GRADE WALL AREA			
ELEMENT	CONDITION/VALUE		
Skylights (*U*-factor)	1		
Slab or below-grade wall (*R*-value)	R-0		
Windows and glass doors	SHGC		*U*-factor
PF < 0.25	Any		Any
0.25 ≤ PF < 0.50	Any		Any
PF ≥ 0.50	Any		Any
Roof assemblies (*R*-value)	Insulation between framing		Continuous insulation
All-wood joist/truss	R-13		R-11
Metal joist/truss	R-13		R-12
Concrete slab or deck	NA		R-11
Metal purlin with thermal block	R-19		R-12
Metal purlin without thermal block	R-30		R-12
Floors over outdoor air or unconditioned space (*R*-value)	Insulation between framing		Continuous insulation
All-wood joist/truss	R-0		R-0
Metal joist/truss	R-0		R-0
Concrete slab or deck	NA		R-0
Above-grade walls (*R*-value)	No framing	Metal framing	Wood framing
Framed			
R-value cavity	NA	R-0	R-0
R-value continuous	NA	R-0	R-0
CMU, ≥ 8 inches, with integral insulation			
R-value cavity	NA	R-0	R-0
R-value continuous	R-0	R-0	R-0
Other masonry walls			
R-value cavity	NA	R-0	R-0
R-value continuous	R-0	R-0	R-0
WINDOW AND GLAZED DOOR AREA GREATER THAN 10 PERCENT BUT NOT GREATER THAN 25 PERCENT OF ABOVE-GRADE WALL AREA			
ELEMENT	CONDITION/VALUE		
Skylights (*U*-factor)	1		
Slab or below-grade wall (*R*-value)	R-0		
Windows and glass doors	SHGC		*U*-factor
PF < 0.25	0.6		Any
0.25 ≤ PF < 0.50	0.7		Any
PF ≥ 0.50	Any		Any
Roof assemblies (*R*-value)	Insulation between framing		Continuous insulation
All-wood joist/truss	R-19		R-14
Metal joist/truss	R-19		R-15
Concrete slab or deck	NA		R-14
Metal purlin with thermal block	R-25		R-15
Metal purlin without thermal block	X		R-15
Floors over outdoor air or unconditioned space (*R*-value)	Insulation between framing		Continuous insulation
All-wood joist/truss	R-0		R-0
Metal joist/truss	R-0		R-0
Concrete slab or deck	NA		R-0
Above-grade walls (*R*-value)	No framing	Metal framing	Wood framing
Framed			
R-value cavity	NA	R-0	R-0
R-value continuous	NA	R-0	R-0
CMU, ≥ 8 inches, with integral insulation			
R-value cavity	NA	R-0	R-0
R-value continuous	R-0	R-0	R-0
Other masonry walls			
R-value cavity	NA	R-0	R-0
R-value continuous	R-0	R-0	R-0

(continued)

ACCEPTABLE PRACTICE FOR COMMERCIAL BUILDING

TABLE 802.2(5)—continued
BUILDING ENVELOPE REQUIREMENTS[a through e] - CLIMATE ZONE 1a

WINDOW AND GLAZED DOOR AREA GREATER THAN 25 PERCENT BUT NOT GREATER THAN 40 PERCENT OF ABOVE-GRADE WALL AREA			
ELEMENT	**CONDITION/VALUE**		
Skylights (U-factor)	1		
Slab or below-grade wall (R-value)	R-0		
Windows and glass doors	**SHGC**		***U*-factor**
PF < 0.25	0.4		0.7
0.25 ≤ PF < 0.50	0.5		0.7
PF ≥ 0.50	0.6		0.7
Roof assemblies (R-value)	**Insulation between framing**		**Continuous insulation**
All-wood joist/truss	R-19		R-16
Metal joist/truss	R-25		R-17
Concrete slab or deck	NA		R-16
Metal purlin with thermal block	R-25		R-17
Metal purlin without thermal block	X		R-17
Floors over outdoor air or unconditioned space (R-value)	**Insulation between framing**		**Continuous insulation**
All-wood joist/truss	R-0		R-0
Metal joist/truss	R-0		R-0
Concrete slab or deck	NA		R-0
Above-grade walls (R-value)	**No framing**	**Metal framing**	**Wood framing**
Framed			
R-value cavity	NA	R-0	R-0
R-value continuous	NA	R-0	R-0
CMU, ≥ 8 inches, with integral insulation			
R-value cavity	NA	R-0	R-0
R-value continuous	R-0	R-0	R-0
Other masonry walls			
R-value cavity	NA	R-0	R-0
R-value continuous	R-0	R-0	R-0
WINDOW AND GLAZED DOOR AREA GREATER THAN 40 PERCENT BUT NOT GREATER THAN 50 PERCENT OF ABOVE-GRADE WALL AREA			
ELEMENT	**CONDITION/VALUE**		
Skylights (U-factor)	1		
Slab or below-grade wall (R-value)	R-0		
Windows and glass doors	**SHGC**		***U*-factor**
PF < 0.25	0.3		0.7
0.25 ≤ PF < 0.50	0.4		0.7
PF ≥ 0.50	0.5		0.7
Roof assemblies (R-value)	**Insulation between framing**		**Continuous insulation**
All-wood joist/truss	R-19		R-16
Metal joist/truss	R-25		R-17
Concrete slab or deck	NA		R-16
Metal purlin with thermal block	R-25		R-17
Metal purlin without thermal block	R-30		R-17
Floors over outdoor air or unconditioned space (R-value)	**Insulation between framing**		**Continuous insulation**
All-wood joist/truss	R-0		R-0
Metal joist/truss	R-0		R-0
Concrete slab or deck	NA		R-0
Above-grade walls (R-value)	**No framing**	**Metal framing**	**Wood framing**
Framed			
R-value cavity	NA	R-0	R-0
R-value continuous	NA	R-0	R-0
CMU, ≥ 8 inches, with integral insulation			
R-value cavity	NA	R-0	R-0
R-value continuous	R-0	R-0	R-0
Other masonry walls			
R-value cavity	NA	R-0	R-0
R-value continuous	R-0	R-0	R-0

For SI: 1 inch = 25.4 mm.

a. Values from Tables 802.2(5) through 802.2(37) shall be used for the purpose of the completion of Tables 802.2(1) through 802.2(4), as applicable based on window and glazed door area.
b. "NA" indicates the condition is not applicable.
c. An R-value of zero indicates no insulation is required.
d. "Any" indicates any available product will comply.
e. "X" indicates no complying option exists for this condition.

TABLE 802.2(6)
BUILDING ENVELOPE REQUIREMENTS[a through e] - CLIMATE ZONE 1b

WINDOW AND GLAZED DOOR AREA 10 PERCENT OR LESS OF ABOVE-GRADE WALL AREA			
ELEMENT	CONDITION/VALUE		
Skylights (U-factor)	1		
Slab or below-grade wall (R-value)	R-0		
Windows and glass doors	SHGC		U-factor
PF < 0.25	Any		Any
0.25 ≤ PF < 0.50	Any		Any
PF ≥ 0.50	Any		Any
Roof assemblies (R-value)	Insulation between framing		Continuous insulation
All-wood joist/truss	R-19		R-14
Metal joist/truss	R-19		R-15
Concrete slab or deck	NA		R-14
Metal purlin with thermal block	R-25		R-15
Metal purlin without thermal block	X		R-15
Floors over outdoor air or unconditioned space (R-value)	Insulation between framing		Continuous insulation
All-wood joist/truss	R-0		R-0
Metal joist/truss	R-0		R-0
Concrete slab or deck	NA		R-0
Above-grade walls (R-value)	No framing	Metal framing	Wood framing
Framed			
R-value cavity	NA	R-0	R-0
R-value continuous	NA	R-0	R-0
CMU, ≥ 8 inches, with integral insulation			
R-value cavity	NA	R-0	R-0
R-value continuous	R-0	R-0	R-0
Other masonry walls			
R-value cavity	NA	R-0	R-0
R-value continuous	R-0	R-0	R-0
WINDOW AND GLAZED DOOR AREA GREATER THAN 10 PERCENT BUT NOT GREATER THAN 25 PERCENT OF ABOVE-GRADE WALL AREA			
ELEMENT	CONDITION/VALUE		
Skylights (U-factor)	1		
Slab or below-grade wall (R-value)	R-0		
Windows and glass doors	SHGC		U-factor
PF < 0.25	0.6		Any
0.25 ≤ PF < 0.50	0.7		Any
PF ≥ 0.50	Any		Any
Roof assemblies (R-value)	Insulation between framing		Continuous insulation
All-wood joist/truss	R-19		R-14
Metal joist/truss	R-19		R-15
Concrete slab or deck	NA		R-14
Metal purlin with thermal block	R-25		R-15
Metal purlin without thermal block	X		R-15
Floors over outdoor air or unconditioned space (R-value)	Insulation between framing		Continuous insulation
All-wood joist/truss	R-0		R-0
Metal joist/truss	R-0		R-0
Concrete slab or deck	NA		R-0
Above-grade walls (R-value)	No framing	Metal framing	Wood framing
Framed			
R-value cavity	NA	R-0	R-0
R-value continuous	NA	R-0	R-0
CMU, ≥ 8 inches, with integral insulation			
R-value cavity	NA	R-0	R-0
R-value continuous	R-0	R-0	R-0
Other masonry walls			
R-value cavity	NA	R-0	R-0
R-value continuous	R-0	R-0	R-0

(continued)

TABLE 802.2(6)—continued
BUILDING ENVELOPE REQUIREMENTS[a through e] - CLIMATE ZONE 1b

WINDOW AND GLAZED DOOR AREA GREATER THAN 25 PERCENT BUT NOT GREATER THAN 40 PERCENT OF ABOVE-GRADE WALL AREA			
ELEMENT	**CONDITION/VALUE**		
Skylights (U-factor)	1		
Slab or below-grade wall (R-value)	R-0		
Windows and glass doors	**SHGC**		**U-factor**
PF < 0.25	0.5		0.7
0.25 ≤ PF < 0.50	0.6		0.7
PF ≥ 0.50	0.7		0.7
Roof assemblies (R-value)	**Insulation between framing**		**Continuous insulation**
All-wood joist/truss	R-19		R-14
Metal joist/truss	R-19		R-15
Concrete slab or deck	NA		R-14
Metal purlin with thermal block	R-25		R-15
Metal purlin without thermal block	X		R-15
Floors over outdoor air or unconditioned space (R-value)	**Insulation between framing**		**Continuous insulation**
All-wood joist/truss	R-0		R-0
Metal joist/truss	R-0		R-0
Concrete slab or deck	NA		R-0
Above-grade walls (R-value)	**No framing**	**Metal framing**	**Wood framing**
Framed			
R-value cavity	NA	R-0	R-0
R-value continuous	NA	R-0	R-0
CMU, ≥ 8 inches, with integral insulation			
R-value cavity	NA	R-0	R-0
R-value continuous	R-0	R-0	R-0
Other masonry walls			
R-value cavity	NA	R-0	R-0
R-value continuous	R-0	R-0	R-0
WINDOW AND GLAZED DOOR AREA GREATER THAN 40 PERCENT BUT NOT GREATER THAN 50 PERCENT OF ABOVE-GRADE WALL AREA			
ELEMENT	**CONDITION/VALUE**		
Skylights (U-factor)	1		
Slab or below-grade wall (R-value)	R-0		
Windows and glass doors	**SHGC**		**U-factor**
PF < 0.25	0.4		0.7
0.25 ≤ PF < 0.50	0.5		0.7
PF ≥ 0.50	0.7		0.7
Roof assemblies (R-value)	**Insulation between framing**		**Continuous insulation**
All-wood joist/truss	R-19		R-14
Metal joist/truss	R-19		R-15
Concrete slab or deck	NA		R-14
Metal purlin with thermal block	R-25		R-15
Metal purlin without thermal block	R-30		R-15
Floors over outdoor air or unconditioned space (R-value)	**Insulation between framing**		**Continuous insulation**
All-wood joist/truss	R-0		R-0
Metal joist/truss	R-0		R-0
Concrete slab or deck	NA		R-0
Above-grade walls (R-value)	**No framing**	**Metal framing**	**Wood framing**
Framed			
R-value cavity	NA	R-0	R-0
R-value continuous	NA	R-0	R-0
CMU, ≥ 8 inches, with integral insulation			
R-value cavity	NA	R-0	R-0
R-value continuous	R-0	R-0	R-0
Other masonry walls			
R-value cavity	NA	R-0	R-0
R-value continuous	R-0	R-0	R-0

For SI: 1 inch = 25.4 mm.

a. Values from Tables 802.2(5) through 802.2(37) shall be used for the purpose of the completion of Tables 802.2(1) through 802.2(4), as applicable based on window and glazed door area.
b. "NA" indicates the condition is not applicable.
c. An R-value of zero indicates no insulation is required.
d. "Any" indicates any available product will comply.
e. "X" indicates no complying option exists for this condition.

TABLE 802.2(7)
BUILDING ENVELOPE REQUIREMENTS[a through e] - CLIMATE ZONE 2a

WINDOW AND GLAZED DOOR AREA 10 PERCENT OR LESS OF ABOVE-GRADE WALL AREA			
ELEMENT	CONDITION/VALUE		
Skylights (U-factor)	1		
Slab or below-grade wall (R-value)	R-0		
Windows and glass doors	SHGC		U-factor
PF < 0.25	Any		Any
0.25 ≤ PF < 0.50	Any		Any
PF ≥ 0.50	Any		Any
Roof assemblies (R-value)	Insulation between framing		Continuous insulation
All-wood joist/truss	R-19		R-13
Metal joist/truss	R-19		R-14
Concrete slab or deck	NA		R-13
Metal purlin with thermal block	R-19		R-14
Metal purlin without thermal block	X		R-14
Floors over outdoor air or unconditioned space (R-value)	Insulation between framing		Continuous insulation
All-wood joist/truss	R-0		R-0
Metal joist/truss	R-11		R-4
Concrete slab or deck	NA		R-0
Above-grade walls (R-value)	No framing	Metal framing	Wood framing
Framed			
R-value cavity	NA	R-0	R-0
R-value continuous	NA	R-0	R-0
CMU, ≥ 8 inches, with integral insulation			
R-value cavity	NA	R-0	R-0
R-value continuous	R-0	R-0	R-0
Other masonry walls			
R-value cavity	NA	R-0	R-0
R-value continuous	R-0	R-0	R-0
WINDOW AND GLAZED DOOR AREA GREATER THAN 10 PERCENT BUT NOT GREATER THAN 25 PERCENT OF ABOVE-GRADE WALL AREA			
ELEMENT	CONDITION/VALUE		
Skylights (U-factor)	1		
Slab or below-grade wall (R-value)	R-0		
Windows and glass doors	SHGC		U-factor
PF < 0.25	0.5		Any
0.25 ≤ PF < 0.50	0.6		Any
PF ≥ 0.50	0.7		Any
Roof assemblies (R-value)	Insulation between framing		Continuous insulation
All-wood joist/truss	R-19		R-13
Metal joist/truss	R-19		R-14
Concrete slab or deck	NA		R-13
Metal purlin with thermal block	R-19		R-14
Metal purlin without thermal block	X		R-14
Floors over outdoor air or unconditioned space (R-value)	Insulation between framing		Continuous insulation
All-wood joist/truss	R-0		R-0
Metal joist/truss	R-11		R-4
Concrete slab or deck	NA		R-0
Above-grade walls (R-value)	No framing	Metal framing	Wood framing
Framed			
R-value cavity	NA	R-0	R-0
R-value continuous	NA	R-0	R-0
CMU, ≥ 8 inches, with integral insulation			
R-value cavity	NA	R-0	R-0
R-value continuous	R-0	R-0	R-0
Other masonry walls			
R-value cavity	NA	R-0	R-0
R-value continuous	R-0	R-0	R-0

(continued)

TABLE 802.2(7)—continued
BUILDING ENVELOPE REQUIREMENTS[a through e] - CLIMATE ZONE 2a

WINDOW AND GLAZED DOOR AREA GREATER THAN 25 PERCENT BUT NOT GREATER THAN 40 PERCENT OF ABOVE-GRADE WALL AREA			
ELEMENT	CONDITION/VALUE		
Skylights (*U*-factor)	1		
Slab or below-grade wall (*R*-value)	R-0		
Windows and glass doors	SHGC		*U*-factor
PF < 0.25	0.4		0.7
0.25 ≤ PF < 0.50	0.5		0.7
PF ≥ 0.50	0.6		0.7
Roof assemblies (*R*-value)	Insulation between framing		Continuous insulation
All-wood joist/truss	R-19		R-13
Metal joist/truss	R-19		R-14
Concrete slab or deck	NA		R-13
Metal purlin with thermal block	R-19		R-14
Metal purlin without thermal block	X		R-14
Floors over outdoor air or unconditioned space (*R*-value)	Insulation between framing		Continuous insulation
All-wood joist/truss	R-0		R-0
Metal joist/truss	R-11		R-4
Concrete slab or deck	NA		R-0
Above-grade walls (*R*-value)	No framing	Metal framing	Wood framing
Framed			
R-value cavity	NA	R-0	R-0
R-value continuous	NA	R-0	R-0
CMU, ≥ 8 inches, with integral insulation			
R-value cavity	NA	R-0	R-0
R-value continuous	R-0	R-0	R-0
Other masonry walls			
R-value cavity	NA	R-0	R-0
R-value continuous	R-0	R-0	R-0
WINDOW AND GLAZED DOOR AREA GREATER THAN 40 PERCENT BUT NOT GREATER THAN 50 PERCENT OF ABOVE-GRADE WALL AREA			
ELEMENT	CONDITION/VALUE		
Skylights (*U*-factor)	1		
Slab or below-grade wall (*R*-value)	R-0		
Windows and glass doors	SHGC		*U*-factor
PF < 0.25	0.4		0.7
0.25 ≤ PF < 0.50	0.5		0.7
PF ≥ 0.50	0.6		0.7
Roof assemblies (*R*-value)	Insulation between framing		Continuous insulation
All-wood joist/truss	R-19		R-13
Metal joist/truss	R-19		R-14
Concrete slab or deck	NA		R-13
Metal purlin with thermal block	R-19		R-14
Metal purlin without thermal block	R-25		R-14
Floors over outdoor air or unconditioned space (*R*-value)	Insulation between framing		Continuous insulation
All-wood joist/truss	R-0		R-0
Metal joist/truss	R-11		R-4
Concrete slab or deck	NA		R-0
Above-grade walls (*R*-value)	No framing	Metal framing	Wood framing
Framed			
R-value cavity	NA	R-7	R-7
R-value continuous	NA	R-0	R-0
CMU, ≥ 8 inches, with integral insulation			
R-value cavity	NA	R-0	R-0
R-value continuous	R-0	R-0	R-0
Other masonry walls			
R-value cavity	NA	R-0	R-0
R-value continuous	R-0	R-0	R-0

For SI: 1 inch = 25.4 mm.

a. Values from Tables 802.2(5) through 802.2(37) shall be used for the purpose of the completion of Tables 802.2(1) through 802.2(4), as applicable based on window and glazed door area.
b. "NA" indicates the condition is not applicable.
c. An *R*-value of zero indicates no insulation is required.
d. "Any" indicates any available product will comply.
e. "X" indicates no complying option exists for this condition.

ACCEPTABLE PRACTICE FOR COMMERCIAL BUILDINGS

TABLE 802.2(8)
BUILDING ENVELOPE REQUIREMENTS[a through e] - CLIMATE ZONE 2b

WINDOW AND GLAZED DOOR AREA 10 PERCENT OR LESS OF ABOVE-GRADE WALL AREA			
ELEMENT	**CONDITION/VALUE**		
Skylights (*U*-factor)	1		
Slab or below-grade wall (*R*-value)	R-0		
Windows and glass doors	SHGC		*U*-factor
PF < 0.25	Any		Any
0.25 ≤ PF < 0.50	Any		Any
PF ≥ 0.50	Any		Any
Roof assemblies (*R*-value)	Insulation between framing		Continuous insulation
All-wood joist/truss	R-19		R-16
Metal joist/truss	R-25		R-17
Concrete slab or deck	NA		R-16
Metal purlin with thermal block	R-25		R-17
Metal purlin without thermal block	X		R-17
Floors over outdoor air or unconditioned space (*R*-value)	Insulation between framing		Continuous insulation
All-wood joist/truss	R-11		R-4
Metal joist/truss	R-11		R-4
Concrete slab or deck	NA		R-1
Above-grade walls (*R*-value)	No framing	Metal framing	Wood framing
Framed			
R-value cavity	NA	R-0	R-0
R-value continuous	NA	R-0	R-0
CMU, ≥ 8 inches, with integral insulation			
R-value cavity	NA	R-0	R-0
R-value continuous	R-0	R-0	R-0
Other masonry walls			
R-value cavity	NA	R-0	R-0
R-value continuous	R-0	R-0	R-0
WINDOW AND GLAZED DOOR AREA GREATER THAN 10 PERCENT BUT NOT GREATER THAN 25 PERCENT OF ABOVE-GRADE WALL AREA			
ELEMENT	**CONDITION/VALUE**		
Skylights (*U*-factor)	1		
Slab or below-grade wall (*R*-value)	R-0		
Windows and glass doors	SHGC		*U*-factor
PF < 0.25	0.5		Any
0.25 ≤ PF < 0.50	0.6		Any
PF ≥ 0.50	0.7		Any
Roof assemblies (*R*-value)	Insulation between framing		Continuous insulation
All-wood joist/truss	R-19		R-16
Metal joist/truss	R-25		R-17
Concrete slab or deck	NA		R-16
Metal purlin with thermal block	R-25		R-17
Metal purlin without thermal block	X		R-17
Floors over outdoor air or unconditioned space (*R*-value)	Insulation between framing		Continuous insulation
All-wood joist/truss	R-11		R-4
Metal joist/truss	R-11		R-4
Concrete slab or deck	NA		R-1
Above-grade walls (*R*-value)	No framing	Metal framing	Wood framing
Framed			
R-value cavity	NA	R-0	R-0
R-value continuous	NA	R-0	R-0
CMU, ≥ 8 inches, with integral insulation			
R-value cavity	NA	R-0	R-0
R-value continuous	R-0	R-0	R-0
Other masonry walls			
R-value cavity	NA	R-0	R-0
R-value continuous	R-0	R-0	R-0

(continued)

ACCEPTABLE PRACTICE FOR COMMERCIAL BUILDING

TABLE 802.2(8)—continued
BUILDING ENVELOPE REQUIREMENTS[a through e] - CLIMATE ZONE 2b

WINDOW AND GLAZED DOOR AREA GREATER THAN 25 PERCENT BUT NOT GREATER THAN 40 PERCENT OF ABOVE-GRADE WALL AREA			
ELEMENT	**CONDITION/VALUE**		
Skylights (U-factor)	1		
Slab or below-grade wall (R-value)	R-0		
Windows and glass doors	SHGC		U-factor
PF < 0.25	0.5		0.7
0.25 ≤ PF < 0.50	0.6		0.7
PF ≥ 0.50	0.7		0.7
Roof assemblies (R-value)	Insulation between framing		Continuous insulation
All-wood joist/truss	R-19		R-16
Metal joist/truss	R-25		R-17
Concrete slab or deck	NA		R-16
Metal purlin with thermal block	R-25		R-17
Metal purlin without thermal block	X		R-17
Floors over outdoor air or unconditioned space (R-value)	Insulation between framing		Continuous insulation
All-wood joist/truss	R-11		R-4
Metal joist/truss	R-11		R-4
Concrete slab or deck	NA		R-1
Above-grade walls (R-value)	No framing	Metal framing	Wood framing
Framed			
R-value cavity	NA	R-11	R-11
R-value continuous	NA	R-0	R-0
CMU, ≥ 8 inches, with integral insulation			
R-value cavity	NA	R-0	R-0
R-value continuous	R-0	R-0	R-0
Other masonry walls			
R-value cavity	NA	R-11	R-11
R-value continuous	R-5	R-0	R-0
WINDOW AND GLAZED DOOR AREA GREATER THAN 40 PERCENT BUT NOT GREATER THAN 50 PERCENT OF ABOVE-GRADE WALL AREA			
ELEMENT	**CONDITION/VALUE**		
Skylights (U-factor)	1		
Slab or below-grade wall (R-value)	R-0		
Windows and glass doors	SHGC		U-factor
PF < 0.25	0.4		0.7
0.25 ≤ PF < 0.50	0.5		0.7
PF ≥ 0.50	0.7		0.7
Roof assemblies (R-value)	Insulation between framing		Continuous insulation
All-wood joist/truss	R-19		R-16
Metal joist/truss	R-25		R-17
Concrete slab or deck	NA		R-16
Metal purlin with thermal block	R-25		R-17
Metal purlin without thermal block	R-30		R-17
Floors over outdoor air or unconditioned space (R-value)	Insulation between framing		Continuous insulation
All-wood joist/truss	R-11		R-4
Metal joist/truss	R-11		R-4
Concrete slab or deck	NA		R-1
Above-grade walls (R-value)	No framing	Metal framing	Wood framing
Framed			
R-value cavity	NA	R-11	R-11
R-value continuous	NA	R-0	R-0
CMU, ≥ 8 inches, with integral insulation			
R-value cavity	NA	R-0	R-0
R-value continuous	R-0	R-0	R-0
Other masonry walls			
R-value cavity	NA	R-11	R-11
R-value continuous	R-5	R-0	R-0

For SI: 1 inch = 25.4 mm.

a. Values from Tables 802.2(5) through 802.2(37) shall be used for the purpose of the completion of Tables 802.2(1) through 802.2(4), as applicable based on window and glazed door area.
b. "NA" indicates the condition is not applicable.
c. An R-value of zero indicates no insulation is required.
d. "Any" indicates any available product will comply.
e. "X" indicates no complying option exists for this condition.

TABLE 802.2(9)
BUILDING ENVELOPE REQUIREMENTS[a through e] - CLIMATE ZONE 3a

WINDOW AND GLAZED DOOR AREA 10 PERCENT OR LESS OF ABOVE-GRADE WALL AREA			
ELEMENT	CONDITION/VALUE		
Skylights (U-factor)	1		
Slab or below-grade wall (R-value)	R-0		
Windows and glass doors	SHGC		U-factor
PF < 0.25	Any		Any
0.25 ≤ PF < 0.50	Any		Any
PF ≥ 0.50	Any		Any
Roof assemblies (R-value)	Insulation between framing		Continuous insulation
All-wood joist/truss	R-11		R-9
Metal joist/truss	R-11		R-10
Concrete slab or deck	NA		R-9
Metal purlin with thermal block	R-13		R-10
Metal purlin without thermal block	R-25		R-10
Floors over outdoor air or unconditioned space (R-value)	Insulation between framing		Continuous insulation
All-wood joist/truss	R-11		R-4
Metal joist/truss	R-11		R-4
Concrete slab or deck	NA		R-2
Above-grade walls (R-value)	No framing	Metal framing	Wood framing
Framed			
R-value cavity	NA	R-0	R-0
R-value continuous	NA	R-0	R-0
CMU, ≥ 8 inches, with integral insulation			
R-value cavity	NA	R-0	R-0
R-value continuous	R-0	R-0	R-0
Other masonry walls			
R-value cavity	NA	R-0	R-0
R-value continuous	R-0	R-0	R-0
WINDOW AND GLAZED DOOR AREA GREATER THAN 10 PERCENT BUT NOT GREATER THAN 25 PERCENT OF ABOVE-GRADE WALL AREA			
ELEMENT	CONDITION/VALUE		
Skylights (U-factor)	1		
Slab or below-grade wall (R-value)	R-0		
Windows and glass doors	SHGC		U-factor
PF < 0.25	0.6		Any
0.25 ≤ PF < 0.50	0.7		Any
PF ≥ 0.50	Any		Any
Roof assemblies (R-value)	Insulation between framing		Continuous insulation
All-wood joist/truss	R-19		R-12
Metal joist/truss	R-19		R-13
Concrete slab or deck	NA		R-12
Metal purlin with thermal block	R-19		R-13
Metal purlin without thermal block	R-30		R-13
Floors over outdoor air or unconditioned space (R-value)	Insulation between framing		Continuous insulation
All-wood joist/truss	R-11		R-4
Metal joist/truss	R-11		R-4
Concrete slab or deck	NA		R-2
Above-grade walls (R-value)	No framing	Metal framing	Wood framing
Framed			
R-value cavity	NA	R-0	R-0
R-value continuous	NA	R-0	R-0
CMU, ≥ 8 inches, with integral insulation			
R-value cavity	NA	R-0	R-0
R-value continuous	R-0	R-0	R-0
Other masonry walls			
R-value cavity	NA	R-0	R-0
R-value continuous	R-0	R-0	R-0

(continued)

TABLE 802.2(9)—continued
BUILDING ENVELOPE REQUIREMENTS[a through e] **- CLIMATE ZONE 3a**

WINDOW AND GLAZED DOOR AREA GREATER THAN 25 PERCENT BUT NOT GREATER THAN 40 PERCENT OF ABOVE-GRADE WALL AREA			
ELEMENT	CONDITION/VALUE		
Skylights (*U*-factor)	1		
Slab or below-grade wall (*R*-value)	R-0		
Windows and glass doors	SHGC	*U*-factor	
PF < 0.25	0.5	0.7	
0.25 ≤ PF < 0.50	0.6	0.7	
PF ≥ 0.50	0.7	0.7	
Roof assemblies (*R*-value)	Insulation between framing	Continuous insulation	
All-wood joist/truss	R-19	R-12	
Metal joist/truss	R-19	R-13	
Concrete slab or deck	NA	R-12	
Metal purlin with thermal block	R-19	R-13	
Metal purlin without thermal block	R-30	R-13	
Floors over outdoor air or unconditioned space (*R*-value)	Insulation between framing	Continuous insulation	
All-wood joist/truss	R-11	R-4	
Metal joist/truss	R-11	R-4	
Concrete slab or deck	NA	R-2	
Above-grade walls (*R*-value)	No framing	Metal framing	Wood framing
Framed			
R-value cavity	NA	R-11	R-11
R-value continuous	NA	R-0	R-0
CMU, ≥ 8 inches, with integral insulation			
R-value cavity	NA	R-0	R-0
R-value continuous	R-0	R-0	R-0
Other masonry walls			
R-value cavity	NA	R-0	R-0
R-value continuous	R-0	R-0	R-0
WINDOW AND GLAZED DOOR AREA GREATER THAN 40 PERCENT BUT NOT GREATER THAN 50 PERCENT OF ABOVE-GRADE WALL AREA			
ELEMENT	CONDITION/VALUE		
Skylights (*U*-factor)	1		
Slab or below-grade wall (*R*-value)	R-0		
Windows and glass doors	SHGC	*U*-factor	
PF < 0.25	0.4	0.7	
0.25 ≤ PF < 0.50	0.5	0.7	
PF ≥ 0.50	0.7	0.7	
Roof assemblies (*R*-value)	Insulation between framing	Continuous insulation	
All-wood joist/truss	R-19	R-12	
Metal joist/truss	R-19	R-13	
Concrete slab or deck	NA	R-12	
Metal purlin with thermal block	R-19	R-13	
Metal purlin without thermal block	R-30	R-13	
Floors over outdoor air or unconditioned space (*R*-value)	Insulation between framing	Continuous insulation	
All-wood joist/truss	R-11	R-4	
Metal joist/truss	R-11	R-4	
Concrete slab or deck	NA	R-2	
Above-grade walls (*R*-value)	No framing	Metal framing	Wood framing
Framed			
R-value cavity	NA	R-11	R-11
R-value continuous	NA	R-0	R-0
CMU, ≥ 8 inches, with integral insulation			
R-value cavity	NA	R-0	R-0
R-value continuous	R-0	R-0	R-0
Other masonry walls			
R-value cavity	NA	R-0	R-0
R-value continuous	R-0	R-0	R-0

For SI: 1 inch = 25.4 mm.

a. Values from Tables 802.2(5) through 802.2(37) shall be used for the purpose of the completion of Tables 802.2(1) through 802.2(4), as applicable based on window and glazed door area.
b. "NA" indicates the condition is not applicable.
c. An *R*-value of zero indicates no insulation is required.
d. "Any" indicates any available product will comply.
e. "X" indicates no complying option exists for this condition.

TABLE 802.2(10)
BUILDING ENVELOPE REQUIREMENTS[a through e] – CLIMATE ZONE 3b

WINDOW AND GLAZED DOOR AREA 10 PERCENT OR LESS OF ABOVE-GRADE WALL AREA			
ELEMENT	**CONDITION/VALUE**		
Skylights (*U*-factor)	1		
Slab or below-grade wall (*R*-value)	R-0		
Windows and glass doors	**SHGC**	**U-factor**	
PF < 0.25	Any	Any	
0.25 ≤ PF < 0.50	Any	Any	
PF ≥ 0.50	Any	Any	
Roof assemblies (*R*-value)	**Insulation between framing**	**Continuous insulation**	
All-wood joist/truss	R-19	R-12	
Metal joist/truss	R-19	R-13	
Concrete slab or deck	NA	R-12	
Metal purlin with thermal block	R-19	R-13	
Metal purlin without thermal block	R-30	R-13	
Floors over outdoor air or unconditioned space (*R*-value)	**Insulation between framing**	**Continuous insulation**	
All-wood joist/truss	R-11	R-4	
Metal joist/truss	R-11	R-4	
Concrete slab or deck	NA	R-2	
Above-grade walls (*R*-value)	**No framing**	**Metal framing**	**Wood framing**
Framed			
R-value cavity	NA	R-0	R-0
R-value continuous	NA	R-0	R-0
CMU, ≥ 8 inches, with integral insulation			
R-value cavity	NA	R-0	R-0
R-value continuous	R-0	R-0	R-0
Other masonry walls			
R-value cavity	NA	R-0	R-0
R-value continuous	R-0	R-0	R-0
WINDOW AND GLAZED DOOR AREA GREATER THAN 10 PERCENT BUT NOT GREATER THAN 25 PERCENT OF ABOVE-GRADE WALL AREA			
ELEMENT	**CONDITION/VALUE**		
Skylights (*U*-factor)	1		
Slab or below-grade wall (*R*-value)	R-0		
Windows and glass doors	**SHGC**	**U-factor**	
PF < 0.25	0.5	Any	
0.25 ≤ PF < 0.50	0.6	Any	
PF ≥ 0.50	0.7	Any	
Roof assemblies (*R*-value)	**Insulation between framing**	**Continuous insulation**	
All-wood joist/truss	R-19	R-12	
Metal joist/truss	R-19	R-13	
Concrete slab or deck	NA	R-12	
Metal purlin with thermal block	R-19	R-13	
Metal purlin without thermal block	R-30	R-13	
Floors over outdoor air or unconditioned space (*R*-value)	**Insulation between framing**	**Continuous insulation**	
All-wood joist/truss	R-11	R-4	
Metal joist/truss	R-11	R-4	
Concrete slab or deck	NA	R-2	
Above-grade walls (*R*-value)	**No framing**	**Metal framing**	**Wood framing**
Framed			
R-value cavity	NA	R-0	R-0
R-value continuous	NA	R-0	R-0
CMU, ≥ 8 inches, with integral insulation			
R-value cavity	NA	R-0	R-0
R-value continuous	R-0	R-0	R-0
Other masonry walls			
R-value cavity	NA	R-0	R-0
R-value continuous	R-0	R-0	R-0

(continued)

TABLE 802.2(10)—continued
BUILDING ENVELOPE REQUIREMENTS[a through e] - CLIMATE ZONE 3b

WINDOW AND GLAZED DOOR AREA GREATER THAN 25 PERCENT BUT NOT GREATER THAN 40 PERCENT OF ABOVE-GRADE WALL AREA			
ELEMENT	CONDITION/VALUE		
Skylights (*U*-factor)	1		
Slab or below-grade wall (*R*-value)	R-0		
Windows and glass doors	SHGC	*U*-factor	
PF < 0.25	0.4	0.7	
0.25 ≤ PF < 0.50	0.5	0.7	
PF ≥ 0.50	0.6	0.7	
Roof assemblies (*R*-value)	Insulation between framing	Continuous insulation	
All-wood joist/truss	R-19	R-12	
Metal joist/truss	R-19	R-13	
Concrete slab or deck	NA	R-12	
Metal purlin with thermal block	R-19	R-13	
Metal purlin without thermal block	R-30	R-13	
Floors over outdoor air or unconditioned space (*R*-value)	Insulation between framing	Continuous insulation	
All-wood joist/truss	R-11	R-4	
Metal joist/truss	R-11	R-4	
Concrete slab or deck	NA	R-2	
Above-grade walls (*R*-value)	No framing	Metal framing	Wood framing
Framed			
R-value cavity	NA	R-0	R-0
R-value continuous	NA	R-0	R-0
CMU, ≥ 8 inches, with integral insulation			
R-value cavity	NA	R-0	R-0
R-value continuous	R-0	R-0	R-0
Other masonry walls			
R-value cavity	NA	R-0	R-0
R-value continuous	R-0	R-0	R-0
WINDOW AND GLAZED DOOR AREA GREATER THAN 40 PERCENT BUT NOT GREATER THAN 50 PERCENT OF ABOVE-GRADE WALL AREA			
ELEMENT	CONDITION/VALUE		
Skylights (*U*-factor)	1		
Slab or below-grade wall (*R*-value)	R-0		
Windows and glass doors	SHGC	*U*-factor	
PF < 0.25	0.4	0.7	
0.25 ≤ PF < 0.50	0.5	0.7	
PF ≥ 0.50	0.6	0.7	
Roof assemblies (*R*-value)	Insulation between framing	Continuous insulation	
All-wood joist/truss	R-19	R-12	
Metal joist/truss	R-19	R-13	
Concrete slab or deck	NA	R-12	
Metal purlin with thermal block	R-19	R-13	
Metal purlin without thermal block	R-30	R-13	
Floors over outdoor air or unconditioned space (*R*-value)	Insulation between framing	Continuous insulation	
All-wood joist/truss	R-11	R-4	
Metal joist/truss	R-11	R-4	
Concrete slab or deck	NA	R-2	
Above-grade walls (*R*-value)	No framing	Metal framing	Wood framing
Framed			
R-value cavity	NA	R-7	R-7
R-value continuous	NA	R-0	R-0
CMU, ≥ 8 inches, with integral insulation			
R-value cavity	NA	R-0	R-0
R-value continuous	R-0	R-0	R-0
Other masonry walls			
R-value cavity	NA	R-0	R-0
R-value continuous	R-0	R-0	R-0

For SI: 1 inch = 25.4 mm.

a. Values from Tables 802.2(5) through 802.2(37) shall be used for the purpose of the completion of Tables 802.2(1) through 802.2(4), as applicable based on window and glazed door area.
b. "NA" indicates the condition is not applicable.
c. An *R*-value of zero indicates no insulation is required.
d. "Any" indicates any available product will comply.
e. "X" indicates no complying option exists for this condition.

ACCEPTABLE PRACTICE FOR COMMERCIAL BUILDINGS

TABLE 802.2(11)
BUILDING ENVELOPE REQUIREMENTS[a through e] - CLIMATE ZONE 3c

WINDOW AND GLAZED DOOR AREA 10 PERCENT OR LESS OF ABOVE-GRADE WALL AREA			
ELEMENT	CONDITION/VALUE		
Skylights (U-factor)	1		
Slab or below-grade wall (R-value)	R-0		
Windows and glass doors	SHGC		U-factor
PF < 0.25	Any		Any
0.25 ≤ PF < 0.50	Any		Any
PF ≥ 0.50	Any		Any
Roof assemblies (R-value)	Insulation between framing		Continuous insulation
All-wood joist/truss	R-19		R-16
Metal joist/truss	R-25		R-17
Concrete slab or deck	NA		R-16
Metal purlin with thermal block	R-25		R-17
Metal purlin without thermal block	X		R-17
Floors over outdoor air or unconditioned space (R-value)	Insulation between framing		Continuous insulation
All-wood joist/truss	R-11		R-4
Metal joist/truss	R-11		R-4
Concrete slab or deck	NA		R-2
Above-grade walls (R-value)	No framing	Metal framing	Wood framing
Framed			
R-value cavity	NA	R-0	R-0
R-value continuous	NA	R-0	R-0
CMU, ≥ 8 inches, with integral insulation			
R-value cavity	NA	R-0	R-0
R-value continuous	R-0	R-0	R-0
Other masonry walls			
R-value cavity	NA	R-0	R-0
R-value continuous	R-0	R-0	R-0
WINDOW AND GLAZED DOOR AREA GREATER THAN 10 PERCENT BUT NOT GREATER THAN 25 PERCENT OF ABOVE-GRADE WALL AREA			
ELEMENT	CONDITION/VALUE		
Skylights (U-factor)	1		
Slab or below-grade wall (R-value)	R-0		
Windows and glass doors	SHGC		U-factor
PF < 0.25	0.5		Any
0.25 ≤ PF < 0.50	0.6		Any
PF ≥ 0.50	0.7		Any
Roof assemblies (R-value)	Insulation between framing		Continuous insulation
All-wood joist/truss	R-25		R-19
Metal joist/truss	R-25		R-20
Concrete slab or deck	NA		R-19
Metal purlin with thermal block	R-30		R-20
Metal purlin without thermal block	X		R-20
Floors over outdoor air or unconditioned space (R-value)	Insulation between framing		Continuous insulation
All-wood joist/truss	R-11		R-4
Metal joist/truss	R-11		R-4
Concrete slab or deck	NA		R-2
Above-grade walls (R-value)	No framing	Metal framing	Wood framing
Framed			
R-value cavity	NA	R-11	R-0
R-value continuous	NA	R-0	R-0
CMU, ≥ 8 inches, with integral insulation			
R-value cavity	NA	R-0	R-0
R-value continuous	R-0	R-0	R-0
Other masonry walls			
R-value cavity	NA	R-0	R-0
R-value continuous	R-0	R-0	R-0

(continued)

TABLE 802.2(11)—continued
BUILDING ENVELOPE REQUIREMENTS[a through e] **- CLIMATE ZONE 3c**

WINDOW AND GLAZED DOOR AREA GREATER THAN 25 PERCENT BUT NOT GREATER THAN 40 PERCENT OF ABOVE-GRADE WALL AREA			
ELEMENT	CONDITION/VALUE		
Skylights (*U*-factor)	1		
Slab or below-grade wall (*R*-value)	R-0		
Windows and glass doors	SHGC	*U*-factor	
PF < 0.25	0.4	0.7	
0.25 ≤ PF < 0.50	0.5	0.7	
PF ≥ 0.50	0.6	0.7	
Roof assemblies (*R*-value)	Insulation between framing	Continuous insulation	
All-wood joist/truss	R-25	R-19	
Metal joist/truss	R-25	R-20	
Concrete slab or deck	NA	R-19	
Metal purlin with thermal block	R-30	R-20	
Metal purlin without thermal block	X	R-20	
Floors over outdoor air or unconditioned space (*R*-value)	Insulation between framing	Continuous insulation	
All-wood joist/truss	R-11	R-4	
Metal joist/truss	R-11	R-4	
Concrete slab or deck	NA	R-2	
Above-grade walls (*R*-value)	No framing	Metal framing	Wood framing
Framed			
R-value cavity	NA	R-11	R-11
R-value continuous	NA	R-0	R-0
CMU, ≥ 8 inches, with integral insulation			
R-value cavity	NA	R-0	R-0
R-value continuous	R-0	R-0	R-0
Other masonry walls			
R-value cavity	NA	R-11	R-11
R-value continuous	R-5	R-0	R-0
WINDOW AND GLAZED DOOR AREA GREATER THAN 40 PERCENT BUT NOT GREATER THAN 50 PERCENT OF ABOVE-GRADE WALL AREA			
ELEMENT	CONDITION/VALUE		
Skylights (*U*-factor)	1		
Slab or below-grade wall (*R*-value)	R-0		
Windows and glass doors	SHGC	*U*-factor	
PF < 0.25	0.4	0.7	
0.25 ≤ PF < 0.50	0.5	0.7	
PF ≥ 0.50	0.6	0.7	
Roof assemblies (*R*-value)	Insulation between framing	Continuous insulation	
All-wood joist/truss	R-25	R-19	
Metal joist/truss	R-25	R-20	
Concrete slab or deck	NA	R-19	
Metal purlin with thermal block	R-30	R-20	
Metal purlin without thermal block	R-38	R-20	
Floors over outdoor air or unconditioned space (*R*-value)	Insulation between framing	Continuous insulation	
All-wood joist/truss	R-11	R-4	
Metal joist/truss	R-11	R-4	
Concrete slab or deck	NA	R-2	
Above-grade walls (*R*-value)	No framing	Metal framing	Wood framing
Framed			
R-value cavity	NA	R-11	R-11
R-value continuous	NA	R-0	R-0
CMU, ≥ 8 inches, with integral insulation			
R-value cavity	NA	R-0	R-0
R-value continuous	R-0	R-0	R-0
Other masonry walls			
R-value cavity	NA	R-11	R-11
R-value continuous	R-5	R-0	R-0

For SI: 1 inch = 25.4 mm.

a. Values from Tables 802.2(5) through 802.2(37) shall be used for the purpose of the completion of Tables 802.2(1) through 802.2(4), as applicable based on window and glazed door area.
b. "NA" indicates the condition is not applicable.
c. An *R*-value of zero indicates no insulation is required.
d. "Any" indicates any available product will comply.
e. "X" indicates no complying option exists for this condition.

ACCEPTABLE PRACTICE FOR COMMERCIAL BUILDINGS

TABLE 802.2(12)
BUILDING ENVELOPE REQUIREMENTS[a through e] - CLIMATE ZONE 4a

WINDOW AND GLAZED DOOR AREA 10 PERCENT OR LESS OF ABOVE-GRADE WALL AREA			
ELEMENT	CONDITION/VALUE		
Skylights (U-factor)	1		
Slab or below-grade wall (R-value)	R-0		
Windows and glass doors	SHGC		U-factor
PF < 0.25	Any		Any
0.25 ≤ PF < 0.50	Any		Any
PF ≥ 0.50	Any		Any
Roof assemblies (R-value)	Insulation between framing		Continuous insulation
All-wood joist/truss	R-13		R-11
Metal joist/truss	R-13		R-12
Concrete slab or deck	NA		R-11
Metal purlin with thermal block	R-19		R-12
Metal purlin without thermal block	R-30		R-12
Floors over outdoor air or unconditioned space (R-value)	Insulation between framing		Continuous insulation
All-wood joist/truss	R-11		R-4
Metal joist/truss	R-11		R-4
Concrete slab or deck	NA		R-3
Above-grade walls (R-value)	No framing	Metal framing	Wood framing
Framed			
R-value cavity	NA	R-0	R-0
R-value continuous	NA	R-0	R-0
CMU, ≥ 8 inches, with integral insulation			
R-value cavity	NA	R-0	R-0
R-value continuous	R-0	R-0	R-0
Other masonry walls			
R-value cavity	NA	R-0	R-0
R-value continuous	R-0	R-0	R-0
WINDOW AND GLAZED DOOR AREA GREATER THAN 10 PERCENT BUT NOT GREATER THAN 25 PERCENT OF ABOVE-GRADE WALL AREA			
ELEMENT	CONDITION/VALUE		
Skylights (U-factor)	1		
Slab or below-grade wall (R-value)	R-0		
Windows and glass doors	SHGC		U-factor
PF < 0.25	0.6		Any
0.25 ≤ PF < 0.50	0.7		Any
PF ≥ 0.50	Any		Any
Roof assemblies (R-value)	Insulation between framing		Continuous insulation
All-wood joist/truss	R-19		R-12
Metal joist/truss	R-19		R-13
Concrete slab or deck	NA		R-12
Metal purlin with thermal block	R-19		R-13
Metal purlin without thermal block	R-30		R-13
Floors over outdoor air or unconditioned space (R-value)	Insulation between framing		Continuous insulation
All-wood joist/truss	R-11		R-4
Metal joist/truss	R-11		R-4
Concrete slab or deck	NA		R-3
Above-grade walls (R-value)	No framing	Metal framing	Wood framing
Framed			
R-value cavity	NA	R-11	R-11
R-value continuous	NA	NA	NA
CMU, ≥ 8 inches, with integral insulation			
R-value cavity	NA	R-0	R-0
R-value continuous	R-0	R-0	R-0
Other masonry walls			
R-value cavity	NA	R-0	R-0
R-value continuous	R-0	R-0	R-0

(continued)

ACCEPTABLE PRACTICE FOR COMMERCIAL BUILDING

TABLE 802.2(12)—continued
BUILDING ENVELOPE REQUIREMENTS[a through e] - CLIMATE ZONE 4a

WINDOW AND GLAZED DOOR AREA GREATER THAN 25 PERCENT BUT NOT GREATER THAN 40 PERCENT OF ABOVE-GRADE WALL AREA			
ELEMENT	CONDITION/VALUE		
Skylights (U-factor)	1		
Slab or below-grade wall (R-value)	R-0		
Windows and glass doors	SHGC		U-factor
PF < 0.25	0.5		0.7
0.25 ≤ PF < 0.50	0.6		0.7
PF ≥ 0.50	0.7		0.7
Roof assemblies (R-value)	Insulation between framing		Continuous insulation
All-wood joist/truss	R-19		R-12
Metal joist/truss	R-19		R-13
Concrete slab or deck	NA		R-12
Metal purlin with thermal block	R-19		R-13
Metal purlin without thermal block	R-30		R-13
Floors over outdoor air or unconditioned space (R-value)	Insulation between framing		Continuous insulation
All-wood joist/truss	R-11		R-4
Metal joist/truss	R-11		R-4
Concrete slab or deck	NA		R-3
Above-grade walls (R-value)	No framing	Metal framing	Wood framing
Framed			
R-value cavity	NA	R-11	R-11
R-value continuous	NA	R-0	R-0
CMU, ≥ 8 inches, with integral insulation			
R-value cavity	NA	R-0	R-0
R-value continuous	R-0	R-0	R-0
Other masonry walls			
R-value cavity	NA	R-11	R-11
R-value continuous	R-5	R-0	R-0
WINDOW AND GLAZED DOOR AREA GREATER THAN 40 PERCENT BUT NOT GREATER THAN 50 PERCENT OF ABOVE-GRADE WALL AREA			
ELEMENT	CONDITION/VALUE		
Skylights (U-factor)	1		
Slab or below-grade wall (R-value)	R-0		
Windows and glass doors	SHGC		U-factor
PF < 0.25	0.4		0.7
0.25 ≤ PF < 0.50	0.5		0.7
PF ≥ 0.50	0.7		0.7
Roof assemblies (R-value)	Insulation between framing		Continuous insulation
All-wood joist/truss	R-19		R-12
Metal joist/truss	R-19		R-13
Concrete slab or deck	NA		R-12
Metal purlin with thermal block	R-19		R-13
Metal purlin without thermal block	R-30		R-13
Floors over outdoor air or unconditioned space (R-value)	Insulation between framing		Continuous insulation
All-wood joist/truss	R-11		R-4
Metal joist/truss	R-11		R-4
Concrete slab or deck	NA		R-3
Above-grade walls (R-value)	No framing	Metal framing	Wood framing
Framed			
R-value cavity	NA	R-11	R-11
R-value continuous	NA	R-0	R-0
CMU, ≥ 8 inches, with integral insulation			
R-value cavity	NA	R-0	R-0
R-value continuous	R-0	R-0	R-0
Other masonry walls			
R-value cavity	NA	R-11	R-11
R-value continuous	R-5	R-0	R-0

For SI: 1 inch = 25.4 mm.

a. Values from Tables 802.2(5) through 802.2(37) shall be used for the purpose of the completion of Tables 802.2(1) through 802.2(4), as applicable based on window and glazed door area.
b. "NA" indicates the condition is not applicable.
c. An R-value of zero indicates no insulation is required.
d. "Any" indicates any available product will comply.
e. "X" indicates no complying option exists for this condition.

ACCEPTABLE PRACTICE FOR COMMERCIAL BUILDINGS

TABLE 802.2(13)
BUILDING ENVELOPE REQUIREMENTS[a through e] - CLIMATE ZONE 4b

WINDOW AND GLAZED DOOR AREA 10 PERCENT OR LESS OF ABOVE-GRADE WALL AREA			
ELEMENT	CONDITION/VALUE		
Skylights (*U*-factor)	1		
Slab or below-grade wall (*R*-value)	R-0		
Windows and glass doors	SHGC		*U*-factor
PF < 0.25	Any		Any
0.25 ≤ PF < 0.50	Any		Any
PF ≥ 0.50	Any		Any
Roof assemblies (*R*-value)	Insulation between framing		Continuous insulation
All-wood joist/truss	R-19		R-16
Metal joist/truss	R-25		R-17
Concrete slab or deck	NA		R-16
Metal purlin with thermal block	R-25		R-17
Metal purlin without thermal block	X		R-17
Floors over outdoor air or unconditioned space (*R*-value)	Insulation between framing		Continuous insulation
All-wood joist/truss	R-11		R-4
Metal joist/truss	R-11		R-4
Concrete slab or deck	NA		R-4
Above-grade walls (*R*-value)	No framing	Metal framing	Wood framing
Framed			
R-value cavity	NA	R-11	R-11
R-value continuous	NA	R-0	R-0
CMU, ≥ 8 inches, with integral insulation			
R-value cavity	NA	R-0	R-0
R-value continuous	R-0	R-0	R-0
Other masonry walls			
R-value cavity	NA	R-0	R-0
R-value continuous	R-0	R-0	R-0
WINDOW AND GLAZED DOOR AREA GREATER THAN 10 PERCENT BUT NOT GREATER THAN 25 PERCENT OF ABOVE-GRADE WALL AREA			
ELEMENT	CONDITION/VALUE		
Skylights (*U*-factor)	1		
Slab or below-grade wall (*R*-value)	R-0		
Windows and glass doors	SHGC		*U*-factor
PF < 0.25	0.6		Any
0.25 ≤ PF < 0.50	0.7		Any
PF ≥ 0.50	Any		Any
Roof assemblies (*R*-value)	Insulation between framing		Continuous insulation
All-wood joist/truss	R-19		R-16
Metal joist/truss	R-25		R-17
Concrete slab or deck	NA		R-16
Metal purlin with thermal block	R-25		R-17
Metal purlin without thermal block	X		R-17
Floors over outdoor air or unconditioned space (*R*-value)	Insulation between framing		Continuous insulation
All-wood joist/truss	R-11		R-4
Metal joist/truss	R-11		R-4
Concrete slab or deck	NA		R-4
Above-grade walls (*R*-value)	No framing	Metal framing	Wood framing
Framed			
R-value cavity	NA	R-11	R-11
R-value continuous	NA	R-0	R-0
CMU, ≥ 8 inches, with integral insulation			
R-value cavity	NA	R-0	R-0
R-value continuous	R-0	R-0	R-0
Other masonry walls			
R-value cavity	NA	R-11	R-11
R-value continuous	R-5	R-0	R-0

(continued)

TABLE 802.2(13)—continued
BUILDING ENVELOPE REQUIREMENTS[a through e] - CLIMATE ZONE 4b

WINDOW AND GLAZED DOOR AREA GREATER THAN 25 PERCENT BUT NOT GREATER THAN 40 PERCENT OF ABOVE-GRADE WALL AREA			
ELEMENT	CONDITION/VALUE		
Skylights (*U*-factor)	1		
Slab or below-grade wall (*R*-value)	R-0		
Windows and glass doors	SHGC		*U*-factor
PF < 0.25	0.4		0.7
0.25 ≤ PF < 0.50	0.5		0.7
PF ≥ 0.50	0.6		0.7
Roof assemblies (*R*-value)	Insulation between framing		Continuous insulation
All-wood joist/truss	R-19		R-16
Metal joist/truss	R-25		R-17
Concrete slab or deck	NA		R-16
Metal purlin with thermal block	R-25		R-17
Metal purlin without thermal block	X		R-17
Floors over outdoor air or unconditioned space (*R*-value)	Insulation between framing		Continuous insulation
All-wood joist/truss	R-11		R-4
Metal joist/truss	R-11		R-4
Concrete slab or deck	NA		R-4
Above-grade walls (*R*-value)	No framing	Metal framing	Wood framing
Framed			
R-value cavity	NA	R-11	R-11
R-value continuous	NA	R-0	R-0
CMU, ≥ 8 inches, with integral insulation			
R-value cavity	NA	R-0	R-0
R-value continuous	R-0	R-0	R-0
Other masonry walls			
R-value cavity	NA	R-11	R-11
R-value continuous	R-5	R-0	R-0
WINDOW AND GLAZED DOOR AREA GREATER THAN 40 PERCENT BUT NOT GREATER THAN 50 PERCENT OF ABOVE-GRADE WALL AREA			
ELEMENT	CONDITION/VALUE		
Skylights (*U*-factor)	1		
Slab or below-grade wall (*R*-value)	R-0		
Windows and glass doors	SHGC		*U*-factor
PF < 0.25	0.4		0.7
0.25 ≤ PF < 0.50	0.5		0.7
PF ≥ 0.50	0.6		0.7
Roof assemblies (*R*-value)	Insulation between framing		Continuous insulation
All-wood joist/truss	R-19		R-16
Metal joist/truss	R-25		R-17
Concrete slab or deck	NA		R-16
Metal purlin with thermal block	R-25		R-17
Metal purlin without thermal block	R-30		R-17
Floors over outdoor air or unconditioned space (*R*-value)	Insulation between framing		Continuous insulation
All-wood joist/truss	R-11		R-4
Metal joist/truss	R-11		R-4
Concrete slab or deck	NA		R-4
Above-grade walls (*R*-value)	No framing	Metal framing	Wood framing
Framed			
R-value cavity	NA	R-13	R-11
R-value continuous	NA	R-3	R-0
CMU, ≥ 8 inches, with integral insulation			
R-value cavity	NA	R-0	R-0
R-value continuous	R-0	R-0	R-0
Other masonry walls			
R-value cavity	NA	R-11	R-11
R-value continuous	R-5	R-0	R-0

For SI: 1 inch = 25.4 mm.

a. Values from Tables 802.2(5) through 802.2(37) shall be used for the purpose of the completion of Tables 802.2(1) through 802.2(4), as applicable based on window and glazed door area.
b. "NA" indicates the condition is not applicable.
c. An *R*-value of zero indicates no insulation is required.
d. "Any" indicates any available product will comply.
e. "X" indicates no complying option exists for this condition.

TABLE 802.2(14)
BUILDING ENVELOPE REQUIREMENTS[a through e] - CLIMATE ZONE 5a

WINDOW AND GLAZED DOOR AREA 10 PERCENT OR LESS OF ABOVE-GRADE WALL AREA			
ELEMENT	**CONDITION/VALUE**		
Skylights (U-factor)	1		
Slab or below-grade wall (R-value)	R-0		
Windows and glass doors	**SHGC**		**U-factor**
PF < 0.25	Any		Any
0.25 ≤ PF < 0.50	Any		Any
PF ≥ 0.50	Any		Any
Roof assemblies (R-value)	**Insulation between framing**		**Continuous insulation**
All-wood joist/truss	R-19		R-14
Metal joist/truss	R-19		R-15
Concrete slab or deck	NA		R-14
Metal purlin with thermal block	R-25		R-15
Metal purlin without thermal block	X		R-15
Floors over outdoor air or unconditioned space (R-value)	**Insulation between framing**		**Continuous insulation**
All-wood joist/truss	R-11		R-5
Metal joist/truss	R-11		R-6
Concrete slab or deck	NA		R-5
Above-grade walls (R-value)	**No framing**	**Metal framing**	**Wood framing**
Framed			
R-value cavity	NA	R-11	R-11
R-value continuous	NA	R-0	R-0
CMU, ≥ 8 inches, with integral insulation			
R-value cavity	NA	R-0	R-0
R-value continuous	R-0	R-0	R-0
Other masonry walls			
R-value cavity	NA	R-0	R-0
R-value continuous	R-0	R-0	R-0
WINDOW AND GLAZED DOOR AREA GREATER THAN 10 PERCENT BUT NOT GREATER THAN 25 PERCENT OF ABOVE-GRADE WALL AREA			
ELEMENT	**CONDITION/VALUE**		
Skylights (U-factor)	1		
Slab or below-grade wall (R-value)	R-0		
Windows and glass doors	**SHGC**		**U-factor**
PF < 0.25	0.6		Any
0.25 ≤ PF < 0.50	0.7		Any
PF ≥ 0.50	Any		Any
Roof assemblies (R-value)	**Insulation between framing**		**Continuous insulation**
All-wood joist/truss	R-19		R-16
Metal joist/truss	R-25		R-17
Concrete slab or deck	NA		R-16
Metal purlin with thermal block	R-25		R-17
Metal purlin without thermal block	X		R-17
Floors over outdoor air or unconditioned space (R-value)	**Insulation between framing**		**Continuous insulation**
All-wood joist/truss	R-11		R-5
Metal joist/truss	R-11		R-6
Concrete slab or deck	NA		R-5
Above-grade walls (R-value)	**No framing**	**Metal framing**	**Wood framing**
Framed			
R-value cavity	NA	R-11	R-11
R-value continuous	NA	R-0	R-0
CMU, ≥ 8 inches, with integral insulation			
R-value cavity	NA	R-0	R-0
R-value continuous	R-0	R-0	R-0
Other masonry walls			
R-value cavity	NA	R-11	R-11
R-value continuous	R-5	R-0	R-0

(continued)

TABLE 802.2(14)—continued
BUILDING ENVELOPE REQUIREMENTS[a through e] - CLIMATE ZONE 5a

WINDOW AND GLAZED DOOR AREA GREATER THAN 25 PERCENT BUT NOT GREATER THAN 40 PERCENT OF ABOVE-GRADE WALL AREA			
ELEMENT	CONDITION/VALUE		
Skylights (*U*-factor)	1		
Slab or below-grade wall (*R*-value)	R-0		
Windows and glass doors	SHGC	*U*-factor	
PF < 0.25	0.5	0.7	
0.25 ≤ PF < 0.50	0.6	0.7	
PF ≥ 0.50	0.7	0.7	
Roof assemblies (*R*-value)	Insulation between framing	Continuous insulation	
All-wood joist/truss	R-19	R-16	
Metal joist/truss	R-25	R-17	
Concrete slab or deck	NA	R-16	
Metal purlin with thermal block	R-25	R-17	
Metal purlin without thermal block	X	R-17	
Floors over outdoor air or unconditioned space (*R*-value)	Insulation between framing	Continuous insulation	
All-wood joist/truss	R-11	R-5	
Metal joist/truss	R-11	R-6	
Concrete slab or deck	NA	R-5	
Above-grade walls (*R*-value)	No framing	Metal framing	Wood framing
Framed			
R-value cavity	NA	R-11	R-11
R-value continuous	NA	R-0	R-0
CMU, ≥ 8 inches, with integral insulation			
R-value cavity	NA	R-0	R-0
R-value continuous	R-0	R-0	R-0
Other masonry walls			
R-value cavity	NA	R-11	R-11
R-value continuous	R-5	R-0	R-0
WINDOW AND GLAZED DOOR AREA GREATER THAN 40 PERCENT BUT NOT GREATER THAN 50 PERCENT OF ABOVE-GRADE WALL AREA			
ELEMENT	CONDITION/VALUE		
Skylights (*U*-factor)	1		
Slab or below-grade wall (*R*-value)	R-0		
Windows and glass doors	SHGC	*U*-factor	
PF < 0.25	0.4	0.7	
0.25 ≤ PF < 0.50	0.5	0.7	
PF ≥ 0.50	0.7	0.7	
Roof assemblies (*R*-value)	Insulation between framing	Continuous insulation	
All-wood joist/truss	R-19	R-16	
Metal joist/truss	R-25	R-17	
Concrete slab or deck	NA	R-16	
Metal purlin with thermal block	R-25	R-17	
Metal purlin without thermal block	R-30	R-17	
Floors over outdoor air or unconditioned space (*R*-value)	Insulation between framing	Continuous insulation	
All-wood joist/truss	R-11	R-5	
Metal joist/truss	R-11	R-6	
Concrete slab or deck	NA	R-5	
Above-grade walls (*R*-value)	No framing	Metal framing	Wood framing
Framed			
R-value cavity	NA	R-11	R-11
R-value continuous	NA	R-0	R-0
CMU, ≥ 8 inches, with integral insulation			
R-value cavity	NA	R-0	R-0
R-value continuous	R-0	R-0	R-0
Other masonry walls			
R-value cavity	NA	R-11	R-11
R-value continuous	R-5	R-0	R-0

For SI: 1 inch = 25.4 mm.

a. Values from Tables 802.2(5) through 802.2(37) shall be used for the purpose of the completion of Tables 802.2(1) through 802.2(4), as applicable based on window and glazed door area.
b. "NA" indicates the condition is not applicable.
c. An *R*-value of zero indicates no insulation is required.
d. "Any" indicates any available product will comply.
e. "X" indicates no complying option exists for this condition.

ACCEPTABLE PRACTICE FOR COMMERCIAL BUILDINGS

TABLE 802.2(15)
BUILDING ENVELOPE REQUIREMENTS[a through e] - CLIMATE ZONE 5b

WINDOW AND GLAZED DOOR AREA 10 PERCENT OR LESS OF ABOVE-GRADE WALL AREA			
ELEMENT	CONDITION/VALUE		
Skylights (U-factor)	1		
Slab or below-grade wall (R-value)	R-0		
Windows and glass doors	SHGC		U-factor
PF < 0.25	Any		Any
0.25 ≤ PF < 0.50	Any		Any
PF ≥ 0.50	Any		Any
Roof assemblies (R-value)	Insulation between framing		Continuous insulation
All-wood joist/truss	R-19		R-14
Metal joist/truss	R-19		R-15
Concrete slab or deck	NA		R-14
Metal purlin with thermal block	R-25		R-15
Metal purlin without thermal block	X		R-15
Floors over outdoor air or unconditioned space (R-value)	Insulation between framing		Continuous insulation
All-wood joist/truss	R-11		R-5
Metal joist/truss	R-11		R-6
Concrete slab or deck	NA		R-5
Above-grade walls (R-value)	No framing	Metal framing	Wood framing
Framed			
R-value cavity	NA	R-11	R-11
R-value continuous	NA	R-0	R-0
CMU, ≥ 8 inches, with integral insulation			
R-value cavity	NA	R-0	R-0
R-value continuous	R-0	R-0	R-0
Other masonry walls			
R-value cavity	NA	R-0	R-0
R-value continuous	R-0	R-0	R-0
WINDOW AND GLAZED DOOR AREA GREATER THAN 10 PERCENT BUT NOT GREATER THAN 25 PERCENT OF ABOVE-GRADE WALL AREA			
ELEMENT	CONDITION/VALUE		
Skylights (U-factor)	1		
Slab or below-grade wall (R-value)	R-0		
Windows and glass doors	SHGC		U-factor
PF < 0.25	0.6		Any
0.25 ≤ PF < 0.50	0.7		Any
PF ≥ 0.50	Any		Any
Roof assemblies (R-value)	Insulation between framing		Continuous insulation
All-wood joist/truss	R-25		R-19
Metal joist/truss	R-25		R-20
Concrete slab or deck	NA		R-19
Metal purlin with thermal block	R-30		R-20
Metal purlin without thermal block	X		R-20
Floors over outdoor air or unconditioned space (R-value)	Insulation between framing		Continuous insulation
All-wood joist/truss	R-11		R-5
Metal joist/truss	R-11		R-6
Concrete slab or deck	NA		R-5
Above-grade walls (R-value)	No framing	Metal framing	Wood framing
Framed			
R-value cavity	NA	R-11	R-11
R-value continuous	NA	R-0	R-0
CMU, ≥ 8 inches, with integral insulation			
R-value cavity	NA	R-11	R-11
R-value continuous	R-5	R-0	R-0
Other masonry walls			
R-value cavity	NA	R-11	R-11
R-value continuous	R-5	R-0	R-0

(continued)

TABLE 802.2(15)—continued
BUILDING ENVELOPE REQUIREMENTS[a through e] - CLIMATE ZONE 5b

WINDOW AND GLAZED DOOR AREA GREATER THAN 25 PERCENT BUT NOT GREATER THAN 40 PERCENT OF ABOVE-GRADE WALL AREA			
ELEMENT	CONDITION/VALUE		
Skylights (*U*-factor)	1		
Slab or below-grade wall (*R*-value)	R-0		
Windows and glass doors	SHGC		*U*-factor
PF < 0.25	0.4		0.7
0.25 ≤ PF < 0.50	0.5		0.7
PF ≥ 0.50	0.6		0.7
Roof assemblies (*R*-value)	Insulation between framing		Continuous insulation
All-wood joist/truss	R-25		R-19
Metal joist/truss	R-25		R-20
Concrete slab or deck	NA		R-19
Metal purlin with thermal block	R-30		R-20
Metal purlin without thermal block	X		R-20
Floors over outdoor air or unconditioned space (*R*-value)	Insulation between framing		Continuous insulation
All-wood joist/truss	R-11		R-5
Metal joist/truss	R-11		R-6
Concrete slab or deck	NA		R-5
Above-grade walls (*R*-value)	No framing	Metal framing	Wood framing
Framed			
R-value cavity	NA	R-11	R-11
R-value continuous	NA	R-0	R-0
CMU, ≥ 8 inches, with integral insulation			
R-value cavity	NA	R-11	R-11
R-value continuous	R-5	R-0	R-0
Other masonry walls			
R-value cavity	NA	R-11	R-11
R-value continuous	R-5	R-0	R-0
WINDOW AND GLAZED DOOR AREA GREATER THAN 40 PERCENT BUT NOT GREATER THAN 50 PERCENT OF ABOVE-GRADE WALL AREA			
ELEMENT	CONDITION/VALUE		
Skylights (*U*-factor)	1		
Slab or below-grade wall (*R*-value)	R-0		
Windows and glass doors	SHGC		*U*-factor
PF < 0.25	0.4		0.7
0.25 ≤ PF < 0.50	0.5		0.7
PF ≥ 0.50	0.6		0.7
Roof assemblies (*R*-value)	Insulation between framing		Continuous insulation
All-wood joist/truss	R-25		R-19
Metal joist/truss	R-25		R-20
Concrete slab or deck	NA		R-19
Metal purlin with thermal block	R-30		R-20
Metal purlin without thermal block	R-38		R-20
Floors over outdoor air or unconditioned space (*R*-value)	Insulation between framing		Continuous insulation
All-wood joist/truss	R-11		R-5
Metal joist/truss	R-11		R-6
Concrete slab or deck	NA		R-5
Above-grade walls (*R*-value)	No framing	Metal framing	Wood framing
Framed			
R-value cavity	NA	R-13	R-11
R-value continuous	NA	R-3	R-0
CMU, ≥ 8 inches, with integral insulation			
R-value cavity	NA	R-11	R-11
R-value continuous	R-5	R-0	R-0
Other masonry walls			
R-value cavity	NA	R-11	R-11
R-value continuous	R-5	R-0	R-0

For SI: 1 inch = 25.4 mm.

a. Values from Tables 802.2(5) through 802.2(37) shall be used for the purpose of the completion of Tables 802.2(1) through 802.2(4), as applicable based on window and glazed door area.
b. "NA" indicates the condition is not applicable.
c. An *R*-value of zero indicates no insulation is required.
d. "Any" indicates any available product will comply.
e. "X" indicates no complying option exists for this condition.

TABLE 802.2(16)
BUILDING ENVELOPE REQUIREMENTS[a through e] - CLIMATE ZONE 6a

WINDOW AND GLAZED DOOR AREA 10 PERCENT OR LESS OF ABOVE-GRADE WALL AREA			
ELEMENT	CONDITION/VALUE		
Skylights (*U*-factor)	0.8		
Slab or below-grade wall (*R*-value)	R-0		
Windows and glass doors	SHGC		*U*-factor
PF < 0.25	Any		Any
0.25 ≤ PF < 0.50	Any		Any
PF ≥ 0.50	Any		Any
Roof assemblies (*R*-value)	Insulation between framing		Continuous insulation
All-wood joist/truss	R-11		R-10
Metal joist/truss	R-13		R-11
Concrete slab or deck	NA		R-10
Metal purlin with thermal block	R-19		R-11
Metal purlin without thermal block	R-25		R-11
Floors over outdoor air or unconditioned space (*R*-value)	Insulation between framing		Continuous insulation
All-wood joist/truss	R-11		R-7
Metal joist/truss	R-11		R-8
Concrete slab or deck	NA		R-7
Above-grade walls (*R*-value)	No framing	Metal framing	Wood framing
Framed			
R-value cavity	NA	R-11	R-11
R-value continuous	NA	R-0	R-0
CMU, ≥ 8 inches, with integral insulation			
R-value cavity	NA	R-0	R-0
R-value continuous	R-0	R-0	R-0
Other masonry walls			
R-value cavity	NA	R-0	R-0
R-value continuous	R-0	R-0	R-0
WINDOW AND GLAZED DOOR AREA GREATER THAN 10 PERCENT BUT NOT GREATER THAN 25 PERCENT OF ABOVE-GRADE WALL AREA			
ELEMENT	CONDITION/VALUE		
Skylights (*U*-factor)	0.8		
Slab or below-grade wall (*R*-value)	R-0		
Windows and glass doors	SHGC		*U*-factor
PF < 0.25	0.7		Any
0.25 ≤ PF < 0.50	Any		Any
PF ≥ 0.50	Any		Any
Roof assemblies (*R*-value)	Insulation between framing		Continuous insulation
All-wood joist/truss	R-25		R-19
Metal joist/truss	R-25		R-20
Concrete slab or deck	NA		R-19
Metal purlin with thermal block	R-30		R-20
Metal purlin without thermal block	X		R-20
Floors over outdoor air or unconditioned space (*R*-value)	Insulation between framing		Continuous insulation
All-wood joist/truss	R-11		R-7
Metal joist/truss	R-11		R-8
Concrete slab or deck	NA		R-7
Above-grade walls (*R*-value)	No framing	Metal framing	Wood framing
Framed			
R-value cavity	NA	R-11	R-11
R-value continuous	NA	R-0	R-0
CMU, ≥ 8 inches, with integral insulation			
R-value cavity	NA	R-0	R-0
R-value continuous	R-0	R-0	R-0
Other masonry walls			
R-value cavity	NA	R-11	R-11
R-value continuous	R-5	R-0	R-0

(continued)

ACCEPTABLE PRACTICE FOR COMMERCIAL BUILDING

TABLE 802.2(16)—continued
BUILDING ENVELOPE REQUIREMENTS[a through e] - CLIMATE ZONE 6a

WINDOW AND GLAZED DOOR AREA GREATER THAN 25 PERCENT BUT NOT GREATER THAN 40 PERCENT OF ABOVE-GRADE WALL AREA			
ELEMENT	**CONDITION/VALUE**		
Skylights (*U*-factor)	0.8		
Slab or below-grade wall (*R*-value)	R-0		
Windows and glass doors	SHGC		*U*-factor
PF < 0.25	0.5		0.7
0.25 ≤ PF < 0.50	0.6		0.7
PF ≥ 0.50	0.7		0.7
Roof assemblies (*R*-value)	Insulation between framing		Continuous insulation
All-wood joist/truss	R-25		R-19
Metal joist/truss	R-25		R-20
Concrete slab or deck	NA		R-19
Metal purlin with thermal block	R-30		R-20
Metal purlin without thermal block	X		R-20
Floors over outdoor air or unconditioned space (*R*-value)	Insulation between framing		Continuous insulation
All-wood joist/truss	R-11		R-7
Metal joist/truss	R-11		R-8
Concrete slab or deck	NA		R-7
Above-grade walls (*R*-value)	No framing	Metal framing	Wood framing
Framed			
R-value cavity	NA	R-11	R-11
R-value continuous	NA	R-0	R-0
CMU, ≥ 8 inches, with integral insulation			
R-value cavity	NA	R-0	R-0
R-value continuous	R-0	R-0	R-0
Other masonry walls			
R-value cavity	NA	R-11	R-11
R-value continuous	R-5	R-0	R-0

WINDOW AND GLAZED DOOR AREA GREATER THAN 40 PERCENT BUT NOT GREATER THAN 50 PERCENT OF ABOVE-GRADE WALL AREA			
ELEMENT	**CONDITION/VALUE**		
Skylights (*U*-factor)	1		
Slab or below-grade wall (*R*-value)	R-0		
Windows and glass doors	SHGC		*U*-factor
PF < 0.25	0.4		0.7
0.25 ≤ PF < 0.50	0.5		0.7
PF ≥ 0.50	0.7		0.7
Roof assemblies (*R*-value)	Insulation between framing		Continuous insulation
All-wood joist/truss	R-25		R-19
Metal joist/truss	R-25		R-20
Concrete slab or deck	NA		R-19
Metal purlin with thermal block	R-30		R-20
Metal purlin without thermal block	R-38		R-20
Floors over outdoor air or unconditioned space (*R*-value)	Insulation between framing		Continuous insulation
All-wood joist/truss	R-11		R-7
Metal joist/truss	R-11		R-8
Concrete slab or deck	NA		R-7
Above-grade walls (*R*-value)	No framing	Metal framing	Wood framing
Framed			
R-value cavity	NA	R-13	R-11
R-value continuous	NA	R-3	R-0
CMU, ≥ 8 inches, with integral insulation			
R-value cavity	NA	R-0	R-0
R-value continuous	R-0	R-0	R-0
Other masonry walls			
R-value cavity	NA	R-11	R-11
R-value continuous	R-5	R-0	R-0

For SI: 1 inch = 25.4 mm.

a. Values from Tables 802.2(5) through 802.2(37) shall be used for the purpose of the completion of Tables 802.2(1) through 802.2(4), as applicable based on window and glazed door area.
b. "NA" indicates the condition is not applicable.
c. An *R*-value of zero indicates no insulation is required.
d. "Any" indicates any available product will comply.
e. "X" indicates no complying option exists for this condition.

ACCEPTABLE PRACTICE FOR COMMERCIAL BUILDINGS

TABLE 802.2(17)
BUILDING ENVELOPE REQUIREMENTS[a through e] - CLIMATE ZONE 6b

WINDOW AND GLAZED DOOR AREA 10 PERCENT OR LESS OF ABOVE-GRADE WALL AREA			
ELEMENT	CONDITION/VALUE		
Skylights (*U*-factor)	1		
Slab or below-grade wall (*R*-value)	R-0		
Windows and glass doors	SHGC		*U*-factor
PF < 0.25	Any		Any
0.25 ≤ PF < 0.50	Any		Any
PF ≥ 0.50	Any		Any
Roof assemblies (*R*-value)	Insulation between framing		Continuous insulation
All-wood joist/truss	R-19		R-16
Metal joist/truss	R-25		R-17
Concrete slab or deck	NA		R-16
Metal purlin with thermal block	R-25		R-17
Metal purlin without thermal block	X		R-17
Floors over outdoor air or unconditioned space (*R*-value)	Insulation between framing		Continuous insulation
All-wood joist/truss	R-11		R-6
Metal joist/truss	R-11		R-6
Concrete slab or deck	NA		R-6
Above-grade walls (*R*-value)	No framing	Metal framing	Wood framing
Framed			
R-value cavity	NA	R-11	R-11
R-value continuous	NA	R-0	R-0
CMU, ≥ 8 inches, with integral insulation			
R-value cavity	NA	R-0	R-0
R-value continuous	R-0	R-0	R-0
Other masonry walls			
R-value cavity	NA	R-0	R-0
R-value continuous	R-0	R-0	R-0
WINDOW AND GLAZED DOOR AREA GREATER THAN 10 PERCENT BUT NOT GREATER THAN 25 PERCENT OF ABOVE-GRADE WALL AREA			
ELEMENT	CONDITION/VALUE		
Skylights (*U*-factor)	1		
Slab or below-grade wall (*R*-value)	R-0		
Windows and glass doors	SHGC		*U*-factor
PF < 0.25	0.6		Any
0.25 ≤ PF < 0.50	0.7		Any
PF ≥ 0.50	Any		Any
Roof assemblies (*R*-value)	Insulation between framing		Continuous insulation
All-wood joist/truss	R-19		R-16
Metal joist/truss	R-25		R-17
Concrete slab or deck	NA		R-16
Metal purlin with thermal block	R-25		R-17
Metal purlin without thermal block	X		R-17
Floors over outdoor air or unconditioned space (*R*-value)	Insulation between framing		Continuous insulation
All-wood joist/truss	R-11		R-6
Metal joist/truss	R-11		R-6
Concrete slab or deck	NA		R-6
Above-grade walls (*R*-value)	No framing	Metal framing	Wood framing
Framed			
R-value cavity	NA	R-11	R-11
R-value continuous	NA	R-0	R-0
CMU, ≥ 8 inches, with integral insulation			
R-value cavity	NA	R-0	R-0
R-value continuous	R-0	R-0	R-0
Other masonry walls			
R-value cavity	NA	R-11	R-11
R-value continuous	R-5	R-0	R-0

(continued)

TABLE 802.2(17)—continued
BUILDING ENVELOPE REQUIREMENTS[a through e] - CLIMATE ZONE 6b

WINDOW AND GLAZED DOOR AREA GREATER THAN 25 PERCENT BUT NOT GREATER THAN 40 PERCENT OF ABOVE-GRADE WALL AREA			
ELEMENT	CONDITION/VALUE		
Skylights (*U*-factor)	1		
Slab or below-grade wall (*R*-value)	R-0		
Windows and glass doors	SHGC		*U*-factor
PF < 0.25	0.5		0.7
0.25 ≤ PF < 0.50	0.6		0.7
PF ≥ 0.50	0.7		0.7
Roof assemblies (*R*-value)	Insulation between framing		Continuous insulation
All-wood joist/truss	R-25		R-19
Metal joist/truss	R-25		R-20
Concrete slab or deck	NA		R-19
Metal purlin with thermal block	R-30		R-20
Metal purlin without thermal block	X		R-20
Floors over outdoor air or unconditioned space (*R*-value)	Insulation between framing		Continuous insulation
All-wood joist/truss	R-11		R-6
Metal joist/truss	R-11		R-6
Concrete slab or deck	NA		R-6
Above-grade walls (*R*-value)	No framing	Metal framing	Wood framing
Framed			
R-value cavity	NA	R-11	R-11
R-value continuous	NA	R-0	R-0
CMU, ≥ 8 inches, with integral insulation			
R-value cavity	NA	R-0	R-0
R-value continuous	R-0	R-0	R-0
Other masonry walls			
R-value cavity	NA	R-11	R-11
R-value continuous	R-5	R-0	R-0
WINDOW AND GLAZED DOOR AREA GREATER THAN 40 PERCENT BUT NOT GREATER THAN 50 PERCENT OF ABOVE-GRADE WALL AREA			
ELEMENT	CONDITION/VALUE		
Skylights (*U*-factor)	1		
Slab or below-grade wall (*R*-value)	R-0		
Windows and glass doors	SHGC		*U*-factor
PF < 0.25	0.4		0.7
0.25 ≤ PF < 0.50	0.5		0.7
PF ≥ 0.50	0.7		0.7
Roof assemblies (*R*-value)	Insulation between framing		Continuous insulation
All-wood joist/truss	R-25		R-19
Metal joist/truss	R-25		R-20
Concrete slab or deck	NA		R-19
Metal purlin with thermal block	R-30		R-20
Metal purlin without thermal block	R-38		R-20
Floors over outdoor air or unconditioned space (*R*-value)	Insulation between framing		Continuous insulation
All-wood joist/truss	R-11		R-6
Metal joist/truss	R-11		R-6
Concrete slab or deck	NA		R-6
Above-grade walls (*R*-value)	No framing	Metal framing	Wood framing
Framed			
R-value cavity	NA	R-11	R-11
R-value continuous	NA	R-0	R-0
CMU, ≥ 8 inches, with integral insulation			
R-value cavity	NA	R-0	R-0
R-value continuous	R-0	R-0	R-0
Other masonry walls			
R-value cavity	NA	R-11	R-11
R-value continuous	R-5	R-0	R-0

For SI: 1 inch = 25.4 mm.

a. Values from Tables 802.2(5) through 802.2(37) shall be used for the purpose of the completion of Tables 802.2(1) through 802.2(4), as applicable based on window and glazed door area.
b. "NA" indicates the condition is not applicable.
c. An *R*-value of zero indicates no insulation is required.
d. "Any" indicates any available product will comply.
e. "X" indicates no complying option exists for this condition.

ACCEPTABLE PRACTICE FOR COMMERCIAL BUILDINGS

TABLE 802.2(18)
BUILDING ENVELOPE REQUIREMENTS[b through f] - CLIMATE ZONE 7a

WINDOW AND GLAZED DOOR AREA 10 PERCENT OR LESS OF ABOVE-GRADE WALL AREA			
ELEMENT	CONDITION/VALUE		
Skylights (*U*-factor)	0.8		
Slab or below-grade wall (*R*-value)	R-0		
Windows and glass doors	SHGC		*U*-factor
PF < 0.25	Any		Any
0.25 ≤ PF < 0.50	Any		Any
PF ≥ 0.50	Any		Any
Roof assemblies (*R*-value)	Insulation between framing		Continuous insulation
All-wood joist/truss	R-19		R-14
Metal joist/truss	R-19		R-15
Concrete slab or deck	NA		R-14
Metal purlin with thermal block	R-25		R-15
Metal purlin without thermal block	X		R-15
Floors over outdoor air or unconditioned space (*R*-value)	Insulation between framing		Continuous insulation
All-wood joist/truss	R-11		R-8
Metal joist/truss	R-11		R-9
Concrete slab or deck	NA		R-8
Above-grade walls (*R*-value)	No framing	Metal framing	Wood framing
Framed			
R-value cavity	NA	R-11	R-11
R-value continuous	NA	R-0	R-0
CMU, ≥ 8 inches, with integral insulation			
R-value cavity	NA	R-0	R-0
R-value continuous	R-0	R-0	R-0
Other masonry walls			
R-value cavity	NA	R-11	R-11
R-value continuous	R-5	R-0	R-0
WINDOW AND GLAZED DOOR AREA GREATER THAN 10 PERCENT BUT NOT GREATER THAN 25 PERCENT OF ABOVE-GRADE WALL AREA			
ELEMENT	CONDITION/VALUE		
Skylights (*U*-factor)	0.8		
Slab or below-grade wall (*R*-value)	R-0		
Windows and glass doors	SHGC		*U*-factor
PF < 0.25	0.5		0.7
0.25 ≤ PF < 0.50	0.6		0.7
PF ≥ 0.50	0.7		0.7
Roof assemblies (*R*-value)	Insulation between framing		Continuous insulation
All-wood joist/truss	R-25		R-19
Metal joist/truss	R-25		R-20
Concrete slab or deck	NA		R-19
Metal purlin with thermal block	R-30		R-20
Metal purlin without thermal block	X		R-20
Floors over outdoor air or unconditioned space (*R*-value)	Insulation between framing		Continuous insulation
All-wood joist/truss	R-11		R-8
Metal joist/truss	R-11		R-9
Concrete slab or deck	NA		R-8
Above-grade walls (*R*-value)	No framing	Metal framing	Wood framing
Framed			
R-value cavity	NA	R-11	R-11
R-value continuous	NA	R-0	R-0
CMU, ≥ 8 inches, with integral insulation			
R-value cavity	NA	R-0	R-0
R-value continuous	R-0	R-0	R-0
Other masonry walls			
R-value cavity	NA	R-11	R-11
R-value continuous	R-5	R-0	R-0

(continued)

TABLE 802.2(18)—continued
BUILDING ENVELOPE REQUIREMENTS[b through f] - CLIMATE ZONE 7a

WINDOW AND GLAZED DOOR AREA GREATER THAN 25 PERCENT BUT NOT GREATER THAN 40 PERCENT OF ABOVE-GRADE WALL AREA			
ELEMENT	CONDITION/VALUE		
Skylights (*U*-factor)	0.8		
Slab or below-grade wall (*R*-value)	R-0		
Windows and glass doors	SHGC		*U*-factor[a]
PF < 0.25	0.4		0.7
0.25 ≤ PF < 0.50	0.5		0.7
PF ≥ 0.50	0.6		0.7
Roof assemblies (*R*-value)	Insulation between framing		Continuous insulation
All-wood joist/truss	R-25		R-19
Metal joist/truss	R-25		R-20
Concrete slab or deck	NA		R-19
Metal purlin with thermal block	R-30		R-20
Metal purlin without thermal block	X		R-20
Floors over outdoor air or unconditioned space (*R*-value)	Insulation between framing		Continuous insulation
All-wood joist/truss	R-11		R-8
Metal joist/truss	R-11		R-9
Concrete slab or deck	NA		R-8
Above-grade walls (*R*-value)	No framing	Metal framing	Wood framing
Framed			
R-value cavity	NA	R-11	R-11
R-value continuous	NA	R-0	R-0
CMU, ≥ 8 inches, with integral insulation			
R-value cavity	NA	R-11	R-11
R-value continuous	R-5	R-0	R-0
Other masonry walls			
R-value cavity	NA	R-11	R-11
R-value continuous	R-5	R-0	R-0
WINDOW AND GLAZED DOOR AREA GREATER THAN 40 PERCENT BUT NOT GREATER THAN 50 PERCENT OF ABOVE-GRADE WALL AREA			
ELEMENT	CONDITION/VALUE		
Skylights (*U*-factor)	0.8		
Slab or below-grade wall (*R*-value)	R-0		
Windows and glass doors	SHGC		*U*-factor
PF < 0.25	0.3		0.7
0.25 ≤ PF < 0.50	0.4		0.7
PF ≥ 0.50	0.5		0.7
Roof assemblies (*R*-value)	Insulation between framing		Continuous insulation
All-wood joist/truss	R-25		R-19
Metal joist/truss	R-25		R-20
Concrete slab or deck	NA		R-19
Metal purlin with thermal block	R-30		R-20
Metal purlin without thermal block	R-38		R-20
Floors over outdoor air or unconditioned space (*R*-value)	Insulation between framing		Continuous insulation
All-wood joist/truss	R-11		R-8
Metal joist/truss	R-11		R-9
Concrete slab or deck	NA		R-8
Above-grade walls (*R*-value)	No framing	Metal framing	Wood framing
Framed			
R-value cavity	NA	R-11	R-11
R-value continuous	NA	R-0	R-0
CMU, ≥ 8 inches, with integral insulation			
R-value cavity	NA	R-11	R-11
R-value continuous	R-5	R-0	R-0
Other masonry walls			
R-value cavity	NA	R-11	R-11
R-value continuous	R-5	R-0	R-0

For SI: 1 inch = 25.4 mm.

a. For buildings over three stories in height, the maximum *U*-factor shall be 0.60.
b. Values from Tables 802.2(5) through 802.2(37) shall be used for the purpose of the completion of Tables 802.2(1) through 802.2(4), as applicable based on window and glazed door area.
c. "NA" indicates the condition is not applicable.
d. An *R*-value of zero indicates no insulation is required.
e. "Any" indicates any available product will comply.
f. "X" indicates no complying option exists for this condition.

TABLE 802.2(19)
BUILDING ENVELOPE REQUIREMENTS[b through f] - CLIMATE ZONE 7b

WINDOW AND GLAZED DOOR AREA 10 PERCENT OR LESS OF ABOVE-GRADE WALL AREA			
ELEMENT	CONDITION/VALUE		
Skylights (U-factor)	0.8		
Slab or below-grade wall (R-value)	R-0		
Windows and glass doors	SHGC		U-factor
PF < 0.25	Any		Any
0.25 ≤ PF < 0.50	Any		Any
PF ≥ 0.50	Any		Any
Roof assemblies (R-value)	Insulation between framing		Continuous insulation
All-wood joist/truss	R-19		R-14
Metal joist/truss	R-19		R-15
Concrete slab or deck	NA		R-14
Metal purlin with thermal block	R-25		R-15
Metal purlin without thermal block	X		R-15
Floors over outdoor air or unconditioned space (R-value)	Insulation between framing		Continuous insulation
All-wood joist/truss	R-11		R-7
Metal joist/truss	R-11		R-8
Concrete slab or deck	NA		R-8
Above-grade walls (R-value)	No framing	Metal framing	Wood framing
Framed			
R-value cavity	NA	R-11	R-11
R-value continuous	NA	R-0	R-0
CMU, ≥ 8 inches, with integral insulation			
R-value cavity	NA	R-0	R-0
R-value continuous	R-0	R-0	R-0
Other masonry walls			
R-value cavity	NA	R-11	R-11
R-value continuous	R-5	R-0	R-0
WINDOW AND GLAZED DOOR AREA GREATER THAN 10 PERCENT BUT NOT GREATER THAN 25 PERCENT OF ABOVE-GRADE WALL AREA			
ELEMENT	CONDITION/VALUE		
Skylights (U-factor)	0.8		
Slab or below-grade wall (R-value)	R-0		
Windows and glass doors	SHGC		U-factor
PF < 0.25	0.5		0.7
0.25 ≤ PF < 0.50	0.6		0.7
PF ≥ 0.50	0.7		0.7
Roof assemblies (R-value)	Insulation between framing		Continuous insulation
All-wood joist/truss	R-25		R-19
Metal joist/truss	R-25		R-20
Concrete slab or deck	NA		R-19
Metal purlin with thermal block	R-30		R-20
Metal purlin without thermal block	X		R-20
Floors over outdoor air or unconditioned space (R-value)	Insulation between framing		Continuous insulation
All-wood joist/truss	R-11		R-7
Metal joist/truss	R-11		R-8
Concrete slab or deck	NA		R-8
Above-grade walls (R-value)	No framing	Metal framing	Wood framing
Framed			
R-value cavity	NA	R-11	R-11
R-value continuous	NA	R-0	R-0
CMU, ≥ 8 inches, with integral insulation			
R-value cavity	NA	R-11	R-11
R-value continuous	R-5	R-0	R-0
Other masonry walls			
R-value cavity	NA	R-11	R-11
R-value continuous	R-5	R-0	R-0

(continued)

TABLE 802.2(19)—continued
BUILDING ENVELOPE REQUIREMENTS[b through f] - CLIMATE ZONE 7b

WINDOW AND GLAZED DOOR AREA GREATER THAN 25 PERCENT BUT NOT GREATER THAN 40 PERCENT OF ABOVE-GRADE WALL AREA			
ELEMENT	CONDITION/VALUE		
Skylights (*U*-factor)	0.8		
Slab or below-grade wall (*R*-value)	R-0		
Windows and glass doors	SHGC		*U*-factor[a]
PF < 0.25	0.4		0.7
0.25 ≤ PF < 0.50	0.5		0.7
PF ≥ 0.50	0.6		0.7
Roof assemblies (*R*-value)	Insulation between framing		Continuous insulation
All-wood joist/truss	R-25		R-19
Metal joist/truss	R-25		R-20
Concrete slab or deck	NA		R-19
Metal purlin with thermal block	R-30		R-20
Metal purlin without thermal block	X		R-20
Floors over outdoor air or unconditioned space (*R*-value)	Insulation between framing		Continuous insulation
All-wood joist/truss	R-11		R-7
Metal joist/truss	R-11		R-8
Concrete slab or deck	NA		R-8
Above-grade walls (*R*-value)	No framing	Metal framing	Wood framing
Framed			
R-value cavity	NA	R-13	R-13
R-value continuous	NA	R-0	R-0
CMU, ≥ 8 inches, with integral insulation			
R-value cavity	NA	R-11	R-11
R-value continuous	R-5	R-0	R-0
Other masonry walls			
R-value cavity	NA	R-13	R-11
R-value continuous	R-6	R-0	R-0

WINDOW AND GLAZED DOOR AREA GREATER THAN 40 PERCENT BUT NOT GREATER THAN 50 PERCENT OF ABOVE-GRADE WALL AREA			
ELEMENT	CONDITION/VALUE		
Skylights (*U*-factor)	0.8		
Slab or below-grade wall (*R*-value)	R-0		
Windows and glass doors	SHGC		*U*-factor
PF < 0.25	0.3		0.7
0.25 ≤ PF < 0.50	0.4		0.7
PF ≥ 0.50	0.5		0.7
Roof assemblies (*R*-value)	Insulation between framing		Continuous insulation
All-wood joist/truss	R-25		R-19
Metal joist/truss	R-25		R-20
Concrete slab or deck	NA		R-19
Metal purlin with thermal block	R-30		R-20
Metal purlin without thermal block	R-38		R-20
Floors over outdoor air or unconditioned space (*R*-value)	Insulation between framing		Continuous insulation
All-wood joist/truss	R-11		R-7
Metal joist/truss	R-11		R-8
Concrete slab or deck	NA		R-8
Above-grade walls (*R*-value)	No framing	Metal framing	Wood framing
Framed			
R-value cavity	NA	R-13	R-13
R-value continuous	NA	R-3	R-0
CMU, ≥ 8 inches, with integral insulation			
R-value cavity	NA	R-11	R-11
R-value continuous	R-5	R-0	R-0
Other masonry walls			
R-value cavity	NA	R-13	R-11
R-value continuous	R-6	R-0	R-0

For SI: 1 inch = 25.4 mm.

a. For buildings over three stories in height, the maximum *U*-factor shall be 0.60.
b. Values from Tables 802.2(5) through 802.2(37) shall be used for the purpose of the completion of Tables 802.2(1) through 802.2(4), as applicable based on window and glazed door area.
c. "NA" indicates the condition is not applicable.
d. An *R*-value of zero indicates no insulation is required.
e. "Any" indicates any available product will comply.
f. "X" indicates no complying option exists for this condition.

ACCEPTABLE PRACTICE FOR COMMERCIAL BUILDINGS

TABLE 802.2(20)
BUILDING ENVELOPE REQUIREMENTS[a through e] - CLIMATE ZONE 8

WINDOW AND GLAZED DOOR AREA 10 PERCENT OR LESS OF ABOVE-GRADE WALL AREA			
ELEMENT	CONDITION/VALUE		
Skylights (U-factor)	0.8		
Slab or below-grade wall (R-value)	R-0		
Windows and glass doors	SHGC		U-factor
PF < 0.25	Any		Any
0.25 ≤ PF < 0.50	Any		Any
PF ≥ 0.50	Any		Any
Roof assemblies (R-value)	Insulation between framing		Continuous insulation
All-wood joist/truss	R-19		R-14
Metal joist/truss	R-19		R-15
Concrete slab or deck	NA		R-14
Metal purlin with thermal block	R-25		R-15
Metal purlin without thermal block	X		R-15
Floors over outdoor air or unconditioned space (R-value)	Insulation between framing		Continuous insulation
All-wood joist/truss	R-11		R-9
Metal joist/truss	R-11		R-10
Concrete slab or deck	NA		R-9
Above-grade walls (R-value)	No framing	Metal framing	Wood framing
Framed			
R-value cavity	NA	R-11	R-11
R-value continuous	NA	R-0	R-0
CMU, ≥ 8 inches, with integral insulation			
R-value cavity	NA	R-11	R-11
R-value continuous	R-5	R-0	R-0
Other masonry walls			
R-value cavity	NA	R-11	R-11
R-value continuous	R-5	R-0	R-0
WINDOW AND GLAZED DOOR AREA GREATER THAN 10 PERCENT BUT NOT GREATER THAN 25 PERCENT OF ABOVE-GRADE WALL AREA			
ELEMENT	CONDITION/VALUE		
Skylights (U-factor)	0.8		
Slab or below-grade wall (R-value)	R-0		
Windows and glass doors	SHGC		U-factor
PF < 0.25	0.5		0.7
0.25 ≤ PF < 0.50	0.6		0.7
PF ≥ 0.50	0.7		0.7
Roof assemblies (R-value)	Insulation between framing		Continuous insulation
All-wood joist/truss	R-25		R-19
Metal joist/truss	R-25		R-20
Concrete slab or deck	NA		R-19
Metal purlin with thermal block	R-30		R-20
Metal purlin without thermal block	X		R-20
Floors over outdoor air or unconditioned space (R-value)	Insulation between framing		Continuous insulation
All-wood joist/truss	R-11		R-9
Metal joist/truss	R-11		R-10
Concrete slab or deck	NA		R-9
Above-grade walls (R-value)	No framing	Metal framing	Wood framing
Framed			
R-value cavity	NA	R-13	R-11
R-value continuous	NA	R-0	R-0
CMU, ≥ 8 inches, with integral insulation			
R-value cavity	NA	R-11	R-11
R-value continuous	R-5	R-0	R-0
Other masonry walls			
R-value cavity	NA	R-13	R-11
R-value continuous	R-6	R-0	R-0

(continued)

ACCEPTABLE PRACTICE FOR COMMERCIAL BUILDING

TABLE 802.2(20)—continued
BUILDING ENVELOPE REQUIREMENTS[a through e] - CLIMATE ZONE 8

WINDOW AND GLAZED DOOR AREA GREATER THAN 25 PERCENT BUT NOT GREATER THAN 40 PERCENT OF ABOVE-GRADE WALL AREA			
ELEMENT	CONDITION/VALUE		
Skylights (U-factor)	0.8		
Slab or below-grade wall (R-value)	R-0		
Windows and glass doors	SHGC		U-factor
PF < 0.25	0.4		0.5
0.25 ≤ PF < 0.50	0.5		0.5
PF ≥ 0.50	0.6		0.5
Roof assemblies (R-value)	Insulation between framing		Continuous insulation
All-wood joist/truss	R-25		R-19
Metal joist/truss	R-25		R-20
Concrete slab or deck	NA		R-19
Metal purlin with thermal block	R-30		R-20
Metal purlin without thermal block	X		R-20
Floors over outdoor air or unconditioned space (R-value)	Insulation between framing		Continuous insulation
All-wood joist/truss	R-11		R-9
Metal joist/truss	R-11		R-10
Concrete slab or deck	NA		R-9
Above-grade walls (R-value)	No framing	Metal framing	Wood framing
Framed			
R-value cavity	NA	R-13	R-11
R-value continuous	NA	R-0	R-0
CMU, ≥ 8 inches, with integral insulation			
R-value cavity	NA	R-11	R-11
R-value continuous	R-5	R-0	R-0
Other masonry walls			
R-value cavity	NA	R-13	R-11
R-value continuous	R-6	R-0	R-0
WINDOW AND GLAZED DOOR AREA GREATER THAN 40 PERCENT BUT NOT GREATER THAN 50 PERCENT OF ABOVE-GRADE WALL AREA			
ELEMENT	CONDITION/VALUE		
Skylights (U-factor)	0.8		
Slab or below-grade wall (R-value)	R-0		
Windows and glass doors	SHGC		U-factor
PF < 0.25	0.3		0.5
0.25 ≤ PF < 0.50	0.4		0.5
PF ≥ 0.50	0.5		0.5
Roof assemblies (R-value)	Insulation between framing		Continuous insulation
All-wood joist/truss	R-25		R-19
Metal joist/truss	R-25		R-20
Concrete slab or deck	NA		R-19
Metal purlin with thermal block	R-30		R-20
Metal purlin without thermal block	R-38		R-20
Floors over outdoor air or unconditioned space (R-value)	Insulation between framing		Continuous insulation
All-wood joist/truss	R-11		R-9
Metal joist/truss	R-11		R-10
Concrete slab or deck	NA		R-9
Above-grade walls (R-value)	No framing	Metal framing	Wood framing
Framed			
R-value cavity	NA	R-13	R-11
R-value continuous	NA	R-0	R-0
CMU, ≥ 8 inches, with integral insulation			
R-value cavity	NA	R-11	R-11
R-value continuous	R-5	R-0	R-0
Other masonry walls			
R-value cavity	NA	R-13	R-11
R-value continuous	R-6	R-0	R-0

For SI: 1 inch = 25.4 mm.

a. Values from Tables 802.2(5) through 802.2(37) shall be used for the purpose of the completion of Tables 802.2(1) through 802.2(4), as applicable based on window and glazed door area.
b. "NA" indicates the condition is not applicable.
c. An R-value of zero indicates no insulation is required.
d. "Any" indicates any available product will comply.
e. "X" indicates no complying option exists for this condition.

TABLE 802.2(21)
BUILDING ENVELOPE REQUIREMENTS[b through f] - CLIMATE ZONE 9a

WINDOW AND GLAZED DOOR AREA 10 PERCENT OR LESS OF ABOVE-GRADE WALL AREA			
ELEMENT	**CONDITION/VALUE**		
Skylights (*U*-factor)	0.8		
Slab or below-grade wall (*R*-value)	R-0		
Windows and glass doors	**SHGC**		***U*-factor**
PF < 0.25	Any		Any
0.25 ≤ PF < 0.50	Any		Any
PF ≥ 0.50	Any		Any
Roof assemblies (*R*-value)	**Insulation between framing**		**Continuous insulation**
All-wood joist/truss	R-19		R-13
Metal joist/truss	R-19		R-14
Concrete slab or deck	NA		R-13
Metal purlin with thermal block	R-19		R-14
Metal purlin without thermal block	X		R-14
Floors over outdoor air or unconditioned space (*R*-value)	**Insulation between framing**		**Continuous insulation**
All-wood joist/truss	R-13		R-12
Metal joist/truss	R-13		R-12
Concrete slab or deck	NA		R-12
Above-grade walls (*R*-value)	**No framing**	**Metal framing**	**Wood framing**
Framed			
R-value cavity	NA	R-11	R-11
R-value continuous	NA	R-0	R-0
CMU, ≥ 8 inches, with integral insulation			
R-value cavity	NA	R-0	R-0
R-value continuous	R-0	R-0	R-0
Other masonry walls			
R-value cavity	NA	R-11	R-11
R-value continuous	R-5	R-0	R-0
WINDOW AND GLAZED DOOR AREA GREATER THAN 10 PERCENT BUT NOT GREATER THAN 25 PERCENT OF ABOVE-GRADE WALL AREA			
ELEMENT	**CONDITION/VALUE**		
Skylights (*U*-factor)	0.8		
Slab or below-grade wall (*R*-value)	R-0		
Windows and glass doors	**SHGC**		***U*-factor**
PF < 0.25	Any		0.7
0.25 ≤ PF < 0.50	Any		0.7
PF ≥ 0.50	Any		0.7
Roof assemblies (*R*-value)	**Insulation between framing**		**Continuous insulation**
All-wood joist/truss	R-19		R-16
Metal joist/truss	R-25		R-17
Concrete slab or deck	NA		R-16
Metal purlin with thermal block	R-25		R-17
Metal purlin without thermal block	X		R-17
Floors over outdoor air or unconditioned space (*R*-value)	**Insulation between framing**		**Continuous insulation**
All-wood joist/truss	R-13		R-12
Metal joist/truss	R-13		R-12
Concrete slab or deck	NA		R-12
Above-grade walls (*R*-value)	**No framing**	**Metal framing**	**Wood framing**
Framed			
R-value cavity	NA	R-11	R-11
R-value continuous	NA	R-0	R-0
CMU, ≥ 8 inches, with integral insulation			
R-value cavity	NA	R-11	R-11
R-value continuous	R-5	R-0	R-0
Other masonry walls			
R-value cavity	NA	R-11	R-11
R-value continuous	R-5	R-0	R-0

(continued)

TABLE 802.2(21)—continued
BUILDING ENVELOPE REQUIREMENTS[b through f] - CLIMATE ZONE 9a

WINDOW AND GLAZED DOOR AREA GREATER THAN 25 PERCENT BUT NOT GREATER THAN 40 PERCENT OF ABOVE-GRADE WALL AREA			
ELEMENT	CONDITION/VALUE		
Skylights (*U*-factor)	0.8		
Slab or below-grade wall (*R*-value)	R-0		
Windows and glass doors	SHGC		*U*-factor[a]
PF < 0.25	0.6		0.7
0.25 ≤ PF < 0.50	0.7		0.7
PF ≥ 0.50	Any		0.7
Roof assemblies (*R*-value)	Insulation between framing		Continuous insulation
All-wood joist/truss	R-25		R-19
Metal joist/truss	R-25		R-20
Concrete slab or deck	NA		R-19
Metal purlin with thermal block	R-30		R-20
Metal purlin without thermal block	X		R-20
Floors over outdoor air or unconditioned space (*R*-value)	Insulation between framing		Continuous insulation
All-wood joist/truss	R-13		R-12
Metal joist/truss	R-13		R-12
Concrete slab or deck	NA		R-12
Above-grade walls (*R*-value)	No framing	Metal framing	Wood framing
Framed			
R-value cavity	NA	R-13	R-11
R-value continuous	NA	R-0	R-0
CMU, ≥ 8 inches, with integral insulation			
R-value cavity	NA	R-11	R-11
R-value continuous	R-5	R-0	R-0
Other masonry walls			
R-value cavity	NA	R-11	R-11
R-value continuous	R-5	R-0	R-0
WINDOW AND GLAZED DOOR AREA GREATER THAN 40 PERCENT BUT NOT GREATER THAN 50 PERCENT OF ABOVE-GRADE WALL AREA			
ELEMENT	CONDITION/VALUE		
Skylights (*U*-factor)	0.8		
Slab or below-grade wall (*R*-value)	R-0		
Windows and glass doors	SHGC		*U*-factor
PF < 0.25	0.5		0.5
0.25 ≤ PF < 0.50	0.7		0.5
PF ≥ 0.50	0.8		0.5
Roof assemblies (*R*-value)	Insulation between framing		Continuous insulation
All-wood joist/truss	R-25		R-19
Metal joist/truss	R-25		R-20
Concrete slab or deck	NA		R-19
Metal purlin with thermal block	R-30		R-20
Metal purlin without thermal block	R-38		R-20
Floors over outdoor air or unconditioned space (*R*-value)	Insulation between framing		Continuous insulation
All-wood joist/truss	R-13		R-12
Metal joist/truss	R-13		R-12
Concrete slab or deck	NA		R-12
Above-grade walls (*R*-value)	No framing	Metal framing	Wood framing
Framed			
R-value cavity	NA	R-13	R-13
R-value continuous	NA	R-11	R-5
CMU, ≥ 8 inches, with integral insulation			
R-value cavity	NA	R-11	R-11
R-value continuous	R-5	R-0	R-0
Other masonry walls			
R-value cavity	NA	R-11	R-11
R-value continuous	R-5	R-0	R-0

For SI: 1 inch = 25.4 mm.

a. For buildings over three stories in height, the maximum *U*-factor shall be 0.60.
b. Values from Tables 802.2(5) through 802.2(37) shall be used for the purpose of the completion of Tables 802.2(1) through 802.2(4), as applicable based on window and glazed door area.
c. "NA" indicates the condition is not applicable.
d. An *R*-value of zero indicates no insulation is required.
e. "Any" indicates any available product will comply.
f. "X" indicates no complying option exists for this condition.

TABLE 802.2(22)
BUILDING ENVELOPE REQUIREMENTS[a through e] - CLIMATE ZONE 9b

WINDOW AND GLAZED DOOR AREA 10 PERCENT OR LESS OF ABOVE-GRADE WALL AREA			
ELEMENT	CONDITION/VALUE		
Skylights (U-factor)	0.8		
Slab or below-grade wall (R-value)	R-0		
Windows and glass doors	SHGC		U-factor
PF < 0.25	Any		Any
0.25 ≤ PF < 0.50	Any		Any
PF ≥ 0.50	Any		Any
Roof assemblies (R-value)	Insulation between framing		Continuous insulation
All-wood joist/truss	R-19		R-15
Metal joist/truss	R-19		R-16
Concrete slab or deck	NA		R-15
Metal purlin with thermal block	R-25		R-16
Metal purlin without thermal block	X		R-16
Floors over outdoor air or unconditioned space (R-value)	Insulation between framing		Continuous insulation
All-wood joist/truss	R-13		R-11
Metal joist/truss	R-13		R-12
Concrete slab or deck	NA		R-12
Above-grade walls (R-value)	No framing	Metal framing	Wood framing
Framed			
R-value cavity	NA	R-11	R-11
R-value continuous	NA	R-0	R-0
CMU, ≥ 8 inches, with integral insulation			
R-value cavity	NA	R-11	R-11
R-value continuous	R-5	R-0	R-0
Other masonry walls			
R-value cavity	NA	R-11	R-11
R-value continuous	R-5	R-0	R-0
WINDOW AND GLAZED DOOR AREA GREATER THAN 10 PERCENT BUT NOT GREATER THAN 25 PERCENT OF ABOVE-GRADE WALL AREA			
ELEMENT	CONDITION/VALUE		
Skylights (U-factor)	0.8		
Slab or below-grade wall (R-value)	R-0		
Windows and glass doors	SHGC		U-factor
PF < 0.25	0.5		0.5
0.25 ≤ PF < 0.50	0.6		0.5
PF ≥ 0.50	0.7		0.5
Roof assemblies (R-value)	Insulation between framing		Continuous insulation
All-wood joist/truss	R-25		R-19
Metal joist/truss	R-25		R-20
Concrete slab or deck	NA		R-19
Metal purlin with thermal block	R-30		R-20
Metal purlin without thermal block	X		R-20
Floors over outdoor air or unconditioned space (R-value)	Insulation between framing		Continuous insulation
All-wood joist/truss	R-13		R-11
Metal joist/truss	R-13		R-12
Concrete slab or deck	NA		R-12
Above-grade walls (R-value)	No framing	Metal framing	Wood framing
Framed			
R-value cavity	NA	R-11	R-11
R-value continuous	NA	R-0	R-0
CMU, ≥ 8 inches, with integral insulation			
R-value cavity	NA	R-11	R-11
R-value continuous	R-5	R-0	R-0
Other masonry walls			
R-value cavity	NA	R-11	R-11
R-value continuous	R-5	R-0	R-0

(continued)

ACCEPTABLE PRACTICE FOR COMMERCIAL BUILDING

TABLE 802.2(22)—continued
BUILDING ENVELOPE REQUIREMENTS[a through e] - CLIMATE ZONE 9b

WINDOW AND GLAZED DOOR AREA GREATER THAN 25 PERCENT BUT NOT GREATER THAN 40 PERCENT OF ABOVE-GRADE WALL AREA			
ELEMENT	CONDITION/VALUE		
Skylights (U-factor)	0.8		
Slab or below-grade wall (R-value)	R-0		
Windows and glass doors	SHGC		U-factor
PF < 0.25	0.4		0.5
0.25 ≤ PF < 0.50	0.5		0.5
PF ≥ 0.50	0.6		0.5
Roof assemblies (R-value)	Insulation between framing		Continuous insulation
All-wood joist/truss	R-25		R-19
Metal joist/truss	R-25		R-20
Concrete slab or deck	NA		R-19
Metal purlin with thermal block	R-30		R-20
Metal purlin without thermal block	X		R-20
Floors over outdoor air or unconditioned space (R-value)	Insulation between framing		Continuous insulation
All-wood joist/truss	R-13		R-11
Metal joist/truss	R-13		R-12
Concrete slab or deck	NA		R-12
Above-grade walls (R-value)	No framing	Metal framing	Wood framing
Framed			
R-value cavity	NA	R-13	R-11
R-value continuous	NA	R-0	R-0
CMU, ≥ 8 inches, with integral insulation			
R-value cavity	NA	R-11	R-11
R-value continuous	R-5	R-0	R-0
Other masonry walls			
R-value cavity	NA	R-13	R-11
R-value continuous	R-6	R-0	R-0
WINDOW AND GLAZED DOOR AREA GREATER THAN 40 PERCENT BUT NOT GREATER THAN 50 PERCENT OF ABOVE-GRADE WALL AREA			
ELEMENT	CONDITION/VALUE		
Skylights (U-factor)	0.8		
Slab or below-grade wall (R-value)	R-0		
Windows and glass doors	SHGC		U-factor
PF < 0.25	0.3		0.5
0.25 ≤ PF < 0.50	0.4		0.5
PF ≥ 0.50	0.5		0.5
Roof assemblies (R-value)	Insulation between framing		Continuous insulation
All-wood joist/truss	R-25		R-19
Metal joist/truss	R-25		R-20
Concrete slab or deck	NA		R-19
Metal purlin with thermal block	R-30		R-20
Metal purlin without thermal block	R-38		R-20
Floors over outdoor air or unconditioned space (R-value)	Insulation between framing		Continuous insulation
All-wood joist/truss	R-13		R-11
Metal joist/truss	R-13		R-12
Concrete slab or deck	NA		R-12
Above-grade walls (R-value)	No framing	Metal framing	Wood framing
Framed			
R-value cavity	NA	R-13	R-13
R-value continuous	NA	R-5	R-3
CMU, ≥ 8 inches, with integral insulation			
R-value cavity	NA	R-11	R-11
R-value continuous	R-5	R-0	R-0
Other masonry walls			
R-value cavity	NA	R-13	R-11
R-value continuous	R-6	R-0	R-0

For SI: 1 inch = 25.4 mm.

a. Values from Tables 802.2(5) through 802.2(37) shall be used for the purpose of the completion of Tables 802.2(1) through 802.2(4), as applicable based on window and glazed door area.
b. "NA" indicates the condition is not applicable.
c. An R-value of zero indicates no insulation is required.
d. "Any" indicates any available product will comply.
e. "X" indicates no complying option exists for this condition.

TABLE 802.2(23)
BUILDING ENVELOPE REQUIREMENTS[a through e] - CLIMATE ZONE 10a

WINDOW AND GLAZED DOOR AREA 10 PERCENT OR LESS OF ABOVE-GRADE WALL AREA			
ELEMENT	CONDITION/VALUE		
Skylights (*U*-factor)	0.8		
Slab or below-grade wall (*R*-value)	R-0		
Windows and glass doors	SHGC		*U*-factor
PF < 0.25	Any		Any
0.25 ≤ PF < 0.50	Any		Any
PF ≥ 0.50	Any		Any
Roof assemblies (*R*-value)	Insulation between framing		Continuous insulation
All-wood joist/truss	R-19		R-14
Metal joist/truss	R-19		R-15
Concrete slab or deck	NA		R-14
Metal purlin with thermal block	R-25		R-15
Metal purlin without thermal block	X		R-15
Floors over outdoor air or unconditioned space (*R*-value)	Insulation between framing		Continuous insulation
All-wood joist/truss	R-19		R-13
Metal joist/truss	R-19		R-13
Concrete slab or deck	NA		R-13
Above-grade walls (*R*-value)	No framing	Metal framing	Wood framing
Framed			
R-value cavity	NA	R-11	R-11
R-value continuous	NA	R-0	R-0
CMU, ≥ 8 inches, with integral insulation			
R-value cavity	NA	R-0	R-0
R-value continuous	R-0	R-0	R-0
Other masonry walls			
R-value cavity	NA	R-11	R-11
R-value continuous	R-5	R-0	R-0
WINDOW AND GLAZED DOOR AREA GREATER THAN 10 PERCENT BUT NOT GREATER THAN 25 PERCENT OF ABOVE-GRADE WALL AREA			
ELEMENT	CONDITION/VALUE		
Skylights (*U*-factor)	0.8		
Slab or below-grade wall (*R*-value)	R-0		
Windows and glass doors	SHGC		*U*-factor
PF < 0.25	0.6		0.7
0.2 ≤ PF < 0.50	0.7		0.7
PF ≥ 0.50	Any		0.7
Roof assemblies (*R*-value)	Insulation between framing		Continuous insulation
All-wood joist/truss	R-19		R-16
Metal joist/truss	R-25		R-17
Concrete slab or deck	NA		R-16
Metal purlin with thermal block	R-25		R-17
Metal purlin without thermal block	X		R-17
Floors over outdoor air or unconditioned space (*R*-value)	Insulation between framing		Continuous insulation
All-wood joist/truss	R-19		R-13
Metal joist/truss	R-19		R-13
Concrete slab or deck	NA		R-13
Above-grade walls (*R*-value)	No framing	Metal framing	Wood framing
Framed			
R-value cavity	NA	R-11	R-11
R-value continuous	NA	R-0	R-0
CMU, ≥ 8 inches, with integral insulation			
R-value cavity	NA	R-11	R-11
R-value continuous	R-5	R-0	R-0
Other masonry walls			
R-value cavity	NA	R-11	R-11
R-value continuous	R-5	R-0	R-0

(*continued*)

TABLE 802.2(23)—continued
BUILDING ENVELOPE REQUIREMENTS[a through e] - CLIMATE ZONE 10a

WINDOW AND GLAZED DOOR AREA GREATER THAN 25 PERCENT BUT NOT GREATER THAN 40 PERCENT OF ABOVE-GRADE WALL AREA			
ELEMENT	CONDITION/VALUE		
Skylights (*U*-factor)	0.8		
Slab or below-grade wall (*R*-value)	R-0		
Windows and glass doors	SHGC	*U*-factor	
PF < 0.25	0.5	0.6	
0.25 ≤ PF < 0.50	0.6	0.6	
PF ≥ 0.50	0.7	0.6	
Roof assemblies (*R*-value)	Insulation between framing	Continuous insulation	
All-wood joist/truss	R-25	R-19	
Metal joist/truss	R-25	R-20	
Concrete slab or deck	NA	R-19	
Metal purlin with thermal block	R-30	R-20	
Metal purlin without thermal block	X	R-20	
Floors over outdoor air or unconditioned space (*R*-value)	Insulation between framing	Continuous insulation	
All-wood joist/truss	R-19	R-13	
Metal joist/truss	R-19	R-13	
Concrete slab or deck	NA	R-13	
Above-grade walls (*R*-value)	No framing	Metal framing	Wood framing
Framed			
R-value cavity	NA	R-11	R-11
R-value continuous	NA	R-0	R-0
CMU, ≥ 8 inches, with integral insulation			
R-value cavity	NA	R-11	R-11
R-value continuous	R-5	R-0	R-0
Other masonry walls			
R-value cavity	NA	R-11	R-11
R-value continuous	R-5	R-0	R-0
WINDOW AND GLAZED DOOR AREA GREATER THAN 40 PERCENT BUT NOT GREATER THAN 50 PERCENT OF ABOVE-GRADE WALL AREA			
ELEMENT	CONDITION/VALUE		
Skylights (*U*-factor)	0.8		
Slab or below-grade wall (*R*-value)	R-0		
Windows and glass doors	SHGC	*U*-factor	
PF < 0.25	0.5	0.4	
0.25 ≤ PF < 0.50	0.6	0.4	
PF ≥ 0.50	0.7	0.4	
Roof assemblies (*R*-value)	Insulation between framing	Continuous insulation	
All-wood joist/truss	R-25	R-19	
Metal joist/truss	R-25	R-20	
Concrete slab or deck	NA	R-19	
Metal purlin with thermal block	R-30	R-20	
Metal purlin without thermal block	R-30	R-20	
Floors over outdoor air or unconditioned space (*R*-value)	Insulation between framing	Continuous insulation	
All-wood joist/truss	R-19	R-13	
Metal joist/truss	R-19	R-13	
Concrete slab or deck	NA	R-13	
Above-grade walls (*R*-value)	No framing	Metal framing	Wood framing
Framed			
R-value cavity	NA	R-13	R-11
R-value continuous	NA	R-3	R-0
CMU, ≥ 8 inches, with integral insulation			
R-value cavity	NA	R-11	R-11
R-value continuous	R-5	R-0	R-0
Other masonry walls			
R-value cavity	NA	R-11	R-11
R-value continuous	R-5	R-0	R-0

For SI: 1 inch = 25.4 mm.

a. Values from Tables 802.2(5) through 802.2(37) shall be used for the purpose of the completion of Tables 802.2(1) through 802.2(4), as applicable based on window and glazed door area.
b. "NA" indicates the condition is not applicable.
c. An *R*-value of zero indicates no insulation is required.
d. "Any" indicates any available product will comply.
e. "X" indicates no complying option exists for this condition.

ACCEPTABLE PRACTICE FOR COMMERCIAL BUILDINGS

TABLE 802.2(24)
BUILDING ENVELOPE REQUIREMENTS[a through e] - CLIMATE ZONE 10b

WINDOW AND GLAZED DOOR AREA 10 PERCENT OR LESS OF ABOVE-GRADE WALL AREA			
ELEMENT	CONDITION/VALUE		
Skylights (*U*-factor)	0.8		
Slab or below-grade wall (*R*-value)	R-0		
Windows and glass doors	SHGC		*U*-factor
PF < 0.25	Any		Any
0.25 ≤ PF < 0.50	Any		Any
PF ≥ 0.50	Any		Any
Roof assemblies (*R*-value)	Insulation between framing		Continuous insulation
All-wood joist/truss	R-19		R-17
Metal joist/truss	R-25		R-18
Concrete slab or deck	NA		R-17
Metal purlin with thermal block	R-30		R-18
Metal purlin without thermal block	X		R-18
Floors over outdoor air or unconditioned space (*R*-value)	Insulation between framing		Continuous insulation
All-wood joist/truss	R-19		R-12
Metal joist/truss	R-19		R-13
Concrete slab or deck	NA		R-13
Above-grade walls (*R*-value)	No framing	Metal framing	Wood framing
Framed			
R-value cavity	NA	R-11	R-11
R-value continuous	NA	R-0	R-0
CMU, ≥ 8 inches, with integral insulation			
R-value cavity	NA	R-11	R-11
R-value continuous	R-5	R-0	R-0
Other masonry walls			
R-value cavity	NA	R-11	R-11
R-value continuous	R-5	R-0	R-0
WINDOW AND GLAZED DOOR AREA GREATER THAN 10 PERCENT BUT NOT GREATER THAN 25 PERCENT OF ABOVE-GRADE WALL AREA			
ELEMENT	CONDITION/VALUE		
Skylights (*U*-factor)	0.8		
Slab or below-grade wall (*R*-value)	R-0		
Windows and glass doors	SHGC		*U*-factor
PF < 0.25	0.5		0.6
0.25 ≤ PF < 0.50	0.6		0.6
PF ≥ 0.50	0.7		0.6
Roof assemblies (*R*-value)	Insulation between framing		Continuous insulation
All-wood joist/truss	R-25		R-19
Metal joist/truss	R-25		R-20
Concrete slab or deck	NA		R-19
Metal purlin with thermal block	R-30		R-20
Metal purlin without thermal block	X		R-20
Floors over outdoor air or unconditioned space (*R*-value)	Insulation between framing		Continuous insulation
All-wood joist/truss	R-19		R-12
Metal joist/truss	R-19		R-13
Concrete slab or deck	NA		R-13
Above-grade walls (*R*-value)	No framing	Metal framing	Wood framing
Framed			
R-value cavity	NA	R-11	R-11
R-value continuous	NA	R-0	R-0
CMU, ≥ 8 inches, with integral insulation			
R-value cavity	NA	R-11	R-11
R-value continuous	R-5	R-0	R-0
Other masonry walls			
R-value cavity	NA	R-11	R-11
R-value continuous	R-5	R-0	R-0

(continued)

TABLE 802.2(24)—continued
BUILDING ENVELOPE REQUIREMENTS[a through e] - CLIMATE ZONE 10b

WINDOW AND GLAZED DOOR AREA GREATER THAN 25 PERCENT BUT NOT GREATER THAN 40 PERCENT OF ABOVE-GRADE WALL AREA			
ELEMENT	CONDITION/VALUE		
Skylights (U-factor)	0.8		
Slab or below-grade wall (R-value)	R-0		
Windows and glass doors	SHGC		U-factor
PF < 0.25	0.4		0.5
0.25 ≤ PF < 0.50	0.5		0.5
PF ≥ 0.50	0.6		0.5
Roof assemblies (R-value)	Insulation between framing		Continuous insulation
All-wood joist/truss	R-25		R-19
Metal joist/truss	R-25		R-20
Concrete slab or deck	NA		R-19
Metal purlin with thermal block	R-30		R-20
Metal purlin without thermal block	X		R-20
Floors over outdoor air or unconditioned space (R-value)	Insulation between framing		Continuous insulation
All-wood joist/truss	R-19		R-12
Metal joist/truss	R-19		R-13
Concrete slab or deck	NA		R-13
Above-grade walls (R-value)	No framing	Metal framing	Wood framing
Framed			
R-value cavity	NA	R-11	R-11
R-value continuous	NA	R-0	R-0
CMU, ≥ 8 inches, with integral insulation			
R-value cavity	NA	R-11	R-11
R-value continuous	R-5	R-0	R-0
Other masonry walls			
R-value cavity	NA	R-11	R-11
R-value continuous	R-5	R-0	R-0
WINDOW AND GLAZED DOOR AREA GREATER THAN 40 PERCENT BUT NOT GREATER THAN 50 PERCENT OF ABOVE-GRADE WALL AREA			
ELEMENT	CONDITION/VALUE		
Skylights (U-factor)	0.8		
Slab or below-grade wall (R-value)	R-0		
Windows and glass doors	SHGC		U-factor
PF < 0.25	0.3		0.5
0.25 ≤ PF < 0.50	0.4		0.5
PF ≥ 0.50	0.5		0.5
Roof assemblies (R-value)	Insulation between framing		Continuous insulation
All-wood joist/truss	R-25		R-19
Metal joist/truss	R-25		R-20
Concrete slab or deck	NA		R-19
Metal purlin with thermal block	R-30		R-20
Metal purlin without thermal block	R-30		R-20
Floors over outdoor air or unconditioned space (R-value)	Insulation between framing		Continuous insulation
All-wood joist/truss	R-19		R-12
Metal joist/truss	R-19		R-13
Concrete slab or deck	NA		R-13
Above-grade walls (R-value)	No framing	Metal framing	Wood framing
Framed			
R-value cavity	NA	R-11	R-11
R-value continuous	NA	R-0	R-0
CMU, ≥ 8 inches, with integral insulation			
R-value cavity	NA	R-11	R-11
R-value continuous	R-5	R-0	R-0
Other masonry walls			
R-value cavity	NA	R-11	R-11
R-value continuous	R-5	R-0	R-0

For SI: 1 inch = 25.4 mm.

a. Values from Tables 802.2(5) through 802.2(37) shall be used for the purpose of the completion of Tables 802.2(1) through 802.2(4), as applicable based on window and glazed door area.
b. "NA" indicates the condition is not applicable.
c. An R-value of zero indicates no insulation is required.
d. "Any" indicates any available product will comply.
e. "X" indicates no complying option exists for this condition.

TABLE 802.2(25)
BUILDING ENVELOPE REQUIREMENTS[a through e] - CLIMATE ZONE 11a

WINDOW AND GLAZED DOOR AREA 10 PERCENT OR LESS OF ABOVE-GRADE WALL AREA			
ELEMENT	CONDITION/VALUE		
Skylights (U-factor)	0.8		
Slab or below-grade wall (R-value)	R-0		
Windows and glass doors	SHGC		U-factor
PF < 0.25	Any		Any
0.25 ≤ PF < 0.50	Any		Any
PF ≥ 0.50	Any		Any
Roof assemblies (R-value)	Insulation between framing		Continuous insulation
All-wood joist/truss	R-19		R-14
Metal joist/truss	R-19		R-15
Concrete slab or deck	NA		R-14
Metal purlin with thermal block	R-25		R-15
Metal purlin without thermal block	X		R-15
Floors over outdoor air or unconditioned space (R-value)	Insulation between framing		Continuous insulation
All-wood joist/truss	R-19		R-14
Metal joist/truss	R-19		R-14
Concrete slab or deck	NA		R-14
Above-grade walls (R-value)	No framing	Metal framing	Wood framing
Framed			
R-value cavity	NA	R-11	R-11
R-value continuous	NA	R-0	R-0
CMU, ≥ 8 inches, with integral insulation			
R-value cavity	NA	R-11	R-11
R-value continuous	R-5	R-0	R-0
Other masonry walls			
R-value cavity	NA	R-11	R-11
R-value continuous	R-5	R-0	R-0

WINDOW AND GLAZED DOOR AREA GREATER THAN 10 PERCENT BUT NOT GREATER THAN 25 PERCENT OF ABOVE-GRADE WALL AREA			
ELEMENT	CONDITION/VALUE		
Skylights (U-factor)	0.8		
Slab or below-grade wall (R-value)	R-0		
Windows and glass doors	SHGC		U-factor
PF < 0.25	0.6		0.7
0.25 ≤ PF < 0.50	0.7		0.7
PF ≥ 0.50	Any		0.7
Roof assemblies (R-value)	Insulation between framing		Continuous insulation
All-wood joist/truss	R-19		R-16
Metal joist/truss	R-25		R-17
Concrete slab or deck	NA		R-16
Metal purlin with thermal block	R-25		R-17
Metal purlin without thermal block	X		R-17
Floors over outdoor air or unconditioned space (R-value)	Insulation between framing		Continuous insulation
All-wood joist/truss	R-19		R-14
Metal joist/truss	R-19		R-14
Concrete slab or deck	NA		R-14
Above-grade walls (R-value)	No framing	Metal framing	Wood framing
Framed			
R-value cavity	NA	R-11	R-11
R-value continuous	NA	R-0	R-0
CMU, ≥ 8 inches, with integral insulation			
R-value cavity	NA	R-11	R-11
R-value continuous	R-5	R-0	R-0
Other masonry walls			
R-value cavity	NA	R-11	R-11
R-value continuous	R-5	R-0	R-0

(continued)

TABLE 802.2(25)—continued
BUILDING ENVELOPE REQUIREMENTS[a through e] - CLIMATE ZONE 11a

WINDOW AND GLAZED DOOR AREA GREATER THAN 25 PERCENT BUT NOT GREATER THAN 40 PERCENT OF ABOVE-GRADE WALL AREA			
ELEMENT	CONDITION/VALUE		
Skylights (U-factor)	0.8		
Slab or below-grade wall (R-value)	R-0		
Windows and glass doors	SHGC		U-factor
PF < 0.25	0.5		0.6
0.25 ≤ PF < 0.50	0.6		0.6
PF ≥ 0.50	0.7		0.6
Roof assemblies (R-value)	Insulation between framing		Continuous insulation
All-wood joist/truss	R-25		R-19
Metal joist/truss	R-25		R-20
Concrete slab or deck	NA		R-19
Metal purlin with thermal block	R-30		R-20
Metal purlin without thermal block	X		R-20
Floors over outdoor air or unconditioned space (R-value)	Insulation between framing		Continuous insulation
All-wood joist/truss	R-19		R-14
Metal joist/truss	R-19		R-14
Concrete slab or deck	NA		R-14
Above-grade walls (R-value)	No framing	Metal framing	Wood framing
Framed			
R-value cavity	NA	R-11	R-11
R-value continuous	NA	R-0	R-0
CMU, ≥ 8 inches, with integral insulation			
R-value cavity	NA	R-11	R-11
R-value continuous	R-5	R-0	R-0
Other masonry walls			
R-value cavity	NA	R-11	R-11
R-value continuous	R-5	R-0	R-0

WINDOW AND GLAZED DOOR AREA GREATER THAN 40 PERCENT BUT NOT GREATER THAN 50 PERCENT OF ABOVE-GRADE WALL AREA			
ELEMENT	CONDITION/VALUE		
Skylights (U-factor)	0.8		
Slab or below-grade wall (R-value)	R-0		
Windows and glass doors	SHGC		U-factor
PF < 0.25	0.5		0.4
0.25 ≤ PF < 0.50	0.6		0.4
PF ≥ 0.50	0.7		0.4
Roof assemblies (R-value)	Insulation between framing		Continuous insulation
All-wood joist/truss	R-25		R-19
Metal joist/truss	R-25		R-20
Concrete slab or deck	NA		R-19
Metal purlin with thermal block	R-30		R-20
Metal purlin without thermal block	R-30		R-20
Floors over outdoor air or unconditioned space (R-value)	Insulation between framing		Continuous insulation
All-wood joist/truss	R-19		R-14
Metal joist/truss	R-19		R-14
Concrete slab or deck	NA		R-14
Above-grade walls (R-value)	No framing	Metal framing	Wood framing
Framed			
R-value cavity	NA	R-13	R-11
R-value continuous	NA	R-0	R-0
CMU, ≥ 8 inches, with integral insulation			
R-value cavity	NA	R-11	R-11
R-value continuous	R-5	R-0	R-0
Other masonry walls			
R-value cavity	NA	R-11	R-11
R-value continuous	R-5	R-0	R-0

For SI: 1 inch = 25.4 mm.

a. Values from Tables 802.2(5) through 802.2(37) shall be used for the purpose of the completion of Tables 802.2(1) through 802.2(4), as applicable based on window and glazed door area.
b. "NA" indicates the condition is not applicable.
c. An R-value of zero indicates no insulation is required.
d. "Any" indicates any available product will comply.
e. "X" indicates no complying option exists for this condition.

ACCEPTABLE PRACTICE FOR COMMERCIAL BUILDINGS

TABLE 802.2(26)
BUILDING ENVELOPE REQUIREMENTS[a through e] **- CLIMATE ZONE 11b**

\multicolumn{4}{c}{WINDOW AND GLAZED DOOR AREA 10 PERCENT OR LESS OF ABOVE-GRADE WALL AREA}			
ELEMENT	\multicolumn{3}{c}{CONDITION/VALUE}		
Skylights (U-factor)	\multicolumn{3}{c}{0.8}		
Slab or below-grade wall (R-value)	\multicolumn{3}{c}{R-0}		
Windows and glass doors	\multicolumn{2}{c}{SHGC}	\multicolumn{2}{c}{U-factor}		
PF < 0.25	Any		Any	
0.25 ≤ PF < 0.50	Any		Any	
PF ≥ 0.50	Any		Any	
Roof assemblies (R-value)	Insulation between framing		Continuous insulation	
All-wood joist/truss	R-25		R-18	
Metal joist/truss	R-25		R-19	
Concrete slab or deck	NA		R-18	
Metal purlin with thermal block	R-30		R-19	
Metal purlin without thermal block	X		R-19	
Floors over outdoor air or unconditioned space (R-value)	Insulation between framing		Continuous insulation	
All-wood joist/truss	R-19		R-14	
Metal joist/truss	R-19		R-15	
Concrete slab or deck	NA		R-15	
Above-grade walls (R-value)	No framing	Metal framing	Wood framing
Framed			
R-value cavity	NA	R-11	R-11
R-value continuous	NA	R-0	R-0
CMU, ≥ 8 inches, with integral insulation			
R-value cavity	NA	R-11	R-11
R-value continuous	R-5	R-0	R-0
Other masonry walls			
R-value cavity	NA	R-11	R-11
R-value continuous	R-5	R-0	R-0
\multicolumn{4}{c}{WINDOW AND GLAZED DOOR AREA GREATER THAN 10 PERCENT BUT NOT GREATER THAN 25 PERCENT OF ABOVE-GRADE WALL AREA}			
ELEMENT	\multicolumn{3}{c}{CONDITION/VALUE}		
Skylights (U-factor)	\multicolumn{3}{c}{0.8}		
Slab or below-grade wall (R-value)	\multicolumn{3}{c}{R-0}		
Windows and glass doors	SHGC		U-factor	
PF < 0.25	0.5		0.6	
0.25 ≤ PF < 0.50	0.6		0.6	
PF ≥ 0.50	0.7		0.6	
Roof assemblies (R-value)	Insulation between framing		Continuous insulation	
All-wood joist/truss	R-25		R-19	
Metal joist/truss	R-25		R-20	
Concrete slab or deck	NA		R-19	
Metal purlin with thermal block	R-30		R-20	
Metal purlin without thermal block	X		R-20	
Floors over outdoor air or unconditioned space (R-value)	Insulation between framing		Continuous insulation	
All-wood joist/truss	R-19		R-14	
Metal joist/truss	R-19		R-15	
Concrete slab or deck	NA		R-15	
Above-grade walls (R-value)	No framing	Metal framing	Wood framing
Framed			
R-value cavity	NA	R-11	R-11
R-value continuous	NA	R-0	R-0
CMU, ≥ 8 inches, with integral insulation			
R-value cavity	NA	R-11	R-11
R-value continuous	R-5	R-0	R-0
Other masonry walls			
R-value cavity	NA	R-11	R-11
R-value continuous	R-5	R-0	R-0

(continued)

TABLE 802.2(26)—continued
BUILDING ENVELOPE REQUIREMENTS[a through e] - CLIMATE ZONE 11b

WINDOW AND GLAZED DOOR AREA GREATER THAN 25 PERCENT BUT NOT GREATER THAN 40 PERCENT OF ABOVE-GRADE WALL AREA			
ELEMENT	CONDITION/VALUE		
Skylights (*U*-factor)	0.8		
Slab or below-grade wall (*R*-value)	R-8		
Windows and glass doors	SHGC		*U*-factor
PF < 0.25	0.4		0.5
0.25 ≤ PF < 0.50	0.5		0.5
PF ≥ 0.50	0.6		0.5
Roof assemblies (*R*-value)	Insulation between framing		Continuous insulation
All-wood joist/truss	R-30		R-23
Metal joist/truss	R-30		R-24
Concrete slab or deck	NA		R-23
Metal purlin with thermal block	X		R-24
Metal purlin without thermal block	X		R-24
Floors over outdoor air or unconditioned space (*R*-value)	Insulation between framing		Continuous insulation
All-wood joist/truss	R-19		R-14
Metal joist/truss	R-19		R-15
Concrete slab or deck	NA		R-15
Above-grade walls (*R*-value)	No framing	Metal framing	Wood framing
Framed			
R-value cavity	NA	R-11	R-11
R-value continuous	NA	R-0	R-0
CMU, ≥ 8 inches, with integral insulation			
R-value cavity	NA	R-11	R-11
R-value continuous	R-5	R-0	R-0
Other masonry walls			
R-value cavity	NA	R-11	R-11
R-value continuous	R-5	R-0	R-0
WINDOW AND GLAZED DOOR AREA GREATER THAN 40 PERCENT BUT NOT GREATER THAN 50 PERCENT OF ABOVE-GRADE WALL AREA			
ELEMENT	CONDITION/VALUE		
Skylights (*U*-factor)	0.8		
Slab or below-grade wall (*R*-value)	R-8		
Windows and glass doors	SHGC		*U*-factor
PF < 0.25	0.3		0.5
0.25 ≤ PF < 0.50	0.4		0.5
PF ≥ 0.50	0.5		0.5
Roof assemblies (*R*-value)	Insulation between framing		Continuous insulation
All-wood joist/truss	R-30		R-23
Metal joist/truss	R-30		R-24
Concrete slab or deck	NA		R-23
Metal purlin with thermal block	R-30		R-24
Metal purlin without thermal block	R-38		R-24
Floors over outdoor air or unconditioned space (*R*-value)	Insulation between framing		Continuous insulation
All-wood joist/truss	R-19		R-14
Metal joist/truss	R-19		R-15
Concrete slab or deck	NA		R-15
Above-grade walls (*R*-value)	No framing	Metal framing	Wood framing
Framed			
R-value cavity	NA	R-13	R-11
R-value continuous	NA	R-3	R-0
CMU, ≥ 8 inches, with integral insulation			
R-value cavity	NA	R-11	R-11
R-value continuous	R-5	R-0	R-0
Other masonry walls			
R-value cavity	NA	R-11	R-11
R-value continuous	R-5	R-0	R-0

For SI: 1 inch = 25.4 mm.

a. Values from Tables 802.2(5) through 802.2(37) shall be used for the purpose of the completion of Tables 802.2(1) through 802.2(4), as applicable based on window and glazed door area.
b. "NA" indicates the condition is not applicable.
c. An *R*-value of zero indicates no insulation is required.
d. "Any" indicates any available product will comply.
e. "X" indicates no complying option exists for this condition.

TABLE 802.2(27)
BUILDING ENVELOPE REQUIREMENTS[a through e] - CLIMATE ZONE 12a

WINDOW AND GLAZED DOOR AREA 10 PERCENT OR LESS OF ABOVE-GRADE WALL AREA			
ELEMENT	CONDITION/VALUE		
Skylights (U-factor)	0.8		
Slab or below-grade wall (R-value)	R-0		
Windows and glass doors	SHGC		U-factor
PF < 0.25	Any		Any
0.25 ≤ PF < 0.50	Any		Any
PF ≥ 0.50	Any		Any
Roof assemblies (R-value)	Insulation between framing		Continuous insulation
All-wood joist/truss	R-19		R-16
Metal joist/truss	R-25		R-17
Concrete slab or deck	NA		R-16
Metal purlin with thermal block	R-25		R-17
Metal purlin without thermal block	X		R-17
Floors over outdoor air or unconditioned space (R-value)	Insulation between framing		Continuous insulation
All-wood joist/truss	R-19		R-16
Metal joist/truss	R-19		R-16
Concrete slab or deck	NA		R-16
Above-grade walls (R-value)	No framing	Metal framing	Wood framing
Framed			
R-value cavity	NA	R-11	R-11
R-value continuous	NA	R-0	R-0
CMU, ≥ 8 inches, with integral insulation			
R-value cavity	NA	R-11	R-11
R-value continuous	R-5	R-0	R-0
Other masonry walls			
R-value cavity	NA	R-11	R-11
R-value continuous	R-5	R-0	R-0
WINDOW AND GLAZED DOOR AREA GREATER THAN 10 PERCENT BUT NOT GREATER THAN 25 PERCENT OF ABOVE-GRADE WALL AREA			
ELEMENT	CONDITION/VALUE		
Skylights (U-factor)	0.8		
Slab or below-grade wall (R-value)	R-0		
Windows and glass doors	SHGC		U-factor
PF < 0.25	0.6		0.6
0.25 ≤ PF < 0.50	0.7		0.6
PF ≥ 0.50	Any		0.6
Roof assemblies (R-value)	Insulation between framing		Continuous insulation
All-wood joist/truss	R-25		R-19
Metal joist/truss	R-25		R-20
Concrete slab or deck	NA		R-19
Metal purlin with thermal block	R-30		R-20
Metal purlin without thermal block	X		R-20
Floors over outdoor air or unconditioned space (R-value)	Insulation between framing		Continuous insulation
All-wood joist/truss	R-19		R-16
Metal joist/truss	R-19		R-16
Concrete slab or deck	NA		R-16
Above-grade walls (R-value)	No framing	Metal framing	Wood framing
Framed			
R-value cavity	NA	R-11	R-11
R-value continuous	NA	R-0	R-0
CMU, ≥ 8 inches, with integral insulation			
R-value cavity	NA	R-11	R-11
R-value continuous	R-5	R-0	R-0
Other masonry walls			
R-value cavity	NA	R-11	R-11
R-value continuous	R-5	R-0	R-0

(continued)

TABLE 802.2(27)—continued
BUILDING ENVELOPE REQUIREMENTS[a through e] - CLIMATE ZONE 12a

WINDOW AND GLAZED DOOR AREA GREATER THAN 25 PERCENT BUT NOT GREATER THAN 40 PERCENT OF ABOVE-GRADE WALL AREA			
ELEMENT	CONDITION/VALUE		
Skylights (U-factor)	0.8		
Slab or below-grade wall (R-value)	R-8		
Windows and glass doors	SHGC		U-factor
PF < 0.25	0.5		0.5
0.25 ≤ PF < 0.50	0.6		0.5
PF ≥ 0.50	0.7		0.5
Roof assemblies (R-value)	Insulation between framing		Continuous insulation
All-wood joist/truss	R-30		R-23
Metal joist/truss	R-30		R-24
Concrete slab or deck	NA		R-23
Metal purlin with thermal block	X		R-24
Metal purlin without thermal block	X		R-24
Floors over outdoor air or unconditioned space (R-value)	Insulation between framing		Continuous insulation
All-wood joist/truss	R-19		R-16
Metal joist/truss	R-19		R-16
Concrete slab or deck	NA		R-16
Above-grade walls (R-value)	No framing	Metal framing	Wood framing
Framed			
R-value cavity	NA	R-11	R-11
R-value continuous	NA	R-0	R-0
CMU, ≥ 8 inches, with integral insulation			
R-value cavity	NA	R-11	R-11
R-value continuous	R-5	R-0	R-0
Other masonry walls			
R-value cavity	NA	R-11	R-11
R-value continuous	R-5	R-0	R-0
WINDOW AND GLAZED DOOR AREA GREATER THAN 40 PERCENT BUT NOT GREATER THAN 50 PERCENT OF ABOVE-GRADE WALL AREA			
ELEMENT	CONDITION/VALUE		
Skylights (U-factor)	0.8		
Slab or below-grade wall (R-value)	R-8		
Windows and glass doors	SHGC		U-factor
PF < 0.25	0.4		0.4
0.25 ≤ PF < 0.50	0.5		0.4
PF ≥ 0.50	0.7		0.4
Roof assemblies (R-value)	Insulation between framing		Continuous insulation
All-wood joist/truss	R-30		R-23
Metal joist/truss	R-30		R-24
Concrete slab or deck	NA		R-23
Metal purlin with thermal block	R-30		R-24
Metal purlin without thermal block	R-38		R-24
Floors over outdoor air or unconditioned space (R-value)	Insulation between framing		Continuous insulation
All-wood joist/truss	R-19		R-16
Metal joist/truss	R-19		R-16
Concrete slab or deck	NA		R-16
Above-grade walls (R-value)	No framing	Metal framing	Wood framing
Framed			
R-value cavity	NA	R-13	R-11
R-value continuous	NA	R-0	R-0
CMU, ≥ 8 inches, with integral insulation			
R-value cavity	NA	R-11	R-11
R-value continuous	R-5	R-0	R-0
Other masonry walls			
R-value cavity	NA	R-11	R-11
R-value continuous	R-5	R-0	R-0

For SI: 1 inch = 25.4 mm.

a. Values from Tables 802.2(5) through 802.2(37) shall be used for the purpose of the completion of Tables 802.2(1) through 802.2(4), as applicable based on window and glazed door area.
b. "NA" indicates the condition is not applicable.
c. An R-value of zero indicates no insulation is required.
d. "Any" indicates any available product will comply.
e. "X" indicates no complying option exists for this condition.

ACCEPTABLE PRACTICE FOR COMMERCIAL BUILDINGS

TABLE 802.2(28)
BUILDING ENVELOPE REQUIREMENTS[a through e] - CLIMATE ZONE 12b

WINDOW AND GLAZED DOOR AREA 10 PERCENT OR LESS OF ABOVE-GRADE WALL AREA			
ELEMENT	CONDITION/VALUE		
Skylights (*U*-factor)	0.8		
Slab or below-grade wall (*R*-value)	R-0		
Windows and glass doors	SHGC		*U*-factor
PF < 0.25	Any		Any
0.25 ≤ PF < 0.50	Any		Any
PF ≥ 0.50	Any		Any
Roof assemblies (*R*-value)	Insulation between framing		Continuous insulation
All-wood joist/truss	R-19		R-16
Metal joist/truss	R-25		R-17
Concrete slab or deck	NA		R-16
Metal purlin with thermal block	R-25		R-17
Metal purlin without thermal block	X		R-17
Floors over outdoor air or unconditioned space (*R*-value)	Insulation between framing		Continuous insulation
All-wood joist/truss	R-19		R-15
Metal joist/truss	R-19		R-16
Concrete slab or deck	NA		R-16
Above-grade walls (*R*-value)	No framing	Metal framing	Wood framing
Framed			
R-value cavity	NA	R-11	R-11
R-value continuous	NA	R-0	R-0
CMU, ≥ 8 inches, with integral insulation			
R-value cavity	NA	R-11	R-11
R-value continuous	R-5	R-0	R-0
Other masonry walls			
R-value cavity	NA	R-11	R-11
R-value continuous	R-5	R-0	R-0
WINDOW AND GLAZED DOOR AREA GREATER THAN 10 PERCENT BUT NOT GREATER THAN 25 PERCENT OF ABOVE-GRADE WALL AREA			
ELEMENT	CONDITION/VALUE		
Skylights (*U*-factor)	0.8		
Slab or below-grade wall (*R*-value)	R-0		
Windows and glass doors	SHGC		*U*-factor
PF < 0.25	0.5		0.6
0.25 ≤ PF < 0.50	0.6		0.6
PF ≥ 0.50	0.7		0.6
Roof assemblies (*R*-value)	Insulation between framing		Continuous insulation
All-wood joist/truss	R-25		R-19
Metal joist/truss	R-25		R-20
Concrete slab or deck	NA		R-19
Metal purlin with thermal block	R-30		R-20
Metal purlin without thermal block	X		R-20
Floors over outdoor air or unconditioned space (*R*-value)	Insulation between framing		Continuous insulation
All-wood joist/truss	R-19		R-15
Metal joist/truss	R-19		R-16
Concrete slab or deck	NA		R-16
Above-grade walls (*R*-value)	No framing	Metal framing	Wood framing
Framed			
R-value cavity	NA	R-11	R-11
R-value continuous	NA	R-0	R-0
CMU, ≥ 8 inches, with integral insulation			
R-value cavity	NA	R-11	R-11
R-value continuous	R-5	R-0	R-0
Other masonry walls			
R-value cavity	NA	R-11	R-11
R-value continuous	R-5	R-0	R-0

(continued)

TABLE 802.2(28)—continued
BUILDING ENVELOPE REQUIREMENTS^{a through e} - CLIMATE ZONE 12b

WINDOW AND GLAZED DOOR AREA GREATER THAN 25 PERCENT BUT NOT GREATER THAN 40 PERCENT OF ABOVE-GRADE WALL AREA			
ELEMENT	CONDITION/VALUE		
Skylights (*U*-factor)	0.8		
Slab or below-grade wall (*R*-value)	R-8		
Windows and glass doors	SHGC		*U*-factor
PF < 0.25	0.4		0.5
0.25 ≤ PF < 0.50	0.5		0.5
PF ≥ 0.50	0.6		0.5
Roof assemblies (*R*-value)	Insulation between framing		Continuous insulation
All-wood joist/truss	R-30		R-23
Metal joist/truss	R-30		R-24
Concrete slab or deck	NA		R-23
Metal purlin with thermal block	X		R-24
Metal purlin without thermal block	X		R-24
Floors over outdoor air or unconditioned space (*R*-value)	Insulation between framing		Continuous insulation
All-wood joist/truss	R-19		R-15
Metal joist/truss	R-19		R-16
Concrete slab or deck	NA		R-16
Above-grade walls (*R*-value)	No framing	Metal framing	Wood framing
Framed			
R-value cavity	NA	R-11	R-11
R-value continuous	NA	R-0	R-0
CMU, ≥ 8 inches, with integral insulation			
R-value cavity	NA	R-11	R-11
R-value continuous	R-5	R-0	R-0
Other masonry walls			
R-value cavity	NA	R-11	R-11
R-value continuous	R-5	R-0	R-0

WINDOW AND GLAZED DOOR AREA GREATER THAN 40 PERCENT BUT NOT GREATER THAN 50 PERCENT OF ABOVE-GRADE WALL AREA			
ELEMENT	CONDITION/VALUE		
Skylights (*U*-factor)	0.8		
Slab or below-grade wall (*R*-value)	R-8		
Windows and glass doors	SHGC		*U*-factor
PF < 0.25	0.3		0.5
0.25 ≤ PF < 0.50	0.4		0.5
PF ≥ 0.50	0.5		0.5
Roof assemblies (*R*-value)	Insulation between framing		Continuous insulation
All-wood joist/truss	R-30		R-23
Metal joist/truss	R-30		R-24
Concrete slab or deck	NA		R-23
Metal purlin with thermal block	R-38		R-24
Metal purlin without thermal block	R-49		R-24
Floors over outdoor air or unconditioned space (*R*-value)	Insulation between framing		Continuous insulation
All-wood joist/truss	R-19		R-15
Metal joist/truss	R-19		R-16
Concrete slab or deck	NA		R-16
Above-grade walls (*R*-value)	No framing	Metal framing	Wood framing
Framed			
R-value cavity	NA	R-13	R-13
R-value continuous	NA	R-3	R-0
CMU, ≥ 8 inches, with integral insulation			
R-value cavity	NA	R-11	R-11
R-value continuous	R-5	R-0	R-0
Other masonry walls			
R-value cavity	NA	R-11	R-11
R-value continuous	R-5	R-0	R-0

For SI: 1 inch = 25.4 mm.

a. Values from Tables 802.2(5) through 802.2(37) shall be used for the purpose of the completion of Tables 802.2(1) through 802.2(4), as applicable based on window and glazed door area.
b. "NA" indicates the condition is not applicable.
c. An *R*-value of zero indicates no insulation is required.
d. "Any" indicates any available product will comply.
e. "X" indicates no complying option exists for this condition.

ACCEPTABLE PRACTICE FOR COMMERCIAL BUILDINGS

TABLE 802.2(29)
BUILDING ENVELOPE REQUIREMENTS[a through e] - CLIMATE ZONE 13a

WINDOW AND GLAZED DOOR AREA 10 PERCENT OR LESS OF ABOVE-GRADE WALL AREA			
ELEMENT	CONDITION/VALUE		
Skylights (U-factor)	0.8		
Slab or below-grade wall (R-value)	R-0		
Windows and glass doors	SHGC		U-factor
PF < 0.25	Any		0.7
0.25 ≤ PF < 0.50	Any		0.7
PF ≥ 0.50	Any		0.7
Roof assemblies (R-value)	Insulation between framing		Continuous insulation
All-wood joist/truss	R-19		R-14
Metal joist/truss	R-19		R-15
Concrete slab or deck	NA		R-14
Metal purlin with thermal block	R-25		R-15
Metal purlin without thermal block	X		R-15
Floors over outdoor air or unconditioned space (R-value)	Insulation between framing		Continuous insulation
All-wood joist/truss	R-19		R-16
Metal joist/truss	R-25		R-17
Concrete slab or deck	NA		R-17
Above-grade walls (R-value)	No framing	Metal framing	Wood framing
Framed			
R-value cavity	NA	R-13	R-11
R-value continuous	NA	R-0	R-0
CMU, ≥ 8 inches, with integral insulation			
R-value cavity	NA	R-11	R-11
R-value continuous	R-5	R-0	R-0
Other masonry walls			
R-value cavity	NA	R-11	R-11
R-value continuous	R-5	R-0	R-0
WINDOW AND GLAZED DOOR AREA GREATER THAN 10 PERCENT BUT NOT GREATER THAN 25 PERCENT OF ABOVE-GRADE WALL AREA			
ELEMENT	CONDITION/VALUE		
Skylights (U-factor)	0.8		
Slab or below-grade wall (R-value)	R-0		
Windows and glass doors	SHGC		U-factor
PF < 0.25	0.6		0.6
0.25 ≤ PF < 0.50	0.7		0.6
PF ≥ 0.50	Any		0.6
Roof assemblies (R-value)	Insulation between framing		Continuous insulation
All-wood joist/truss	R-25		R-19
Metal joist/truss	R-25		R-20
Concrete slab or deck	NA		R-19
Metal purlin with thermal block	R-30		R-20
Metal purlin without thermal block	X		R-20
Floors over outdoor air or unconditioned space (R-value)	Insulation between framing		Continuous insulation
All-wood joist/truss	R-19		R-16
Metal joist/truss	R-25		R-17
Concrete slab or deck	NA		R-17
Above-grade walls (R-value)	No framing	Metal framing	Wood framing
Framed			
R-value cavity	NA	R-13	R-11
R-value continuous	NA	R-0	R-0
CMU, ≥ 8 inches, with integral insulation			
R-value cavity	NA	R-11	R-11
R-value continuous	R-5	R-0	R-0
Other masonry walls			
R-value cavity	NA	R-11	R-11
R-value continuous	R-5	R-0	R-0

(continued)

TABLE 802.2(29)—continued
BUILDING ENVELOPE REQUIREMENTS[a through e] - CLIMATE ZONE 13a

WINDOW AND GLAZED DOOR AREA GREATER THAN 25 PERCENT BUT NOT GREATER THAN 40 PERCENT OF ABOVE-GRADE WALL AREA			
ELEMENT	CONDITION/VALUE		
Skylights (*U*-factor)	0.8		
Slab or below-grade wall (*R*-value)	R-8		
Windows and glass doors	SHGC		*U*-factor
PF < 0.25	0.5		0.5
0.25 ≤ PF < 0.50	0.6		0.5
PF ≥ 0.50	0.7		0.5
Roof assemblies (*R*-value)	Insulation between framing		Continuous insulation
All-wood joist/truss	R-30		R-23
Metal joist/truss	R-30		R-24
Concrete slab or deck	NA		R-23
Metal purlin with thermal block	X		R-24
Metal purlin without thermal block	X		R-24
Floors over outdoor air or unconditioned space (*R*-value)	Insulation between framing		Continuous insulation
All-wood joist/truss	R-19		R-16
Metal joist/truss	R-25		R-17
Concrete slab or deck	NA		R-17
Above-grade walls (*R*-value)	No framing	Metal framing	Wood framing
Framed			
R-value cavity	NA	R-13	R-11
R-value continuous	NA	R-0	R-0
CMU, ≥ 8 inches, with integral insulation			
R-value cavity	NA	R-11	R-11
R-value continuous	R-5	R-0	R-0
Other masonry walls			
R-value cavity	NA	R-11	R-11
R-value continuous	R-5	R-0	R-0
WINDOW AND GLAZED DOOR AREA GREATER THAN 40 PERCENT BUT NOT GREATER THAN 50 PERCENT OF ABOVE-GRADE WALL AREA			
ELEMENT	CONDITION/VALUE		
Skylights (*U*-factor)	0.8		
Slab or below-grade wall (*R*-value)	R-8		
Windows and glass doors	SHGC		*U*-factor
PF < 0.25	0.4		0.4
0.25 ≤ PF < 0.50	0.5		0.4
PF ≥ 0.50	0.7		0.4
Roof assemblies (*R*-value)	Insulation between framing		Continuous insulation
All-wood joist/truss	R-30		R-23
Metal joist/truss	R-30		R-24
Concrete slab or deck	NA		R-23
Metal purlin with thermal block	R-30		R-24
Metal purlin without thermal block	R-38		R-24
Floors over outdoor air or unconditioned space (*R*-value)	Insulation between framing		Continuous insulation
All-wood joist/truss	R-19		R-16
Metal joist/truss	R-25		R-17
Concrete slab or deck	NA		R-17
Above-grade walls (*R*-value)	No framing	Metal framing	Wood framing
Framed			
R-value cavity	NA	R-13	R-11
R-value continuous	NA	R-0	R-0
CMU, ≥ 8 inches, with integral insulation			
R-value cavity	NA	R-11	R-11
R-value continuous	R-5	R-0	R-0
Other masonry walls			
R-value cavity	NA	R-11	R-11
R-value continuous	R-5	R-0	R-0

For SI: 1 inch = 25.4 mm.

a. Values from Tables 802.2(5) through 802.2(37) shall be used for the purpose of the completion of Tables 802.2(1) through 802.2(4), as applicable based on window and glazed door area.
b. "NA" indicates the condition is not applicable.
c. An *R*-value of zero indicates no insulation is required.
d. "Any" indicates any available product will comply.
e. "X" indicates no complying option exists for this condition.

ACCEPTABLE PRACTICE FOR COMMERCIAL BUILDINGS

TABLE 802.2(30)
BUILDING ENVELOPE REQUIREMENTS[a through e] - CLIMATE ZONE 13b

WINDOW AND GLAZED DOOR AREA 10 PERCENT OR LESS OF ABOVE-GRADE WALL AREA			
ELEMENT	CONDITION/VALUE		
Skylights (*U*-factor)	0.8		
Slab or below-grade wall (*R*-value)	R-0		
Windows and glass doors	SHGC		*U*-factor
PF < 0.25	Any		Any
0.25 ≤ PF < 0.50	Any		Any
PF ≥ 0.50	Any		Any
Roof assemblies (*R*-value)	Insulation between framing		Continuous insulation
All-wood joist/truss	R-25		R-18
Metal joist/truss	R-25		R-19
Concrete slab or deck	NA		R-18
Metal purlin with thermal block	R-30		R-19
Metal purlin without thermal block	NA		R-19
Floors over outdoor air or unconditioned space (*R*-value)	Insulation between framing		Continuous insulation
All-wood joist/truss	R-19		R-17
Metal joist/truss	R-25		R-17
Concrete slab or deck	NA		R-17
Above-grade walls (*R*-value)	No framing	Metal framing	Wood framing
Framed			
R-value cavity	NA	R-13	R-11
R-value continuous	NA	R-0	R-0
CMU, ≥ 8 inches, with integral insulation			
R-value cavity	NA	R-11	R-11
R-value continuous	R-5	R-0	R-0
Other masonry walls			
R-value cavity	NA	R-11	R-11
R-value continuous	R-5	R-0	R-0
WINDOW AND GLAZED DOOR AREA GREATER THAN 10 PERCENT BUT NOT GREATER THAN 25 PERCENT OF ABOVE-GRADE WALL AREA			
ELEMENT	CONDITION/VALUE		
Skylights (*U*-factor)	0.8		
Slab or below-grade wall (*R*-value)	R-0		
Windows and glass doors	SHGC		*U*-factor
PF < 0.25	0.5		0.5
0.25 ≤ PF < 0.50	0.6		0.5
PF ≥ 0.50	0.7		0.5
Roof assemblies (*R*-value)	Insulation between framing		Continuous insulation
All-wood joist/truss	R-25		R-19
Metal joist/truss	R-25		R-20
Concrete slab or deck	NA		R-19
Metal purlin with thermal block	R-30		R-20
Metal purlin without thermal block	X		R-20
Floors over outdoor air or unconditioned space (*R*-value)	Insulation between framing		Continuous insulation
All-wood joist/truss	R-19		R-17
Metal joist/truss	R-25		R-17
Concrete slab or deck	NA		R-17
Above-grade walls (*R*-value)	No framing	Metal framing	Wood framing
Framed			
R-value cavity	NA	R-13	R-11
R-value continuous	NA	R-0	R-0
CMU, ≥ 8 inches, with integral insulation			
R-value cavity	NA	R-11	R-11
R-value continuous	R-5	R-0	R-0
Other masonry walls			
R-value cavity	NA	R-11	R-11
R-value continuous	R-5	R-0	R-0

(continued)

TABLE 802.2(30)—continued
BUILDING ENVELOPE REQUIREMENTS[a through e] - CLIMATE ZONE 13b

WINDOW AND GLAZED DOOR AREA GREATER THAN 25 PERCENT BUT NOT GREATER THAN 40 PERCENT OF ABOVE-GRADE WALL AREA			
ELEMENT	CONDITION/VALUE		
Skylights (*U*-factor)	0.8		
Slab or below-grade wall (*R*-value)	R-8		
Windows and glass doors	SHGC	*U*-factor	
PF < 0.25	0.4	0.5	
0.25 ≤ PF < 0.50	0.5	0.5	
PF ≥ 0.50	0.6	0.5	
Roof assemblies (*R*-value)	Insulation between framing	Continuous insulation	
All-wood joist/truss	R-30	R-23	
Metal joist/truss	R-30	R-24	
Concrete slab or deck	NA	R-23	
Metal purlin with thermal block	X	R-24	
Metal purlin without thermal block	X	R-24	
Floors over outdoor air or unconditioned space (*R*-value)	Insulation between framing	Continuous insulation	
All-wood joist/truss	R-19	R-17	
Metal joist/truss	R-25	R-17	
Concrete slab or deck	NA	R-17	
Above-grade walls (*R*-value)	No framing	Metal framing	Wood framing
Framed			
R-value cavity	NA	R-13	R-11
R-value continuous	NA	R-0	R-0
CMU, ≥ 8 inches, with integral insulation			
R-value cavity	NA	R-11	R-11
R-value continuous	R-5	R-0	R-0
Other masonry walls			
R-value cavity	NA	R-11	R-11
R-value continuous	R-5	R-0	R-0
WINDOW AND GLAZED DOOR AREA GREATER THAN 40 PERCENT BUT NOT GREATER THAN 50 PERCENT OF ABOVE-GRADE WALL AREA			
ELEMENT	CONDITION/VALUE		
Skylights (*U*-factor)	0.8		
Slab or below-grade wall (*R*-value)	R-8		
Windows and glass doors	SHGC	*U*-factor	
PF < 0.25	0.4	0.4	
0.25 ≤ PF < 0.50	0.5	0.4	
PF ≥ 0.50	0.6	0.4	
Roof assemblies (*R*-value)	Insulation between framing	Continuous insulation	
All-wood joist/truss	R-30	R-23	
Metal joist/truss	R-30	R-24	
Concrete slab or deck	NA	R-23	
Metal purlin with thermal block	R-38	R-24	
Metal purlin without thermal block	R-49	R-24	
Floors over outdoor air or unconditioned space (*R*-value)	Insulation between framing	Continuous insulation	
All-wood joist/truss	R-19	R-17	
Metal joist/truss	R-25	R-17	
Concrete slab or deck	NA	R-17	
Above-grade walls (*R*-value)	No framing	Metal framing	Wood framing
Framed			
R-value cavity	NA	R-13	R-13
R-value continuous	NA	R-7	R-3
CMU, ≥ 8 inches, with integral insulation			
R-value cavity	NA	R-11	R-11
R-value continuous	R-5	R-0	R-0
Other masonry walls			
R-value cavity	NA	R-11	R-11
R-value continuous	R-5	R-0	R-0

For SI: 1 inch = 25.4 mm.

a. Values from Tables 802.2(5) through 802.2(37) shall be used for the purpose of the completion of Tables 802.2(1) through 802.2(4), as applicable based on window and glazed door area.
b. "NA" indicates the condition is not applicable.
c. An *R*-value of zero indicates no insulation is required.
d. "Any" indicates any available product will comply.
e. "X" indicates no complying option exists for this condition.

TABLE 802.2(31)
BUILDING ENVELOPE REQUIREMENTS[a through e] - CLIMATE ZONE 14a

WINDOW AND GLAZED DOOR AREA 10 PERCENT OR LESS OF ABOVE-GRADE WALL AREA			
ELEMENT	**CONDITION/VALUE**		
Skylights (*U*-factor)	0.8		
Slab or below-grade wall (*R*-value)	R-0		
Windows and glass doors	**SHGC**		***U*-factor**
PF < 0.25	Any		0.7
0.25 ≤ PF < 0.50	Any		0.7
PF ≥ 0.50	Any		0.7
Roof assemblies (*R*-value)	**Insulation between framing**		**Continuous insulation**
All-wood joist/truss	R-19		R-17
Metal joist/truss	R-25		R-18
Concrete slab or deck	NA		R-17
Metal purlin with thermal block	R-30		R-18
Metal purlin without thermal block	X		R-18
Floors over outdoor air or unconditioned space (*R*-value)	**Insulation between framing**		**Continuous insulation**
All-wood joist/truss	R-25		R-18
Metal joist/truss	R-25		R-19
Concrete slab or deck	NA		R-19
Above-grade walls (*R*-value)	**No framing**	**Metal framing**	**Wood framing**
Framed			
R-value cavity	NA	R-13	R-11
R-value continuous	NA	R-3	R-0
CMU, ≥ 8 inches, with integral insulation			
R-value cavity	NA	R-11	R-11
R-value continuous	R-5	R-0	R-0
Other masonry walls			
R-value cavity	NA	R-11	R-11
R-value continuous	R-5	R-0	R-0
WINDOW AND GLAZED DOOR AREA GREATER THAN 10 PERCENT BUT NOT GREATER THAN 25 PERCENT OF ABOVE-GRADE WALL AREA			
ELEMENT	**CONDITION/VALUE**		
Skylights (*U*-factor)	0.8		
Slab or below-grade wall (*R*-value)	R-8		
Windows and glass doors	**SHGC**		***U*-factor**
PF < 0.25	0.5		0.6
0.25 ≤ PF < 0.50	0.6		0.6
PF ≥ 0.50	0.7		0.6
Roof assemblies (*R*-value)	**Insulation between framing**		**Continuous insulation**
All-wood joist/truss	R-25		R-19
Metal joist/truss	R-25		R-20
Concrete slab or deck	NA		R-19
Metal purlin with thermal block	R-30		R-20
Metal purlin without thermal block	X		R-20
Floors over outdoor air or unconditioned space (*R*-value)	**Insulation between framing**		**Continuous insulation**
All-wood joist/truss	R-25		R-18
Metal joist/truss	R-25		R-19
Concrete slab or deck	NA		R-19
Above-grade walls (*R*-value)	**No framing**	**Metal framing**	**Wood framing**
Framed			
R-value cavity	NA	R-13	R-11
R-value continuous	NA	R-3	R-0
CMU, ≥ 8 inches, with integral insulation			
R-value cavity	NA	R-11	R-11
R-value continuous	R-5	R-0	R-0
Other masonry walls			
R-value cavity	NA	R-11	R-11
R-value continuous	R-5	R-0	R-0

(continued)

TABLE 802.2(31)—continued
BUILDING ENVELOPE REQUIREMENTS[a through e] - CLIMATE ZONE 14a

WINDOW AND GLAZED DOOR AREA GREATER THAN 25 PERCENT BUT NOT GREATER THAN 40 PERCENT OF ABOVE-GRADE WALL AREA			
ELEMENT	CONDITION/VALUE		
Skylights (U-factor)	0.8		
Slab or below-grade wall (R-value)	R-8		
Windows and glass doors	SHGC		U-factor
PF < 0.25	0.4		0.5
0.25 ≤ PF < 0.50	0.5		0.5
PF ≥ 0.50	0.6		0.5
Roof assemblies (R-value)	Insulation between framing		Continuous insulation
All-wood joist/truss	R-30		R-23
Metal joist/truss	R-30		R-24
Concrete slab or deck	NA		R-23
Metal purlin with thermal block	X		R-24
Metal purlin without thermal block	X		R-24
Floors over outdoor air or unconditioned space (R-value)	Insulation between framing		Continuous insulation
All-wood joist/truss	R-25		R-18
Metal joist/truss	R-25		R-19
Concrete slab or deck	NA		R-19
Above-grade walls (R-value)	No framing	Metal framing	Wood framing
Framed			
R-value cavity	NA	R-13	R-11
R-value continuous	NA	R-3	R-0
CMU, ≥ 8 inches, with integral insulation			
R-value cavity	NA	R-11	R-11
R-value continuous	R-5	R-0	R-0
Other masonry walls			
R-value cavity	NA	R-11	R-11
R-value continuous	R-5	R-0	R-0
WINDOW AND GLAZED DOOR AREA GREATER THAN 40 PERCENT BUT NOT GREATER THAN 50 PERCENT OF ABOVE-GRADE WALL AREA			
ELEMENT	CONDITION/VALUE		
Skylights (U-factor)	0.8		
Slab or below-grade wall (R-value)	R-8		
Windows and glass doors	SHGC		U-factor
PF < 0.25	0.4		0.4
0.25 ≤ PF < 0.50	0.5		0.4
PF ≥ 0.50	0.6		0.4
Roof assemblies (R-value)	Insulation between framing		Continuous insulation
All-wood joist/truss	R-30		R-23
Metal joist/truss	R-30		R-24
Concrete slab or deck	NA		R-23
Metal purlin with thermal block	R-38		R-24
Metal purlin without thermal block	R-38		R-24
Floors over outdoor air or unconditioned space (R-value)	Insulation between framing		Continuous insulation
All-wood joist/truss	R-25		R-18
Metal joist/truss	R-25		R-19
Concrete slab or deck	NA		R-19
Above-grade walls (R-value)	No framing	Metal framing	Wood framing
Framed			
R-value cavity	NA	R-13	R-11
R-value continuous	NA	R-3	R-0
CMU, ≥ 8 inches, with integral insulation			
R-value cavity	NA	R-11	R-11
R-value continuous	R-5	R-0	R-0
Other masonry walls			
R-value cavity	NA	R-11	R-11
R-value continuous	R-5	R-0	R-0

For SI: 1 inch = 25.4 mm.

a. Values from Tables 802.2(5) through 802.2(37) shall be used for the purpose of the completion of Tables 802.2(1) through 802.2(4), as applicable based on window and glazed door area.
b. "NA" indicates the condition is not applicable.
c. An R-value of zero indicates no insulation is required.
d. "Any" indicates any available product will comply.
e. "X" indicates no complying option exists for this condition.

**TABLE 802.2(32)
BUILDING ENVELOPE REQUIREMENTS[a through e] - CLIMATE ZONE 14b**

WINDOW AND GLAZED DOOR AREA 10 PERCENT OR LESS OF ABOVE-GRADE WALL AREA			
ELEMENT	**CONDITION/VALUE**		
Skylights (*U*-factor)	0.8		
Slab or below-grade wall (*R*-value)	R-0		
Windows and glass doors	SHGC		*U*-factor
PF < 0.25	Any		0.7
0.25 ≤ PF < 0.50	Any		0.7
PF ≥ 0.50	Any		0.7
Roof assemblies (*R*-value)	Insulation between framing		Continuous insulation
All-wood joist/truss	R-25		R-19
Metal joist/truss	R-25		R-20
Concrete slab or deck	NA		R-19
Metal purlin with thermal block	R-30		R-20
Metal purlin without thermal block	NA		R-20
Floors over outdoor air or unconditioned space (*R*-value)	Insulation between framing		Continuous insulation
All-wood joist/truss	R-25		R-19
Metal joist/truss	R-25		R-19
Concrete slab or deck	NA		R-19
Above-grade walls (*R*-value)	No framing	Metal framing	Wood framing
Framed			
R-value cavity	NA	R-13	R-11
R-value continuous	NA	R-3	R-0
CMU, ≥ 8 inches, with integral insulation			
R-value cavity	NA	R-11	R-11
R-value continuous	R-5	R-0	R-0
Other masonry walls			
R-value cavity	NA	R-11	R-11
R-value continuous	R-5	R-0	R-0
WINDOW AND GLAZED DOOR AREA GREATER THAN 10 PERCENT BUT NOT GREATER THAN 25 PERCENT OF ABOVE-GRADE WALL AREA			
ELEMENT	**CONDITION/VALUE**		
Skylights (*U*-factor)	0.8		
Slab or below-grade wall (*R*-value)	R-8		
Windows and glass doors	SHGC		*U*-factor
PF < 0.25	0.5		0.5
0.25 ≤ PF < 0.50	0.6		0.5
PF ≥ 0.50	0.7		0.5
Roof assemblies (*R*-value)	Insulation between framing		Continuous insulation
All-wood joist/truss	R-25		R-19
Metal joist/truss	R-25		R-20
Concrete slab or deck	NA		R-19
Metal purlin with thermal block	R-30		R-20
Metal purlin without thermal block	X		R-20
Floors over outdoor air or unconditioned space (*R*-value)	Insulation between framing		Continuous insulation
All-wood joist/truss	R-25		R-19
Metal joist/truss	R-25		R-19
Concrete slab or deck	NA		R-19
Above-grade walls (*R*-value)	No framing	Metal framing	Wood framing
Framed			
R-value cavity	NA	R-13	R-11
R-value continuous	NA	R-3	R-0
CMU, ≥ 8 inches, with integral insulation			
R-value cavity	NA	R-11	R-11
R-value continuous	R-5	R-0	R-0
Other masonry walls			
R-value cavity	NA	R-11	R-11
R-value continuous	R-5	R-0	R-0

(continued)

ACCEPTABLE PRACTICE FOR COMMERCIAL BUILDING

TABLE 802.2(32)—continued
BUILDING ENVELOPE REQUIREMENTS[a through e] - CLIMATE ZONE 14b

WINDOW AND GLAZED DOOR AREA GREATER THAN 25 PERCENT BUT NOT GREATER THAN 40 PERCENT OF ABOVE-GRADE WALL AREA			
ELEMENT	CONDITION/VALUE		
Skylights (U-factor)	0.8		
Slab or below-grade wall (R-value)	R-8		
Windows and glass doors	SHGC		U-factor
PF < 0.25	0.4		0.5
0.25 ≤ PF < 0.50	0.5		0.5
PF ≥ 0.50	0.6		0.5
Roof assemblies (R-value)	Insulation between framing		Continuous insulation
All-wood joist/truss	R-30		R-23
Metal joist/truss	R-30		R-24
Concrete slab or deck	NA		R-23
Metal purlin with thermal block	X		R-24
Metal purlin without thermal block	X		R-24
Floors over outdoor air or unconditioned space (R-value)	Insulation between framing		Continuous insulation
All-wood joist/truss	R-25		R-19
Metal joist/truss	R-25		R-19
Concrete slab or deck	NA		R-19
Above-grade walls (R-value)	No framing	Metal framing	Wood framing
Framed			
R-value cavity	NA	R-13	R-11
R-value continuous	NA	R-3	R-0
CMU, ≥ 8 inches, with integral insulation			
R-value cavity	NA	R-11	R-11
R-value continuous	R-5	R-0	R-0
Other masonry walls			
R-value cavity	NA	R-11	R-11
R-value continuous	R-5	R-0	R-0
WINDOW AND GLAZED DOOR AREA GREATER THAN 40 PERCENT BUT NOT GREATER THAN 50 PERCENT OF ABOVE-GRADE WALL AREA			
ELEMENT	CONDITION/VALUE		
Skylights (U-factor)	0.8		
Slab or below-grade wall (R-value)	R-8		
Windows and glass doors	SHGC		U-factor
PF < 0.25	0.4		0.4
0.25 ≤ PF < 0.50	0.5		0.4
PF ≥ 0.50	0.6		0.4
Roof assemblies (R-value)	Insulation between framing		Continuous insulation
All-wood joist/truss	R-30		R-23
Metal joist/truss	R-30		R-24
Concrete slab or deck	NA		R-23
Metal purlin with thermal block	R-38		R-24
Metal purlin without thermal block	R-49		R-24
Floors over outdoor air or unconditioned space (R-value)	Insulation between framing		Continuous insulation
All-wood joist/truss	R-25		R-19
Metal joist/truss	R-25		R-19
Concrete slab or deck	NA		R-19
Above-grade walls (R-value)	No framing	Metal framing	Wood framing
Framed			
R-value cavity	NA	R-13	R-13
R-value continuous	NA	R-7	R-3
CMU, ≥ 8 inches, with integral insulation			
R-value cavity	NA	R-11	R-11
R-value continuous	R-5	R-0	R-0
Other masonry walls			
R-value cavity	NA	R-11	R-11
R-value continuous	R-5	R-0	R-0

For SI: 1 inch = 25.4 mm.

a. Values from Tables 802.2(5) through 802.2(37) shall be used for the purpose of the completion of Tables 802.2(1) through 802.2(4), as applicable based on window and glazed door area.
b. "NA" indicates the condition is not applicable.
c. An R-value of zero indicates no insulation is required.
d. "Any" indicates any available product will comply.
e. "X" indicates no complying option exists for this condition.

TABLE 802.2(33)
BUILDING ENVELOPE REQUIREMENTS[a through e] - CLIMATE ZONE 15

WINDOW AND GLAZED DOOR AREA 10 PERCENT OR LESS OF ABOVE-GRADE WALL AREA			
ELEMENT	CONDITION/VALUE		
Skylights (U-factor)	0.6		
Slab or below-grade wall (R-value)	R-0		
Windows and glass doors	**SHGC**		***U*-factor**
PF < 0.25	Any		0.7
0.25 ≤ PF < 0.50	Any		0.7
PF ≥ 0.50	Any		0.7
Roof assemblies (R-value)	**Insulation between framing**		**Continuous insulation**
All-wood joist/truss	R-25		R-19
Metal joist/truss	R-25		R-20
Concrete slab or deck	NA		R-19
Metal purlin with thermal block	R-30		R-20
Metal purlin without thermal block	X		R-20
Floors over outdoor air or unconditioned space (R-value)	**Insulation between framing**		**Continuous insulation**
All-wood joist/truss	R-25		R-22
Metal joist/truss	R-30		R-23
Concrete slab or deck	NA		R-22
Above-grade walls (R-value)	**No framing**	**Metal framing**	**Wood framing**
Framed			
R-value cavity	NA	R-13	R-11
R-value continuous	NA	R-3	R-0
CMU, ≥ 8 inches, with integral insulation			
R-value cavity	NA	R-11	R-11
R-value continuous	R-5	R-0	R-0
Other masonry walls			
R-value cavity	NA	R-11	R-11
R-value continuous	R-5	R-0	R-0
WINDOW AND GLAZED DOOR AREA GREATER THAN 10 PERCENT BUT NOT GREATER THAN 25 PERCENT OF ABOVE-GRADE WALL AREA			
ELEMENT	CONDITION/VALUE		
Skylights (U-factor)	0.6		
Slab or below-grade wall (R-value)	R-8		
Windows and glass doors	**SHGC**		***U*-factor**
PF < 0.25	0.5		0.5
0.25 ≤ PF < 0.50	0.6		0.5
PF ≥ 0.50	0.7		0.5
Roof assemblies (R-value)	**Insulation between framing**		**Continuous insulation**
All-wood joist/truss	R-25		R-19
Metal joist/truss	R-25		R-20
Concrete slab or deck	NA		R-19
Metal purlin with thermal block	R-30		R-20
Metal purlin without thermal block	X		R-20
Floors over outdoor air or unconditioned space (R-value)	**Insulation between framing**		**Continuous insulation**
All-wood joist/truss	R-25		R-22
Metal joist/truss	R-30		R-23
Concrete slab or deck	NA		R-22
Above-grade walls (R-value)	**No framing**	**Metal framing**	**Wood framing**
Framed			
R-value cavity	NA	R-13	R-11
R-value continuous	NA	R-3	R-0
CMU, ≥ 8 inches, with integral insulation			
R-value cavity	NA	R-11	R-11
R-value continuous	R-5	R-0	R-0
Other masonry walls			
R-value cavity	NA	R-11	R-11
R-value continuous	R-5	R-0	R-0

(continued)

TABLE 802.2(33)—continued
BUILDING ENVELOPE REQUIREMENTS[a through e] - CLIMATE ZONE 15

WINDOW AND GLAZED DOOR AREA GREATER THAN 25 PERCENT BUT NOT GREATER THAN 40 PERCENT OF ABOVE-GRADE WALL AREA			
ELEMENT	CONDITION/VALUE		
Skylights (*U*-factor)	0.6		
Slab or below-grade wall (*R*-value)	R-8		
Windows and glass doors	SHGC		*U*-factor
PF < 0.25	0.5		0.4
0.25 ≤ PF < 0.50	0.6		0.4
PF ≥ 0.50	0.7		0.4
Roof assemblies (*R*-value)	Insulation between framing		Continuous insulation
All-wood joist/truss	R-30		R-23
Metal joist/truss	R-30		R-24
Concrete slab or deck	NA		R-23
Metal purlin with thermal block	X		R-24
Metal purlin without thermal block	X		R-24
Floors over outdoor air or unconditioned space (*R*-value)	Insulation between framing		Continuous insulation
All-wood joist/truss	R-25		R-22
Metal joist/truss	R-30		R-23
Concrete slab or deck	NA		R-22
Above-grade walls (*R*-value)	No framing	Metal framing	Wood framing
Framed			
R-value cavity	NA	R-13	R-11
R-value continuous	NA	R-3	R-0
CMU, ≥ 8 inches, with integral insulation			
R-value cavity	NA	R-11	R-11
R-value continuous	R-5	R-0	R-0
Other masonry walls			
R-value cavity	NA	R-13	R-11
R-value continuous	R-6	R-0	R-0
WINDOW AND GLAZED DOOR AREA GREATER THAN 40 PERCENT BUT NOT GREATER THAN 50 PERCENT OF ABOVE-GRADE WALL AREA			
ELEMENT	CONDITION/VALUE		
Skylights (*U*-factor)	0.6		
Slab or below-grade wall (*R*-value)	R-8		
Windows and glass doors	SHGC		*U*-factor
PF < 0.25	0.4		0.4
0.25 ≤ PF < 0.50	0.5		0.4
PF ≥ 0.50	0.7		0.4
Roof assemblies (*R*-value)	Insulation between framing		Continuous insulation
All-wood joist/truss	R-30		R-23
Metal joist/truss	R-30		R-24
Concrete slab or deck	NA		R-23
Metal purlin with thermal block	R-38		R-24
Metal purlin without thermal block	X		R-24
Floors over outdoor air or unconditioned space (*R*-value)	Insulation between framing		Continuous insulation
All-wood joist/truss	R-25		R-22
Metal joist/truss	R-30		R-23
Concrete slab or deck	NA		R-22
Above-grade walls (*R*-value)	No framing	Metal framing	Wood framing
Framed			
R-value cavity	NA	R-13	R-13
R-value continuous	NA	R-7	R-4
CMU, ≥ 8 inches, with integral insulation			
R-value cavity	NA	R-13	R-11
R-value continuous	R-5	R-0	R-0
Other masonry walls			
R-value cavity	NA	R-13	R-11
R-value continuous	R-6	R-3	R-0

For SI: 1 inch = 25.4 mm.

a. Values from Tables 802.2(5) through 802.2(37) shall be used for the purpose of the completion of Tables 802.2(1) through 802.2(4), as applicable based on window and glazed door area.
b. "NA" indicates the condition is not applicable.
c. An *R*-value of zero indicates no insulation is required.
d. "Any" indicates any available product will comply.
e. "X" indicates no complying option exists for this condition.

TABLE 802.2(34)
BUILDING ENVELOPE REQUIREMENTS[a through e] - CLIMATE ZONE 16

WINDOW AND GLAZED DOOR AREA 10 PERCENT OR LESS OF ABOVE-GRADE WALL AREA			
ELEMENT	CONDITION/VALUE		
Skylights (*U*-factor)	0.6		
Slab or below-grade wall (*R*-value)	R-8		
Windows and glass doors	SHGC		*U*-factor
PF < 0.25	0.7		0.6
0.25 ≤ PF < 0.50	Any		0.6
PF ≥ 0.50	Any		0.6
Roof assemblies (*R*-value)	Insulation between framing		Continuous insulation
All-wood joist/truss	R-25		R-19
Metal joist/truss	R-25		R-20
Concrete slab or deck	NA		R-19
Metal purlin with thermal block	R-30		R-20
Metal purlin without thermal block	X		R-20
Floors over outdoor air or unconditioned space (*R*-value)	Insulation between framing		Continuous insulation
All-wood joist/truss	R-25		R-22
Metal joist/truss	R-30		R-23
Concrete slab or deck	NA		R-22
Above-grade walls (*R*-value)	No framing	Metal framing	Wood framing
Framed			
R-value cavity	NA	R-13	R-11
R-value continuous	NA	R-3	R-0
CMU, ≥ 8 inches, with integral insulation			
R-value cavity	NA	R-11	R-11
R-value continuous	R-5	R-0	R-0
Other masonry walls			
R-value cavity	NA	R-11	R-11
R-value continuous	R-5	R-0	R-0
WINDOW AND GLAZED DOOR AREA GREATER THAN 10 PERCENT BUT NOT GREATER THAN 25 PERCENT OF ABOVE-GRADE WALL AREA			
ELEMENT	CONDITION/VALUE		
Skylights (*U*-factor)	0.6		
Slab or below-grade wall (*R*-value)	R-8		
Windows and glass doors	SHGC		*U*-factor
PF < 0.25	0.7		0.5
0.25 ≤ PF < 0.50	Any		0.5
PF ≥ 0.50	Any		0.5
Roof assemblies (*R*-value)	Insulation between framing		Continuous insulation
All-wood joist/truss	R-30		R-23
Metal joist/truss	R-30		R-24
Concrete slab or deck	NA		R-23
Metal purlin with thermal block	X		R-24
Metal purlin without thermal block	X		R-24
Floors over outdoor air or unconditioned space (*R*-value)	Insulation between framing		Continuous insulation
All-wood joist/truss	R-25		R-22
Metal joist/truss	R-30		R-23
Concrete slab or deck	NA		R-22
Above-grade walls (*R*-value)	No framing	Metal framing	Wood framing
Framed			
R-value cavity	NA	R-13	R-11
R-value continuous	NA	R-3	R-0
CMU, ≥ 8 inches, with integral insulation			
R-value cavity	NA	R-11	R-11
R-value continuous	R-5	R-0	R-0
Other masonry walls			
R-value cavity	NA	R-13	R-11
R-value continuous	R-9	R-3	R-0

(continued)

ACCEPTABLE PRACTICE FOR COMMERCIAL BUILDING

TABLE 802.2(34)—continued
BUILDING ENVELOPE REQUIREMENTS[a through e] - CLIMATE ZONE 16

WINDOW AND GLAZED DOOR AREA GREATER THAN 25 PERCENT BUT NOT GREATER THAN 40 PERCENT OF ABOVE-GRADE WALL AREA			
ELEMENT	**CONDITION/VALUE**		
Skylights (U-factor)	0.6		
Slab or below-grade wall (R-value)	R-8		
Windows and glass doors	SHGC		U-factor
PF < 0.25	0.5		0.4
0.25 ≤ PF < 0.50	0.6		0.4
PF ≥ 0.50	0.7		0.4
Roof assemblies (R-value)	Insulation between framing		Continuous insulation
All-wood joist/truss	R-30		R-23
Metal joist/truss	R-30		R-24
Concrete slab or deck	NA		R-23
Metal purlin with thermal block	X		R-24
Metal purlin without thermal block	X		R-24
Floors over outdoor air or unconditioned space (R-value)	Insulation between framing		Continuous insulation
All-wood joist/truss	R-25		R-22
Metal joist/truss	R-30		R-23
Concrete slab or deck	NA		R-22
Above-grade walls (R-value)	No framing	Metal framing	Wood framing
Framed			
R-value cavity	NA	R-13	R-13
R-value continuous	NA	R-3	R-0
CMU, ≥ 8 inches, with integral insulation			
R-value cavity	NA	R-13	R-11
R-value continuous	R-6	R-0	R-0
Other masonry walls			
R-value cavity	NA	R-13	R-13
R-value continuous	R-9	R-3	R-0
WINDOW AND GLAZED DOOR AREA GREATER THAN 40 PERCENT BUT NOT GREATER THAN 50 PERCENT OF ABOVE-GRADE WALL AREA			
ELEMENT	**CONDITION/VALUE**		
Skylights (U-factor)	0.6		
Slab or below-grade wall (R-value)	R-8		
Windows and glass doors	SHGC		U-factor
PF < 0.25	0.4		0.4
0.25 ≤ PF < 0.50	0.5		0.4
PF ≥ 0.50	0.7		0.4
Roof assemblies (R-value)	Insulation between framing		Continuous insulation
All-wood joist/truss	R-30		R-23
Metal joist/truss	R-30		R-24
Concrete slab or deck	NA		R-23
Metal purlin with thermal block	R-38		R-24
Metal purlin without thermal block	X		R-24
Floors over outdoor air or unconditioned space (R-value)	Insulation between framing		Continuous insulation
All-wood joist/truss	R-25		R-22
Metal joist/truss	R-30		R-23
Concrete slab or deck	NA		R-22
Above-grade walls (R-value)	No framing	Metal framing	Wood framing
Framed			
R-value cavity	NA	R-13	R-13
R-value continuous	NA	R-14	R-7
CMU, ≥ 8 inches, with integral insulation			
R-value cavity	NA	R-13	R-13
R-value continuous	R-10	R-3	R-0
Other masonry walls			
R-value cavity	NA	R-13	R-13
R-value continuous	R-9	R-3	R-3

For SI: 1 inch = 25.4 mm.

a. Values from Tables 802.2(5) through 802.2(37) shall be used for the purpose of the completion of Tables 802.2(1) through 802.2(4), as applicable based on window and glazed door area.
b. "NA" indicates the condition is not applicable.
c. An R-value of zero indicates no insulation is required.
d. "Any" indicates any available product will comply.
e. "X" indicates no complying option exists for this condition.

TABLE 802.2(35)
BUILDING ENVELOPE REQUIREMENTS[b through f] - CLIMATE ZONE 17

WINDOW AND GLAZED DOOR AREA 10 PERCENT OR LESS OF ABOVE-GRADE WALL AREA			
ELEMENT	**CONDITION/VALUE**		
Skylights (*U*-factor)	0.6		
Slab or below-grade wall (*R*-value)	R-8		
Windows and glass doors	**SHGC**	**U-factor**	
PF < 0.25	0.7	0.5	
0.25 ≤ PF < 0.50	Any	0.5	
PF ≥ 0.50	Any	0.5	
Roof assemblies (*R*-value)	**Insulation between framing**	**Continuous insulation**	
All-wood joist/truss	R-30	R-23	
Metal joist/truss	R-30	R-24	
Concrete slab or deck	NA	R-23	
Metal purlin with thermal block	X	R-24	
Metal purlin without thermal block	X	R-24	
Floors over outdoor air or unconditioned space (*R*-value)	**Insulation between framing**	**Continuous insulation**	
All-wood joist/truss	R-25	R-22	
Metal joist/truss	R-30	R-23	
Concrete slab or deck	NA	R-22	
Above-grade walls (*R*-value)	**No framing**	**Metal framing**	**Wood framing**
Framed			
R-value cavity	NA	R-13	R-13
R-value continuous	NA	R-3	R-0
CMU, ≥ 8 inches, with integral insulation			
R-value cavity	NA	R-13	R-11
R-value continuous	R-6	R-0	R-0
Other masonry walls			
R-value cavity	NA	R-13	R-11
R-value continuous	R-6	R-0	R-0
WINDOW AND GLAZED DOOR AREA GREATER THAN 10 PERCENT BUT NOT GREATER THAN 25 PERCENT OF ABOVE-GRADE WALL AREA			
ELEMENT	**CONDITION/VALUE**		
Skylights (*U*-factor)	0.6		
Slab or below-grade wall (*R*-value)	R-8		
Windows and glass doors	**SHGC**	**U-factor**	
PF < 0.25	0.7	0.4	
0.25 ≤ PF < 0.50	Any	0.4	
PF ≥ 0.50	Any	0.4	
Roof assemblies (*R*-value)	**Insulation between framing**	**Continuous insulation**	
All-wood joist/truss	R-30	R-23	
Metal joist/truss	R-30	R-24	
Concrete slab or deck	NA	R-23	
Metal purlin with thermal block	X	R-24	
Metal purlin without thermal block	X	R-24	
Floors over outdoor air or unconditioned space (*R*-value)	**Insulation between framing**	**Continuous insulation**	
All-wood joist/truss	R-25	R-22	
Metal joist/truss	R-30	R-23	
Concrete slab or deck	NA	R-22	
Above-grade walls (*R*-value)	**No framing**	**Metal framing**	**Wood framing**
Framed			
R-value cavity	NA	R-13	R-13
R-value continuous	NA	R-3	R-0
CMU, ≥ 8 inches, with integral insulation			
R-value cavity	NA	R-13	R-11
R-value continuous	R-6	R-0	R-0
Other masonry walls			
R-value cavity	NA	R-13	R-11
R-value continuous	R-9	R-3	R-0

(continued)

TABLE 802.2(35)—continued
BUILDING ENVELOPE REQUIREMENTS[b through f] - CLIMATE ZONE 17

WINDOW AND GLAZED DOOR AREA GREATER THAN 25 PERCENT BUT NOT GREATER THAN 40 PERCENT OF ABOVE-GRADE WALL AREA			
ELEMENT	CONDITION/VALUE		
Skylights (U-factor)	0.6		
Slab or below-grade wall (R-value)	R-8		
Windows and glass doors	SHGC[a]		U-factor
PF < 0.25	0.7		0.4
0.25 ≤ PF < 0.50	Any		0.4
PF ≥ 0.50	Any		0.4
Roof assemblies (R-value)	Insulation between framing		Continuous insulation
All-wood joist/truss	R-30		R-23
Metal joist/truss	R-30		R-24
Concrete slab or deck	NA		R-23
Metal purlin with thermal block	X		R-24
Metal purlin without thermal block	X		R-24
Floors over outdoor air or unconditioned space (R-value)	Insulation between framing		Continuous insulation
All-wood joist/truss	R-25		R-22
Metal joist/truss	R-30		R-23
Concrete slab or deck	NA		R-22
Above-grade walls (R-value)	No framing	Metal framing	Wood framing
Framed			
R-value cavity	NA	R-13	R-13
R-value continuous	NA	R-4	R-3
CMU, ≥ 8 inches, with integral insulation			
R-value cavity	NA	R-13	R-13
R-value continuous	R-10	R-4	R-3
Other masonry walls			
R-value cavity	NA	R-13	R-13
R-value continuous	R-10	R-4	R-3
WINDOW AND GLAZED DOOR AREA GREATER THAN 40 PERCENT BUT NOT GREATER THAN 50 PERCENT OF ABOVE-GRADE WALL AREA			
ELEMENT	CONDITION/VALUE		
Skylights (U-factor)	0.6		
Slab or below-grade wall (R-value)	R-8		
Windows and glass doors	SHGC		U-factor
PF < 0.25	0.4		0.4
0.25 ≤ PF < 0.50	0.5		0.4
PF ≥ 0.50	0.7		0.4
Roof assemblies (R-value)	Insulation between framing		Continuous insulation
All-wood joist/truss	R-30		R-23
Metal joist/truss	R-30		R-24
Concrete slab or deck	NA		R-23
Metal purlin with thermal block	R-38		R-24
Metal purlin without thermal block	X		R-24
Floors over outdoor air or unconditioned space (R-value)	Insulation between framing		Continuous insulation
All-wood joist/truss	R-25		R-22
Metal joist/truss	R-30		R-23
Concrete slab or deck	NA		R-22
Above-grade walls (R-value)	No framing	Metal framing	Wood framing
Framed			
R-value cavity	NA	R-13	R-13
R-value continuous	NA	R-14	R-14
CMU, ≥ 8 inches, with integral insulation			
R-value cavity	NA	R-13	R-13
R-value continuous	R-14	R-10	R-7
Other masonry walls			
R-value cavity	NA	R-13	R-13
R-value continuous	R-14	R-10	R-7

For SI: 1 inch = 25.4 mm.

a. For buildings over three stories in height, the maximum SHGC shall be 0.60.
b. Values from Tables 802.2(5) through 802.2(37) shall be used for the purpose of the completion of Tables 802.2(1) through 802.2(4), as applicable based on window and glazed door area.
c. "NA" indicates the condition is not applicable.
d. An R-value of zero indicates no insulation is required.
e. "Any" indicates any available product will comply.
f. "X" indicates no complying option exists for this condition.

ACCEPTABLE PRACTICE FOR COMMERCIAL BUILDINGS

TABLE 802.2(36)
BUILDING ENVELOPE REQUIREMENTS[a through e] - CLIMATE ZONE 18

\multicolumn{4}{c}{WINDOW AND GLAZED DOOR AREA 10 PERCENT OR LESS OF ABOVE-GRADE WALL AREA}			
ELEMENT	\multicolumn{3}{c}{CONDITION/VALUE}		
Skylights (U-factor)	\multicolumn{3}{c}{0.6}		
Slab or below-grade wall (R-value)	\multicolumn{3}{c}{R-12}		
Windows and glass doors	\multicolumn{2}{c}{SHGC}	\multicolumn{1}{c}{U-factor}		
PF < 0.25	0.7		0.6
0.25 ≤ PF < 0.50	Any		0.6
PF ≥ 0.50	Any		0.6
Roof assemblies (R-value)	\multicolumn{2}{c}{Insulation between framing}	Continuous insulation		
All-wood joist/truss	R-30		R-23
Metal joist/truss	R-30		R-24
Concrete slab or deck	NA		R-23
Metal purlin with thermal block	X		R-24
Metal purlin without thermal block	X		R-24
Floors over outdoor air or unconditioned space (R-value)	\multicolumn{2}{c}{Insulation between framing}	Continuous insulation		
All-wood joist/truss	R-25		R-22
Metal joist/truss	R-30		R-23
Concrete slab or deck	NA		R-22
Above-grade walls (R-value)	No framing	Metal framing	Wood framing
Framed			
R-value cavity	NA	R-13	R-13
R-value continuous	NA	R-4	R-3
CMU, ≥ 8 inches, with integral insulation			
R-value cavity	NA	R-13	R-13
R-value continuous	R-9	R-3	R-3
Other masonry walls			
R-value cavity	NA	R-13	R-13
R-value continuous	R-10	R-4	R-3
\multicolumn{4}{c}{WINDOW AND GLAZED DOOR AREA GREATER THAN 10 PERCENT BUT NOT GREATER THAN 25 PERCENT OF ABOVE-GRADE WALL AREA}			
ELEMENT	\multicolumn{3}{c}{CONDITION/VALUE}		
Skylights (U-factor)	\multicolumn{3}{c}{0.6}		
Slab or below-grade wall (R-value)	\multicolumn{3}{c}{R-12}		
Windows and glass doors	\multicolumn{2}{c}{SHGC}	U-factor		
PF < 0.25	0.7		0.4
0.25 ≤ PF < 0.50	Any		0.4
PF ≥ 0.50	Any		0.4
Roof assemblies (R-value)	\multicolumn{2}{c}{Insulation between framing}	Continuous insulation		
All-wood joist/truss	R-30		R-23
Metal joist/truss	R-30		R-24
Concrete slab or deck	NA		R-23
Metal purlin with thermal block	X		R-24
Metal purlin without thermal block	X		R-24
Floors over outdoor air or unconditioned space (R-value)	\multicolumn{2}{c}{Insulation between framing}	Continuous insulation		
All-wood joist/truss	R-25		R-22
Metal joist/truss	R-30		R-23
Concrete slab or deck	NA		R-22
Above-grade walls (R-value)	No framing	Metal framing	Wood framing
Framed			
R-value cavity	NA	R-13	R-13
R-value continuous	NA	R-4	R-3
CMU, ≥ 8 inches, with integral insulation			
R-value cavity	NA	R-13	R-13
R-value continuous	R-9	R-3	R-3
Other masonry walls			
R-value cavity	NA	R-13	R-13
R-value continuous	R-10	R-4	R-3

(continued)

ACCEPTABLE PRACTICE FOR COMMERCIAL BUILDING

TABLE 802.2(36)—continued
BUILDING ENVELOPE REQUIREMENTS[a through e] - CLIMATE ZONE 18

WINDOW AND GLAZED DOOR AREA GREATER THAN 25 PERCENT BUT NOT GREATER THAN 40 PERCENT OF ABOVE-GRADE WALL AREA			
ELEMENT	CONDITION/VALUE		
Skylights (*U*-factor)	X		
Slab or below-grade wall (*R*-value)	X		
Windows and glass doors	SHGC	*U*-factor	
PF < 0.25	X	X	
0.25 ≤ PF < 0.50	X	X	
PF ≥ 0.50	X	X	
Roof assemblies (*R*-value)	Insulation between framing	Continuous insulation	
All-wood joist/truss	X	X	
Metal joist/truss	X	X	
Concrete slab or deck	X	X	
Metal purlin with thermal block	X	X	
Metal purlin without thermal block	X	X	
Floors over outdoor air or unconditioned space (*R*-value)	Insulation between framing	Continuous insulation	
All-wood joist/truss	X	X	
Metal joist/truss	X	X	
Concrete slab or deck	X	X	
Above-grade walls (*R*-value)	No framing	Metal framing	Wood framing
Framed			
R-value cavity	NA	X	X
R-value continuous	NA	X	X
CMU, ≥ 8 inches, with integral insulation			
R-value cavity	X	X	X
R-value continuous	X	X	X
Other masonry walls			
R-value cavity	X	X	X
R-value continuous	X	X	X
WINDOW AND GLAZED DOOR AREA GREATER THAN 40 PERCENT BUT NOT GREATER THAN 50 PERCENT OF ABOVE-GRADE WALL AREA			
ELEMENT	CONDITION/VALUE		
Skylights (*U*-factor)	X		
Slab or below-grade wall (*R*-value)	X		
Windows and glass doors	SHGC	*U*-factor	
PF < 0.25	X	X	
0.25 ≤ PF < 0.50	X	X	
PF ≥ 0.50	X	X	
Roof assemblies (*R*-value)	Insulation between framing	Continuous insulation	
All-wood joist/truss	X	X	
Metal joist/truss	X	X	
Concrete slab or deck	X	X	
Metal purlin with thermal block	X	X	
Metal purlin without thermal block	X	X	
Floors over outdoor air or unconditioned space (*R*-value)	Insulation between framing	Continuous insulation	
All-wood joist/truss	X	X	
Metal joist/truss	X	X	
Concrete slab or deck	X	X	
Above-grade walls (*R*-value)	No framing	Metal framing	Wood framing
Framed			
R-value cavity	NA	X	X
R-value continuous	NA	X	X
CMU, ≥ 8 inches, with integral insulation			
R-value cavity	X	X	X
R-value continuous	X	X	X
Other masonry walls			
R-value cavity	X	X	X
R-value continuous	X	X	X

For SI: 1 inch = 25.4 mm.

a. Values from Tables 802.2(5) through 802.2(37) shall be used for the purpose of the completion of Tables 802.2(1) through 802.2(4), as applicable based on window and glazed door area.
b. "NA" indicates the condition is not applicable.
c. An *R*-value of zero indicates no insulation is required.
d. "Any" indicates any available product will comply.
e. "X" indicates no complying option exists for this condition.

TABLE 802.2(37)
BUILDING ENVELOPE REQUIREMENTS[a through e] - CLIMATE ZONE 19

WINDOW AND GLAZED DOOR AREA 10 PERCENT OR LESS OF ABOVE-GRADE WALL AREA			
ELEMENT	CONDITION/VALUE		
Skylights (*U*-factor)	0.6		
Slab or below-grade wall (*R*-value)	R-12		
Windows and glass doors	SHGC		*U*-factor
PF < 0.25	0.7		0.5
0.25 ≤ PF < 0.50	Any		0.5
PF ≥ 0.50	Any		0.5
Roof assemblies (*R*-value)	Insulation between framing		Continuous insulation
All-wood joist/truss	R-30		R-23
Metal joist/truss	R-30		R-24
Concrete slab or deck	NA		R-23
Metal purlin with thermal block	X		R-24
Metal purlin without thermal block	X		R-24
Floors over outdoor air or unconditioned space (*R*-value)	Insulation between framing		Continuous insulation
All-wood joist/truss	R-25		R-22
Metal joist/truss	R-30		R-23
Concrete slab or deck	NA		R-22
Above-grade walls (*R*-value)	No framing	Metal framing	Wood framing
Framed			
R-value cavity	NA	R-13	R-13
R-value continuous	NA	R-4	R-3
CMU, ≥ 8 inches, with integral insulation			
R-value cavity	NA	R-13	R-13
R-value continuous	R-9	R-3	R-3
Other masonry walls			
R-value cavity	NA	R-13	R-13
R-value continuous	R-10	R-4	R-3
WINDOW AND GLAZED DOOR AREA GREATER THAN 10 PERCENT BUT NOT GREATER THAN 25 PERCENT OF ABOVE-GRADE WALL AREA			
ELEMENT	CONDITION/VALUE		
Skylights (*U*-factor)	0.6		
Slab or below-grade wall (*R*-value)	R-12		
Windows and glass doors	SHGC		*U*-factor
PF < 0.25	0.7		0.4
0.25 ≤ PF < 0.50	Any		0.4
PF ≥ 0.50	Any		0.4
Roof assemblies (*R*-value)	Insulation between framing		Continuous insulation
All-wood joist/truss	R-30		R-23
Metal joist/truss	R-30		R-24
Concrete slab or deck	NA		R-23
Metal purlin with thermal block	X		R-24
Metal purlin without thermal block	X		R-24
Floors over outdoor air or unconditioned space (*R*-value)	Insulation between framing		Continuous insulation
All-wood joist/truss	R-25		R-22
Metal joist/truss	R-30		R-23
Concrete slab or deck	NA		R-22
Above-grade walls (*R*-value)	No framing	Metal framing	Wood framing
Framed			
R-value cavity	NA	R-13	R-13
R-value continuous	NA	R-4	R-3
CMU, ≥ 8 inches, with integral insulation			
R-value cavity	NA	R-13	R-13
R-value continuous	R-9	R-3	R-3
Other masonry walls			
R-value cavity	NA	R-13	R-13
R-value continuous	R-10	R-4	R-3

(continued)

ACCEPTABLE PRACTICE FOR COMMERCIAL BUILDING

TABLE 802.2(37)—continued
BUILDING ENVELOPE REQUIREMENTS[a through e] - CLIMATE ZONE 19

WINDOW AND GLAZED DOOR AREA GREATER THAN 25 PERCENT BUT NOT GREATER THAN 40 PERCENT OF ABOVE-GRADE WALL AREA			
ELEMENT	**CONDITION/VALUE**		
Skylights (U-factor)	X		
Slab or below-grade wall (R-value)	X		
Windows and glass doors	SHGC		U-factor
PF < 0.25	X		X
0.25 ≤ PF < 0.50	X		X
PF ≥ 0.50	X		X
Roof assemblies (R-value)	Insulation between framing		Continuous insulation
All-wood joist/truss	X		X
Metal joist/truss	X		X
Concrete slab or deck	X		X
Metal purlin with thermal block	X		X
Metal purlin without thermal block	X		X
Floors over outdoor air or unconditioned space (R-value)	Insulation between framing		Continuous insulation
All-wood joist/truss	X		X
Metal joist/truss	X		X
Concrete slab or deck	X		X
Above-grade walls (R-value)	No framing	Metal framing	Wood framing
Framed			
R-value cavity	NA	X	X
R-value continuous	NA	X	X
CMU, ≥ 8 inches, with integral insulation			
R-value cavity	X	X	X
R-value continuous	X	X	X
Other masonry walls			
R-value cavity	X	X	X
R-value continuous	X	X	X
WINDOW AND GLAZED DOOR AREA GREATER THAN 40 PERCENT BUT NOT GREATER THAN 50 PERCENT OF ABOVE-GRADE WALL AREA			
ELEMENT	**CONDITION/VALUE**		
Skylights (U-factor)	X		
Slab or below-grade wall (R-value)	X		
Windows and glass doors	SHGC		U-factor
PF < 0.25	X		X
0.25 ≤ PF < 0.50	X		X
PF ≥ 0.50	X		X
Roof assemblies (R-value)	Insulation between framing		Continuous insulation
All-wood joist/truss	X		X
Metal joist/truss	X		X
Concrete slab or deck	X		X
Metal purlin with thermal block	X		X
Metal purlin without thermal block	X		X
Floors over outdoor air or unconditioned space (R-value)	Insulation between framing		Continuous insulation
All-wood joist/truss	X		X
Metal joist/truss	X		X
Concrete slab or deck	X		X
Above-grade walls (R-value)	No framing	Metal framing	Wood framing
Framed			
R-value cavity	NA	X	X
R-value continuous	NA	X	X
CMU, ≥ 8 inches, with integral insulation			
R-value cavity	X	X	X
R-value continuous	X	X	X
Other masonry walls			
R-value cavity	X	X	X
R-value continuous	X	X	X

For SI: 1 inch = 25.4 mm.

a. Values from Tables 802.2(5) through 802.2(37) shall be used for the purpose of the completion of Tables 802.2(1) through 802.2(4), as applicable based on window and glazed door area.
b. "NA" indicates the condition is not applicable.
c. An R-value of zero indicates no insulation is required.
d. "Any" indicates any available product will comply.
e. "X" indicates no complying option exists for this condition.

CHAPTER 9
CLIMATE MAPS

SECTION 901
GENERAL

901.1 Scope. The criteria of this chapter establish design conditions based on political boundaries for use with climate-dependent requirements in Chapters 5, 6 and 8, as applicable.

SECTION 902
CLIMATE ZONES

902.1 General. The climate zone for use in Table 302.1 shall be selected from the applicable map in Figures 902.1(1) through 902.1(51) corresponding to the state and county of each jurisdiction.

CLIMATE MAPS

Zone	County
6B	Autauga (H)
4B	Baldwin (H)
5A	Barbour (H)
6B	Bibb (H)
7A	Blount
5A	Bullock (H)
5A	Butler (H)
6B	Calhoun (H)
6B	Chambers (H)
7A	Cherokee
6B	Chilton (H)
5A	Choctaw (H)
5A	Clarke (H)
7A	Clay
7A	Cleburne
4B	Coffee (H)
8	Colbert (H)
5A	Conecuh (H)
6B	Coosa (H)
4B	Covington (H)
5A	Crenshaw (H)
7A	Cullman
4B	Dale (H)
5A	Dallas (H)
8	De Kalb
6B	Elmore (H)
4B	Escambia (H)
7A	Etowah
7A	Fayette
8	Franklin
4B	Geneva (H)
5A	Greene (H)
5A	Hale (H)
4B	Henry (H)

Zone	County
4B	Houston (H)
8	Jackson
6B	Jefferson (H)
7A	Lamar
8	Lauderdale
8	Lawrence
6B	Lee (H)
8	Limestone
5A	Lowndes (H)
6B	Macon (H)
8	Madison
5A	Marengo (H)
7A	Marion
8	Marshall
4B	Mobile (H)
5A	Monroe (H)
6B	Montgomery (H)
8	Morgan
5A	Perry (H)
6B	Pickens (H)
5A	Pike (H)
7A	Randolph
5A	Russell (H)
6B	Shelby (H)
6B	St Clair (H)
5A	Sumter (H)
6B	Talladega (H)
6B	Tallapoosa (H)
6B	Tuscaloosa (H)
6B	Walker (H)
5A	Washington (H)
5A	Wilcox (H)
7A	Winston

a. Counties identified with (H) shall be considered "hot and humid climate areas" for purposes of the application of Section 502.1.1.

FIGURE 902.1(1)
ALABAMA[a]

CLIMATE MAPS

Zone	Borough[a]		
16	Adak Region	17	Kenai Peninsula
19	Alaska Gateway	15	Ketchikan Gateway
17	Aleutian Region	16	Kodiak Island
17	Aleutians East	18	Kuspuk
17	Anchorage	17	Lake ard Peninsula
15	Annette Island	18	Lower Kuskokwim
19	Bering Straits	18	Lower Yukon
17	Bristol Bay	17	Matanuska-Susitna
16	Chatham	19	North Slope
17	Chugach	19	Northwest Arctic
18	Copper River	17	Pribilof Islands
18	Delta/Greely	15	Sitka
18	Denali	15	Southeast Island
18	Fairbanks North Star	17	Southwest Region
16	Haines	17	Yakutat
19	Iditarod Area	19	Yukon Flats
16	Juneau	19	Yukon-Koyukuk
18	Kashunamiut	18	Yupiit

a. Borough refers to boroughs, united home rule municipalities and regional education attendance areas.

**FIGURE 902.1(2)
ALASKA**

CLIMATE MAPS

Zone	County
13B	Apache
6B	Cochise
14A	Coconino
8	Gila
6B	Graham
6B	Greenlee
3C	La Paz
3C	Maricopa
7B	Mohave
10B	Navajo
4B	Pima
4B	Pinal
6B	Santa Cruz
10B	Yavapai
3C	Yuma

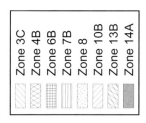

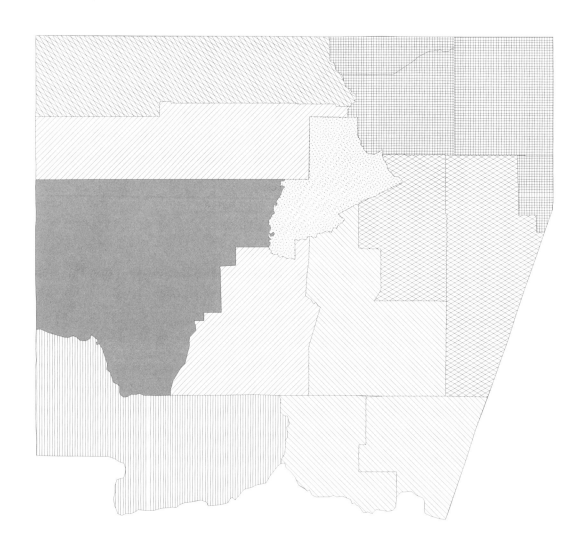

FIGURE 902.1(3)
ARIZONA

CLIMATE MAPS

Zone	County	Zone	County
6B	Arkansas (H)	7B	Lee (H)
6B	Ashley (H)	6B	Lincoln (H)
9B	Baxter	6B	Little River (H)
9B	Benton	7B	Logan (H)
9B	Boone	7B	Lonoke (H)
6B	Bradley (H)	9B	Madison
6B	Calhoun (H)	9B	Marion
9B	Carroll	6B	Miller (H)
6B	Chicot (H)	8	Mississippi
6B	Clark (H)	7B	Monroe (H)
8	Clay	8	Montgomery
8	Cleburne	6B	Nevada (H)
6B	Cleveland (H)	9B	Newton
6B	Columbia (H)	6B	Ouachita (H)
7B	Conway (H)	7B	Perry (H)
8	Craighead	7B	Phillips (H)
8	Crawford	7B	Pike (H)
7B	Crittenden (H)	8	Poinsett
7B	Cross (H)	8	Polk
6B	Dallas (H)	8	Pope
6B	Desha (H)	7B	Prairie (H)
6B	Drew (H)	7B	Pulaski (H)
7B	Faulkner (H)	8	Randolph
8	Franklin	7B	Saline (H)
8	Fulton	9B	Scott (H)
7B	Garland (H)	7B	Searcy
6B	Grant (H)	8	Sebastian
8	Greene	7B	Sevier (H)
7B	Hempstead (H)	8	Sharp
7B	Hot Spring (H)	7B	St Francis (H)
7B	Howard (H)	9B	Stone
8	Independence	6B	Union (H)
8	Izard	8	Van Buren
8	Jackson	9B	Washington
6B	Jefferson (H)	7B	White (H)
8	Johnson	7B	Woodruff (H)
6B	Lafayette (H)	7B	Yell (H)
8	Lawrence		

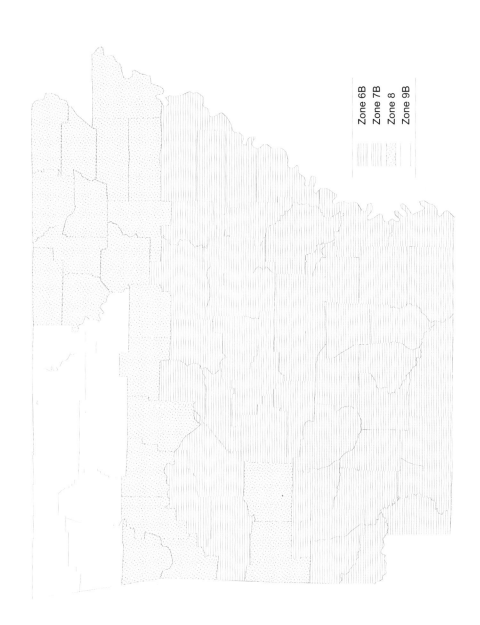

Zone 6B
Zone 7B
Zone 8
Zone 9B

FIGURE 902.1(4)
ARKANSAS[a]

a. Counties identified with (H) shall be considered "hot and humid climate areas" for purposes of the application of Section 502.1.1.

CLIMATE MAPS

Zone	County
6A	Alameda
15	Alpine
8	Amador
6B	Butte
8	Calaveras
6B	Colusa
6A	Contra Costa
9A	Del Norte
8	El Dorado
6B	Fresno
6B	Glenn
9A	Humboldt
3A	Imperial
9B	Inyo
5B	Kern
6B	Kings
8	Lake
13B	Lassen
4A	Los Angeles
6B	Madera
6A	Marin
8	Mariposa
8	Mendocino
6B	Merced
15	Modoc
15	Mono

Zone	County
6A	Monterey
6A	Napa
11A	Nevada
4A	Orange
8	Placer
13B	Plumas
4B	Riverside
6B	Sacramento
6B	San Benito
4B	San Bernardino
3A	San Diego
6A	San Francisco
6A	San Joaquin
6A	San Luis Obispo
6A	San Mateo
5A	Santa Barbara
6A	Santa Clara
6A	Santa Cruz
6B	Shasta
11A	Sierra
11A	Siskiyou
6A	Solano
6A	Sonoma
6B	Stanislaus
6B	Sutter
6B	Tehama
9B	Trinity
6B	Tulare
8	Tuolumne
4A	Ventura
6B	Yolo
6B	Yuba

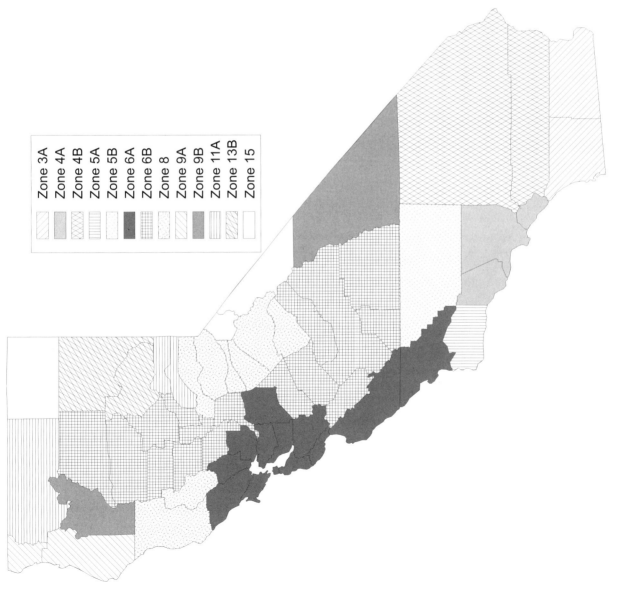

**FIGURE 902.1(5)
CALIFORNIA**

CLIMATE MAPS

Zone	County	Zone	County
13B	Adams	15	La Plata
16	Alamosa	17	Lake
13B	Arapahoe	13B	Larimer
16	Archuleta	11B	Las Animas
11B	Baca	13B	Lincoln
11B	Bent	13B	Logan
13B	Boulder	13B	Mesa
16	Chaffee	17	Mineral
13B	Cheyenne	15	Moffat
17	Clear Creek	15	Montezuma
16	Conejos	13B	Montrose
16	Costilla	13B	Morgan
11B	Crowley	11B	Otero
16	Custer	15	Ouray
13B	Delta	17	Park
13B	Denver	13B	Phillips
15	Dolores	17	Pitkin
13B	Douglas	11B	Prowers
15	Eagle	11B	Pueblo
13B	El Paso	15	Rio Blanco
13B	Elbert	17	Rio Grande
11B	Fremont	17	Routt
15	Garfield	16	Saguache
13B	Gilpin	17	San Juan
17	Grand	15	San Miguel
17	Gunnison	13B	Sedgwick
17	Hinsdale	17	Summit
11B	Huerfano	13B	Teller
17	Jackson	13B	Washington
13B	Jefferson	13B	Weld
13B	Kiowa	13B	Yuma
13B	Kit Carson		

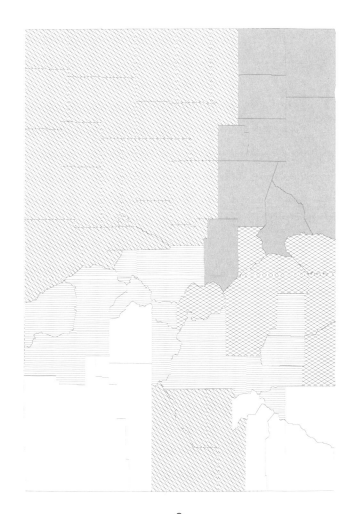

**FIGURE 902.1(6)
COLORADO**

CLIMATE MAPS

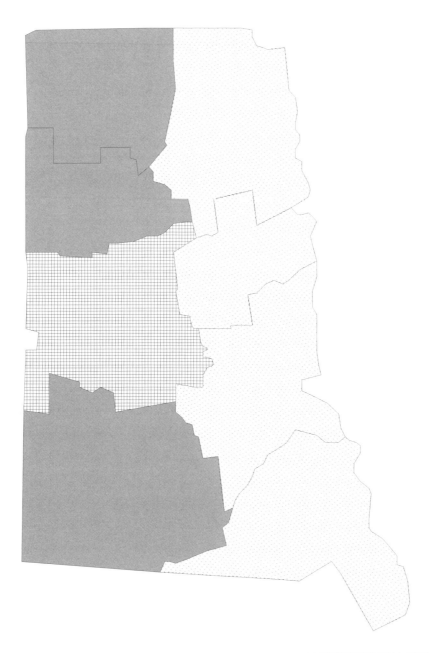

FIGURE 902.1(7)
CONNECTICUT

CLIMATE MAPS

Zone	County
9B	Kent
10B	New Castle
9B	Sussex

FIGURE 902.1(8)
DELAWARE

CLIMATE MAPS

FIGURE 902.1(9)
DISTRICT OF COLUMBIA

CLIMATE MAPS

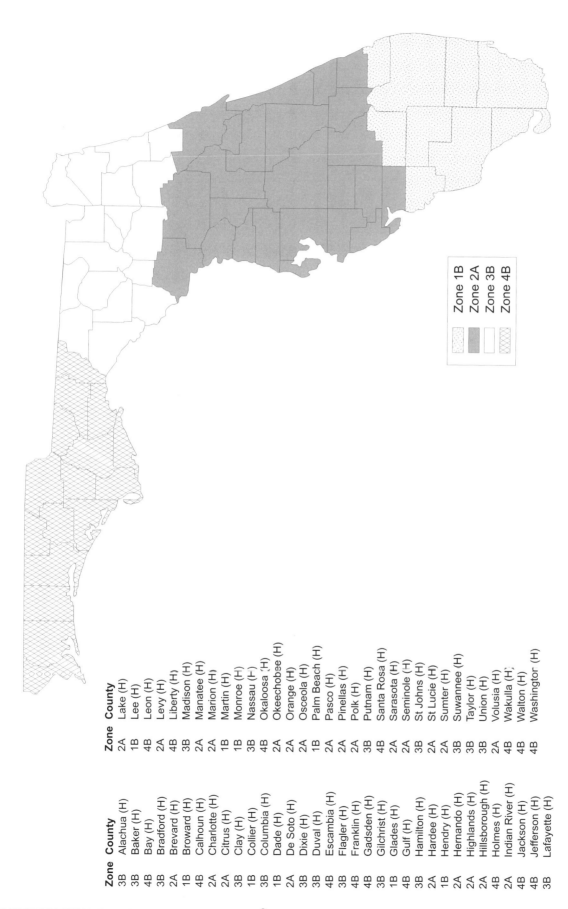

FIGURE 902.1(10)
FLORIDA[a]

a. Counties identified with (H) shall be considered "hot and humid climate areas" for purposes of the application of Section 502.1.1.

CLIMATE MAPS

Zone	County
4B	Appling (H)
4B	Atkinson (H)
4B	Bacon (H)
4B	Baker (H)
6B	Baldwin (H)
7A	Banks
7A	Barrow
7A	Bartow
5A	Ben Hill (H)
4B	Berrien (H)
5A	Bibb (H)
5A	Bleckley (H)
4B	Brantley (H)
4B	Brooks (H)
4B	Bryan (H)
5A	Bulloch (H)
6B	Burke (H)
7A	Butts
5A	Calhoun (H)
4B	Camden (H)
5A	Candler (H)
7A	Carroll
8	Catoosa
4B	Charlton (H)
4B	Chatham (H)
5A	Chattahoochee (H)
8	Chattooga
8	Cherokee
7A	Clarke
5A	Clay (H)
7A	Clayton
4B	Clinch (H)
7A	Cobb
5A	Coffee (H)
4B	Colquitt (H)
6B	Columbia (H)
4B	Cook (H)
7A	Coweta
5A	Crawford (H)
5A	Crisp (H)
8	Dade
8	Dawson
7A	De Kalb
4B	Decatur (H)
5A	Dodge (H)
5A	Dooly (H)
4B	Dougherty (H)
7A	Douglas
5A	Early (H)
4B	Echols (H)
4B	Effingham (H)
7A	Elbert
5A	Emanuel (H)
4B	Evans (H)
8	Fannin
7A	Fayette
7A	Floyd
8	Forsyth
7A	Franklin
7A	Fulton
8	Gilmer
6B	Glascock (H)
4B	Glynn (H)
8	Gordon
4B	Grady (H)
6B	Greene (H)
7A	Gwinnett
8	Habersham
7A	Hall
6B	Hancock (H)
7A	Haralson
6B	Harris (H)
7A	Hart
6B	Heard (H)
7A	Henry
5A	Houston (H)
5A	Irwin (H)
7A	Jackson
6B	Jasper (H)
4B	Jeff Davis (H)
6B	Jefferson (H)
5A	Jenkins (H)
5A	Johnson (H)
6B	Jones (H)
6B	Lamar (H)
4B	Lanier (H)
5A	Laurens (H)
5A	Lee (H)
4B	Liberty (H)
6B	Lincoln (H)
4B	Long (H)
4B	Lowndes (H)
8	Lumpkin
5A	Macon (H)
7A	Madison
5A	Marion (H)
6B	McDuffie (H)
4B	McIntosh (H)
6B	Meriwether (H)
4B	Miller (H)
4B	Mitchell (H)
6B	Monroe (H)
4B	Montgomery (H)
5A	Morgan (H)
6B	Murray
8	Muscogee (H)
5A	Newton
7A	Oconee
7A	Oglethorpe
7A	Paulding
5A	Peach (H)
8	Pickens
4B	Pierce (H)
6B	Pike (H)
7A	Polk
5A	Pulaski (H)
6B	Putnam (H)
5A	Quitman (H)
8	Rabun
5A	Randolph (H)
6B	Richmond (H)
7A	Rockdale
5A	Schley (H)
5A	Screven (H)
5A	Seminole (H)
4B	Spalding
7A	Stephens
5A	Stewart (H)
5A	Sumter (H)
5A	Talbot (H)
5A	Taliaferro (H)
6B	Tattnall (H)
4B	Taylor (H)
5A	Telfair (H)
5A	Terrell (H)
4B	Thomas (H)
5A	Tift (H)
4B	Toombs (H)
8	Towns
5A	Treutlen (H)
5A	Troup (H)
5A	Turner (H)
6B	Twiggs (H)
5A	Union
5A	Upson (H)
8	Walker
5A	Walton
7A	Ware (H)
4B	Warren (H)
6B	Washington (H)
6B	Wayne (H)
4B	Webster (H)
5A	Wheeler (H)
5A	White
8	Whitfield
8	Wilcox (H)
5A	Wilkes
7A	Wilkinson (H)
5A	Worth (H)

Zone 4B
Zone 5A
Zone 6B
Zone 7A
Zone 8

FIGURE 902.1(11)
GEORGIA[a]

a. Counties identified with (H) shall be considered "hot and humid climate areas" for purposes of the application of Section 502.1.1.

CLIMATE MAPS

Zone 1A

Zone	County
1A	Hawaii (H)
1A	Honolulu (H)
1A	Kalawao (H)
1A	Kauai H)
1A	Maui (H)

a. Counties identified with (H) shall be considered "hot and humid climate areas" for purposes of the application of Section 502.1.1.

FIGURE 902.1(12)
HAWAII[a]

CLIMATE MAPS

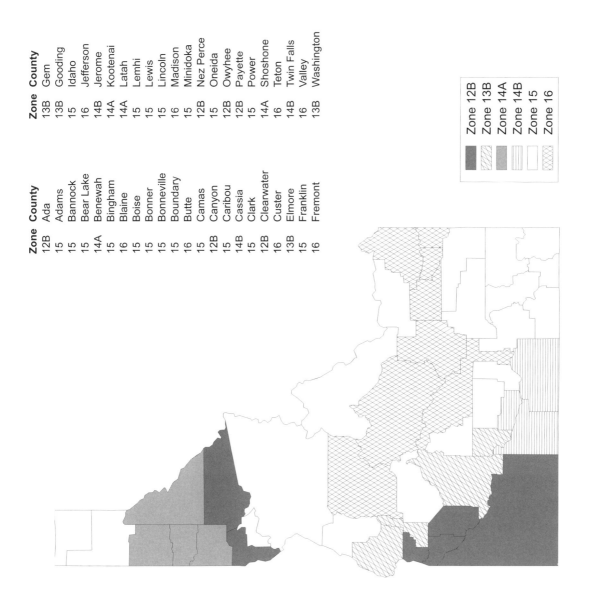

FIGURE 902.1(13)
IDAHO

CLIMATE MAPS

Zone	County	Zone	County	Zone	County
12B	Adams	10B	Hardin	12B	Morgan
10B	Alexander	13B	Henderson	12B	Moultrie
11B	Bond	13B	Henry	14B	Ogle
14B	Boone	13B	Iroquois	13B	Peoria
12B	Brown	10B	Jackson	10B	Perry
13B	Bureau	11B	Jasper	12B	Piatt
11B	Calhoun	11B	Jefferson	12B	Pike
14B	Carroll	10B	Jersey	10B	Pope
12B	Cass	14B	Jo Daviess	10B	Pulaski
12B	Champaign	10B	Johnson	13B	Putnam
11B	Christian	14B	Kane	10B	Randolph
12B	Clark	13B	Kankakee	11B	Richland
11B	Clay	13B	Kendall	13B	Rock Island
10B	Clinton	13B	Knox	10B	Saline
12B	Coles	13B	La Salle	12B	Sangamon
14B	Cook	14B	Lake	12B	Schuyler
11B	Crawford	11B	Lawrence	12B	Scott
12B	Cumberland	14B	Lee	11B	Shelby
14B	De Kalb	13B	Livingston	10B	St Clair
12B	De Witt	12B	Logan	13B	Stark
12B	Douglas	12B	Macon	14B	Stephenson
14B	Du Page	11B	Macoupin	12B	Tazewell
12B	Edgar	10B	Madison	10B	Union
11B	Edwards	12B	Marion	12B	Vermilion
11B	Effingham	13B	Marshall	11B	Wabash
11B	Fayette	12B	Mason	13B	Warren
13B	Ford	10B	Massac	10B	Washington
10B	Franklin	13B	McDonough	11B	Wayne
13B	Fulton	14B	McHenry	10B	White
10B	Gallatin	12B	McLean	14B	Whiteside
11B	Greene	12B	Menard	13B	Will
13B	Grundy	13B	Mercer	10B	Williamson
10B	Hamilton	10B	Monroe	14B	Winnebago
13B	Hancock	11B	Montgomery	13B	Woodford

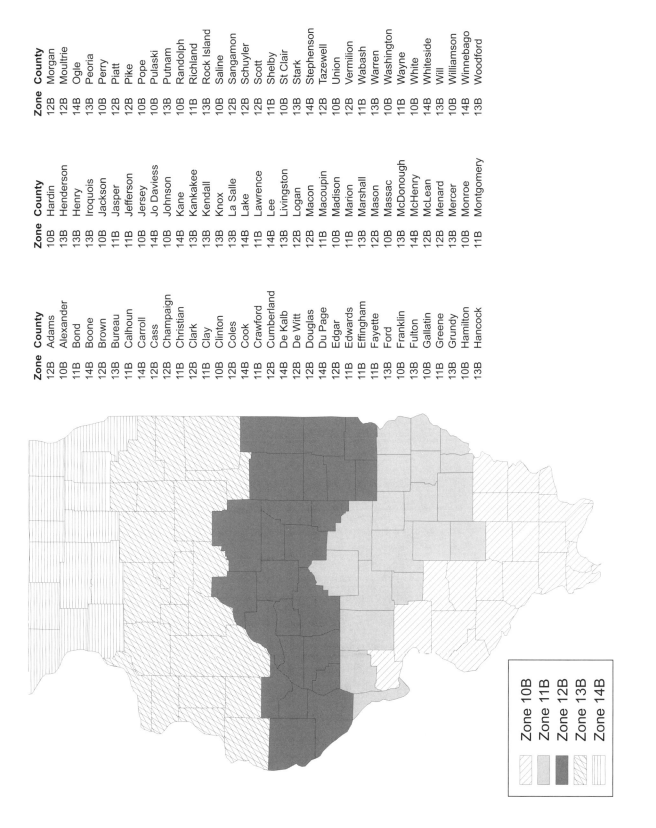

**FIGURE 902.1(14)
ILLINOIS**

CLIMATE MAPS

Zone	County	Zone	County	Zone	County
13B	Adams	12B	Hendricks	11B	Pike
13B	Allen	12B	Henry	13B	Porter
11B	Bartholomew	13B	Howard	10B	Posey
13B	Benton	14A	Huntington	13B	Pulaski
13B	Blackford	11B	Jackson	12B	Putnam
12B	Boone	13B	Jasper	13B	Randolph
11B	Brown	13B	Jay	11B	Ripley
13B	Carroll	10B	Jefferson	12B	Rush
13B	Cass	11B	Jennings	11B	Scott
10B	Clark	12B	Johnson	12B	Shelby
12B	Clay	11B	Knox	10B	Spencer
13B	Clinton	14A	Kosciusko	13B	St Joseph
11B	Crawford	13B	La Porte	13B	Starke
11B	Daviess	14A	Lagrange	14A	Steuben
13B	De Kalb	13B	Lake	11B	Sullivan
11B	Dearborn	11B	Lawrence	10B	Switzerland
12B	Decatur	13B	Madison	13B	Tippecanoe
13B	Delaware	12B	Marion	13B	Tipton
11B	Dubois	13B	Marshall	12B	Union
13B	Elkhart	11B	Martin	10B	Vanderburgh
12B	Fayette	14A	Miami	12B	Vermillion
10B	Floyd	11B	Monroe	12B	Vigo
12B	Fountain	12B	Montgomery	14A	Wabash
12B	Franklin	12B	Morgan	12B	Warren
14A	Fulton	13B	Newton	10B	Warrick
10B	Gibson	14A	Noble	11B	Washington
13B	Grant	11B	Ohio	12B	Wayne
11B	Greene	11B	Orange	13B	Wells
12B	Hamilton	12B	Owen	13B	White
12B	Hancock	12B	Parke	14A	Whitley
10B	Harrison	10B	Perry		

Zone 10B
Zone 11B
Zone 12B
Zone 13B
Zone 14A

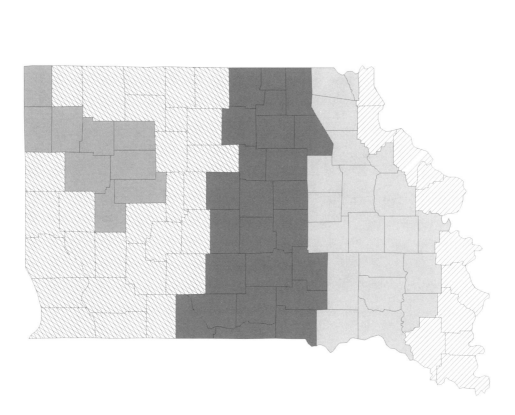

FIGURE 902.1(15)
INDIANA

CLIMATE MAPS

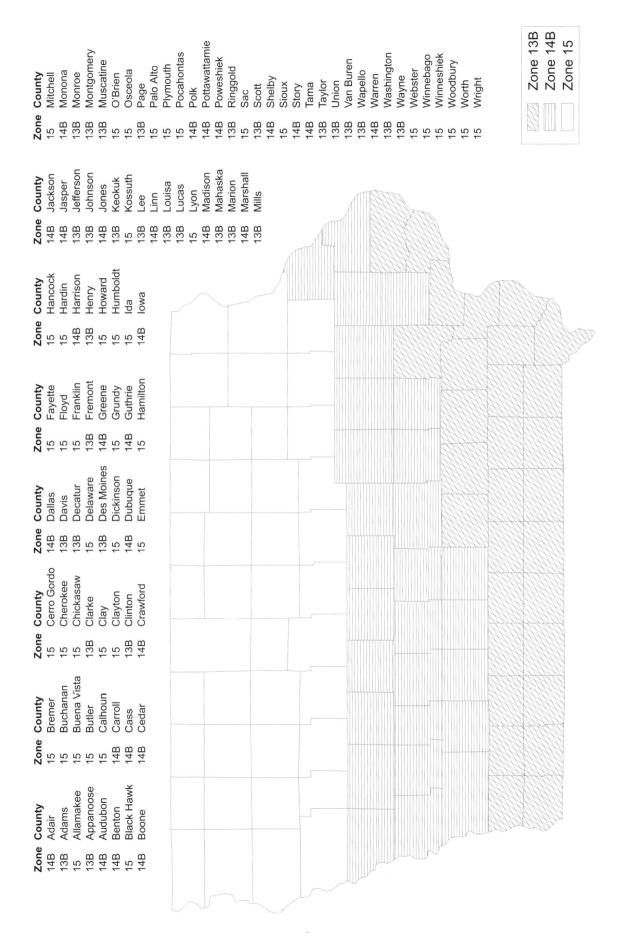

FIGURE 902.1(16)
IOWA

CLIMATE MAPS

Zone	County	Zone	County	Zone	County	Zone	County	Zone	County	Zone	County	Zone	County
10B	Allen	9B	Cherokee	11B	Dickinson	11B	Geary	11B	Haskell	12B	Lane	10B	Miami	12B	Osborne
10B	Anderson	13B	Cheyenne	11B	Doniphan	12B	Gove	11B	Hodgeman	11B	Leavenworth	12B	Mitchell	11B	Ottawa
11B	Atchison	10B	Clark	10B	Douglas	12B	Graham	11B	Jackson	11B	Lincoln	9B	Montgomery	11B	Pawnee
9B	Barber	11B	Clay	11B	Edwards	11B	Grant	11B	Jefferson	10B	Linn	11B	Morris	12B	Phillips
11B	Barton	12B	Cloud	9B	Elk	11B	Gray	12B	Jewell	12B	Logan	10B	Morton	11B	Pottawatomie
10B	Bourbon	10B	Coffey	12B	Ellis	12B	Greeley	11B	Johnson	11B	Lyon	11B	Nemaha	10B	Pratt
11B	Brown	9B	Comanche	11B	Ellsworth	10B	Greenwood	11B	Kearny	11B	Marion	9B	Neosho	13B	Rawlins
10B	Butler	9B	Cowley	11B	Finney	11B	Hamilton	10B	Kingman	12B	Marshall	12B	Ness	11B	Reno
10B	Chase	9B	Crawford	11B	Ford	9B	Harper	10B	Kiowa	11B	McPherson	13B	Norton	12B	Republic
9B	Chautauqua	13B	Decatur	10B	Franklin	11B	Harvey	9B	Labette	10B	Meade	10B	Osage	11B	Rice

Zone	County
11B	Riley
12B	Rooks
11B	Rush
11B	Russell
11B	Saline
12B	Scott
10B	Sedgwick
10B	Seward
11B	Shawnee
12B	Sheridan
13B	Sherman
12B	Smith
11B	Stafford
11B	Stanton
10B	Stevens
9B	Sumner
13B	Thomas
12B	Trego
11B	Wabaunsee
12B	Wallace
12B	Washington
12B	Wichita
9B	Wilson
10B	Woodson
11B	Wyandotte

Zone 9B
Zone 10B
Zone 11B
Zone 12B
Zone 13B

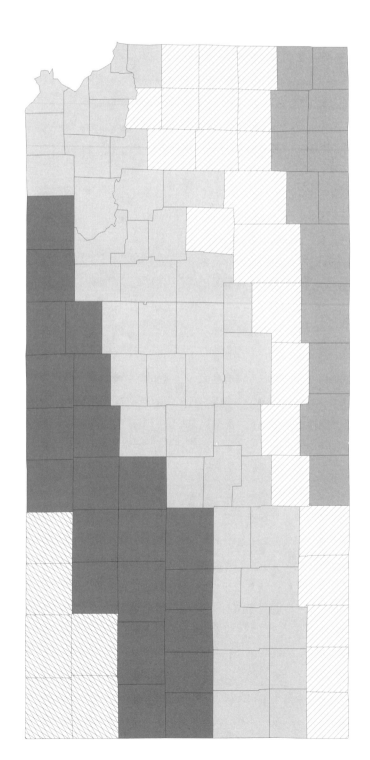

FIGURE 902.1(17)
KANSAS

CLIMATE MAPS

Zone	County	Zone	County	Zone	County	Zone	County	Zone	County	Zone	County	Zone	County
9B	Adair	9B	Caldwell	10B	Estill	11B	Harrison	10B	Lee	9B	McCracken	11B	Nicholas	11B	Scott
9B	Allen	9B	Calloway	10B	Fayette	9B	Hart	10B	Leslie	10B	McCreary	9B	Ohio	10B	Shelby
10B	Anderson	11B	Campbell	11B	Fleming	9B	Henderson	10B	Letcher	9B	McLean	10B	Oldham	9B	Simpson
9B	Ballard	9B	Carlisle	10B	Floyd	10B	Henry	11B	Lewis	9B	Meade	10B	Owen	10B	Spencer
9B	Barren	10B	Carroll	10B	Franklin	9B	Hickman	10B	Lincoln	10B	Menifee	10B	Owsley	9B	Taylor
11B	Bath	11B	Carter	9B	Fulton	9B	Hopkins	9B	Livingston	10B	Mercer	11B	Pendleton	9B	Todd
10B	Bell	10B	Casey	11B	Gallatin	10B	Jackson	9B	Logan	9B	Metcalfe	10B	Perry	9B	Trigg
11B	Boone	9B	Christian	10B	Garrard	10B	Jefferson	9B	Lyon	9B	Monroe	10B	Pike	10B	Trimble
10B	Bourbon	10B	Clark	11B	Grant	10B	Jessamine	10B	Madison	10B	Montgomery	10B	Powell	9B	Union
11B	Boyd	10B	Clay	9B	Graves	11B	Johnson	10B	Magoffin	10B	Morgan	10B	Pulaski	9B	Warren
10B	Boyle	10B	Clinton	9B	Grayson	11B	Kenton	10B	Marion	9B	Muhlenberg	11B	Robertson	10B	Washington
11B	Bracken	9B	Crittenden	11B	Green	10B	Knott	9B	Marshall	10B	Nelson	10B	Rockcastle	10B	Wayne
10B	Breathitt	9B	Cumberland	11B	Greenup	10B	Knox	11B	Martin			11B	Rowan	9B	Webster
9B	Breckenridge	9B	Daviess	9B	Hancock	9B	Larue	11B	Mason			10B	Russell	10B	Whitley
10B	Bullitt	9B	Edmonson	9B	Hardin	10B	Laurel							10B	Wolfe
9B	Butler	11B	Elliot	10B	Harlan	11B	Lawrence							10B	Woodford

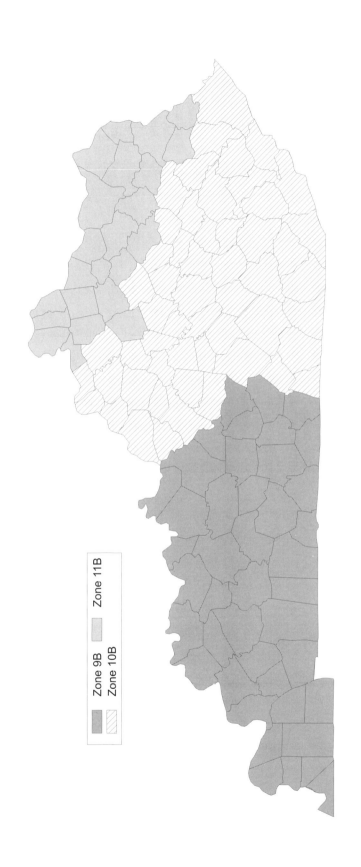

FIGURE 902.1(18)
KENTUCKY

Zone 9B
Zone 10B
Zone 11B

CLIMATE MAPS

Zone	Parish	Zone	Parish	Zone	Parish
4B	Acadia (H)	6B	East Carroll (H)	5A	Natchitoches (H)
4B	Allen (H)	4B	East Feliciana (H)	3B	Orleans (H)
4B	Ascension (H)	4B	Evangeline (H)	6B	Ouachita (H)
3B	Assumption (H)	6B	Franklin (H)	3B	Plaquemines (H)
5A	Avoyelles (H)	5A	Grant (H)	4B	Pointe Coupee (H)
4B	Beauregard (H)	4B	Iberia (H)	5A	Rapides (H)
6B	Bienville (H)	4B	Iberville (H)	5A	Red River (H)
6B	Bossier (H)	6B	Jackson (H)	6B	Richland (H)
6B	Caddo (H)	3B	Jefferson (H)	5A	Sabine (H)
4B	Calcasieu (H)	4B	Jefferson Davis (H)	3B	St Bernard (H)
6B	Caldwell (H)	5A	La Salle (H)	3B	St Charles (H)
4B	Cameron (H)	4B	Lafayette (H)	4B	St Helena (H)
5A	Catahoula (H)	3B	Lafourche (H)	3B	St James (H)
6B	Claiborne (H)	6B	Lincoln (H)	3B	St John The Baptist (H)
5A	Concordia (H)	4B	Livingston (H)	4B	St Landry (H)
5A	De Soto (H)	6B	Madison (H)	4B	St Martin (H)
4B	East Baton Rouge (H)	6B	Morehouse (H)	3B	St Mary (H)
				4B	St Tammany (H)
				4B	Tangipahoa (H)
				5A	Tensas (H)
				3B	Terrebonne (H)
				6B	Union (H)
				4B	Vermilion (H)
				5A	Vernon (H)
				4B	Washington (H)
				6B	Webster (H)
				4B	West Baton Rouge (H)
				6B	West Carroll (H)
				4B	West Feliciana (H)
				5A	Winn (H)

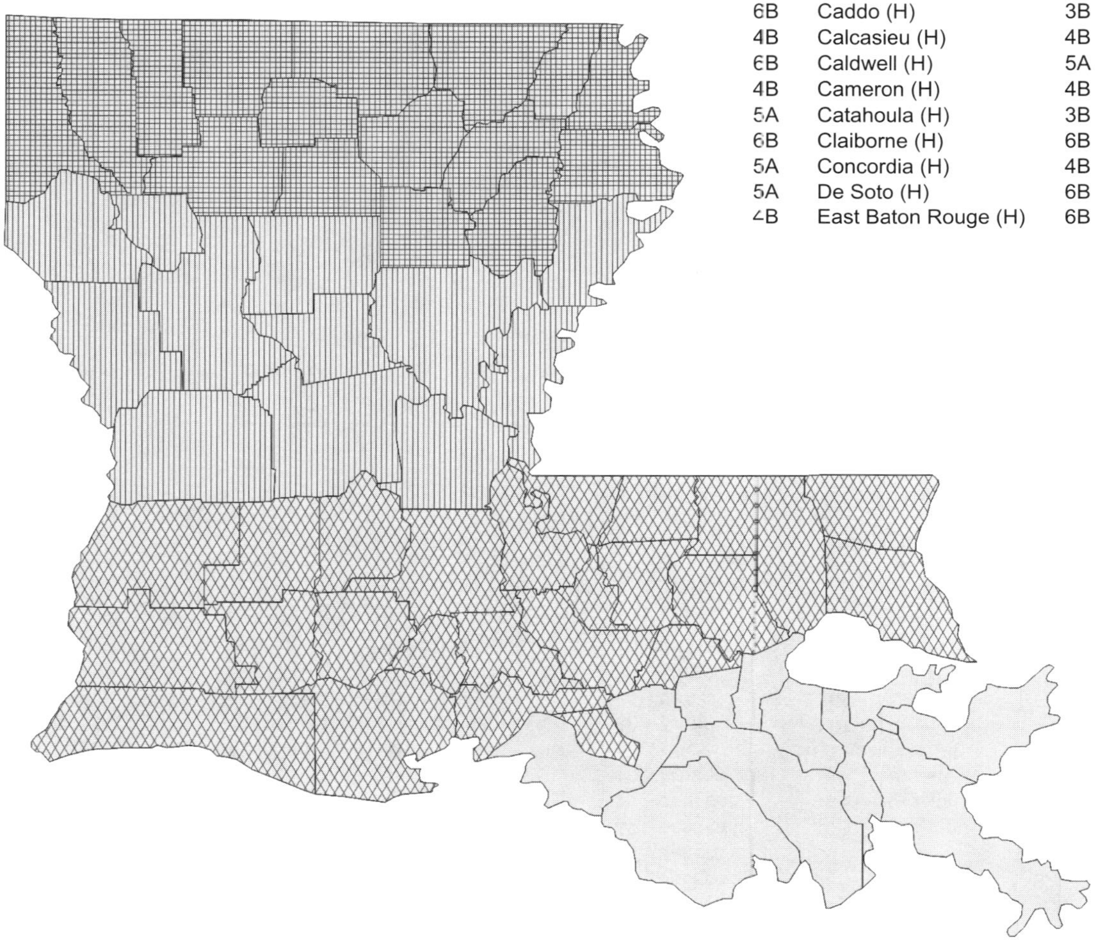

Zone 3B
Zone 4B
Zone 5A
Zone 6B

FIGURE 902.1(19)
LOUISIANA[a]

a. Counties identified with (H) shall be considered "hot and humid climate areas" for purposes of the application of Section 502.1.1.

CLIMATE MAPS

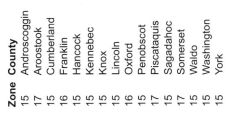

Zone	County
15	Androscoggin
17	Aroostook
15	Cumberland
16	Franklin
15	Hancock
15	Kennebec
15	Knox
15	Lincoln
16	Oxford
15	Penobscot
17	Piscataquis
15	Sagadahoc
17	Somerset
15	Waldo
15	Washington
15	York

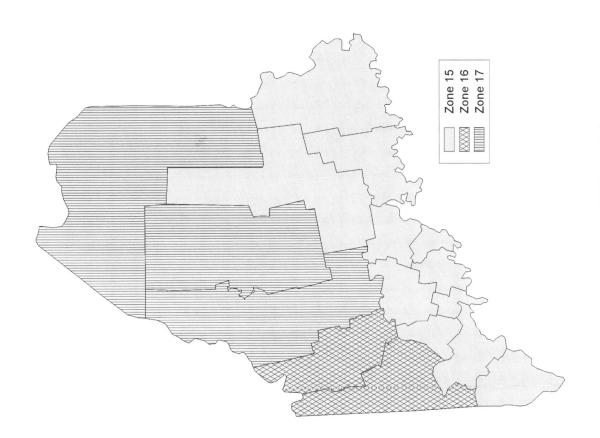

FIGURE 902.1(20)
MAINE

CLIMATE MAPS

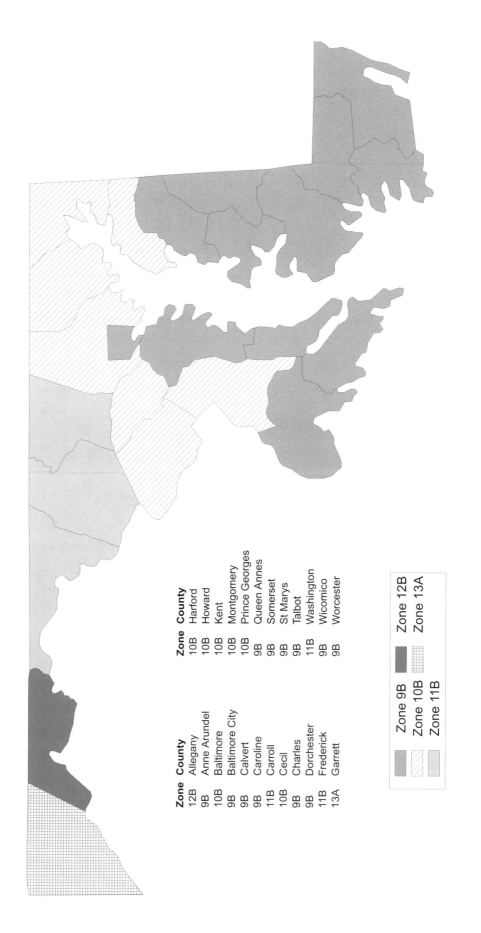

FIGURE 902.1(21)
MARYLAND

CLIMATE MAPS

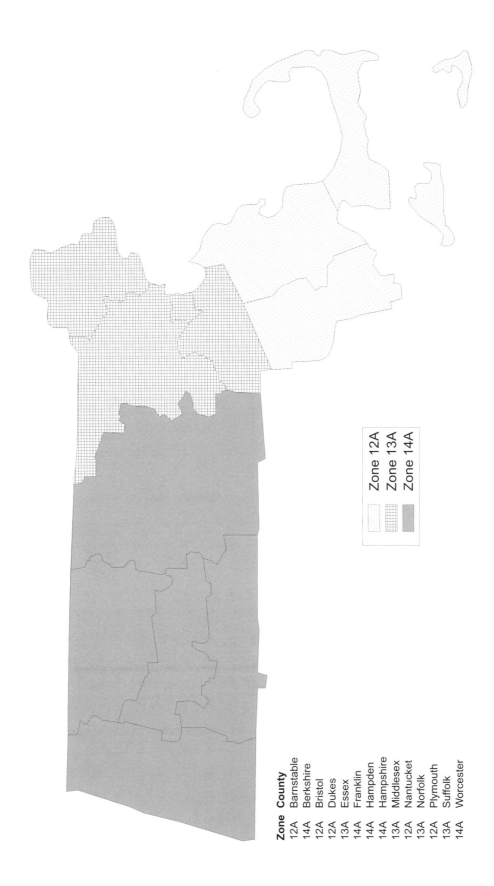

FIGURE 902.1(22)
MASSACHUSETTS

Zone	County
12A	Barnstable
14A	Berkshire
12A	Bristol
12A	Dukes
13A	Essex
14A	Franklin
14A	Hampden
14A	Hampshire
13A	Middlesex
12A	Nantucket
13A	Norfolk
12A	Plymouth
13A	Suffolk
14A	Worcester

CLIMATE MAPS

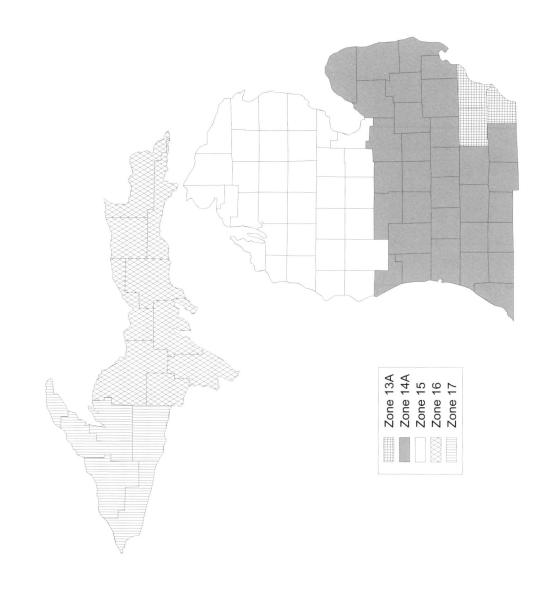

FIGURE 902.1(23)
MICHIGAN

Zone	County	Zone	County
15	Alcona	17	Keweenaw
16	Alger	15	Lake
14A	Allegan	14A	Lapeer
15	Alpena	15	Leelanau
15	Antrim	14A	Lenawee
15	Arenac	14A	Livingston
17	Baraga	16	Luce
14A	Barry	16	Mackinac
15	Bay	14A	Macomb
15	Benzie	15	Manistee
14A	Berrien	16	Marquette
14A	Branch	15	Mason
14A	Calhoun	15	Mecosta
14A	Cass	16	Menominee
15	Charlevoix	15	Midland
15	Cheboygan	15	Missaukee
16	Chippewa	13A	Monroe
15	Clare	14A	Montcalm
14A	Clinton	15	Montmorency
15	Crawford	14A	Muskegon
16	Delta	15	Newaygo
16	Dickinson	14A	Oakland
14A	Eaton	15	Oceana
15	Emmet	15	Ogemaw
14A	Genesee	17	Ontonagon
15	Gladwin	15	Osceola
17	Gogebic	15	Oscoda
15	Grand Traverse	15	Otsego
14A	Gratiot	14A	Ottawa
14A	Hillsdale	15	Presque Isle
17	Houghton	15	Roscommon
14A	Huron	14A	Saginaw
14A	Ingham	14A	Sanilac
14A	Ionia	16	Schoolcraft
15	Iosco	14A	Shiawassee
17	Iron	14A	St Clair
15	Isabella	14A	St Joseph
14A	Jackson	14A	Tuscola
14A	Kalamazoo	14A	Van Buren
15	Kalkaska	13A	Washtenaw
14A	Kent	13A	Wayne
		15	Wexford

CLIMATE MAPS

Zone	County	Zone	County
17	Aitkin	17	Marshall
16	Anoka	15	Martin
17	Becker	15	McLeod
17	Beltrami	16	Meeker
16	Benton	16	Mille Lacs
16	Big Stone	16	Morrison
15	Blue Earth	15	Mower
15	Brown	15	Murray
17	Carlton	15	Nicollet
15	Carver	15	Nobles
17	Cass	17	Norman
16	Chippewa	15	Olmsted
16	Chisago	17	Otter Tail
17	Clay	17	Pennington
17	Clearwater	16	Pine
17	Cook	15	Pipestone
15	Cottonwood	17	Polk
17	Crow Wing	16	Pope
15	Dakota	15	Ramsey
15	Dodge	17	Red Lake
16	Douglas	15	Redwood
15	Faribault	15	Renville
15	Fillmore	15	Rice
15	Freeborn	15	Rock
17	Goodhue	17	Roseau
16	Grant	15	Scott
15	Hennepin	16	Sherburne
15	Houston	15	Sibley
17	Hubbard	17	St Louis
16	Isanti	16	Stearns
17	Itasca	15	Steele
15	Jackson	16	Stevens
16	Kanabec	16	Swift
16	Kandiyohi	16	Todd
17	Kittson	16	Traverse
17	Koochiching	15	Wabasha
15	Lac Qui Parle	17	Wadena
17	Lake	15	Waseca
17	Lake Of The Woods	15	Washington
15	Le Sueur	15	Watonwan
15	Lincoln	17	Wilkin
15	Lyon	15	Winona
17	Mahnomen	16	Wright
		15	Yellow Medicine

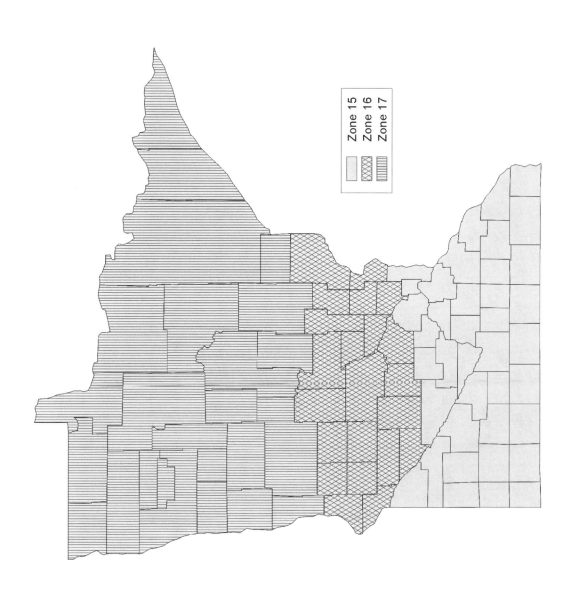

**FIGURE 902.1(24)
MINNESOTA**

CLIMATE MAPS

Zone	County	Zone	County
5A	Adams (H)	6B	Leflore (H)
7B	Alcorn	5A	Lincoln (H)
4B	Amite (H)	6B	Lowndes (H)
6B	Attala (H)	6B	Madison (H)
7B	Benton	4B	Marion (H)
6B	Bolivar (H)	7B	Marshall
6B	Calhoun (H)	6B	Monroe (H)
6B	Carroll (H)	6B	Montgomery (H)
6B	Chickasaw (H)	6B	Neshoba (H)
6B	Choctaw (H)	6B	Newton (H)
5A	Claiborne (H)	6B	Noxubee (H)
5A	Clarke (H)	6B	Oktibbeha (H)
6B	Clay (H)	7B	Panola
7B	Coahoma	4B	Pearl River (H)
5A	Copiah (H)	5A	Perry (H)
5A	Covington (H)	4B	Pike (H)
7B	De Soto	7B	Pontotoc
5A	Forrest (H)	7B	Prentiss
5A	Franklin (H)	7B	Quitman
4B	George (H)	6B	Rankin (H)
5A	Greene (H)	6B	Scott (H)
6B	Grenada (H)	5A	Sharkey (H)
4B	Hancock (H)	5A	Simpson (H)
4B	Harrison (H)	5A	Smith (H)
6B	Hinds (H)	4B	Stone (H)
6B	Holmes (H)	6B	Sunflower (H)
6B	Humphreys (H)	7B	Tallahatchie
6B	Issaquena (H)	7B	Tate
7B	Itawamba	7B	Tippah
4B	Jackson (H)	7B	Tishomingo
5A	Jasper (H)	7B	Tunica
5A	Jefferson (H)	7B	Union
5A	Jefferson Davis (H)	4B	Walthall (H)
5A	Jones (H)	6B	Warren (H)
6B	Kemper (H)	6B	Washington (H)
7B	Lafayette	5A	Wayne (H)
4B	Lamar (H)	6B	Webster (H)
6B	Lauderdale (H)	4B	Wilkinson (H)
5A	Lawrence (H)	6B	Winston (H)
6B	Leake (H)	7B	Yalobusha
7B	Lee	6B	Yazoo (H)

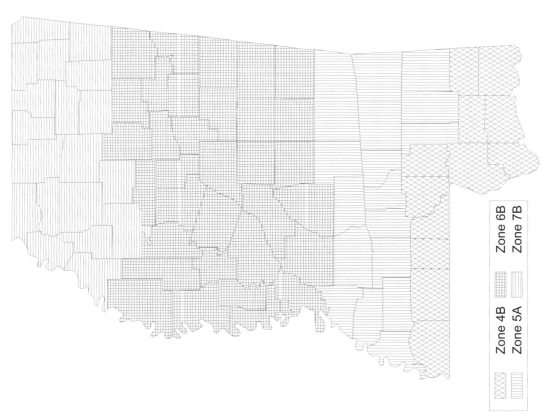

Zone 4B | Zone 6B
Zone 5A | Zone 7B

FIGURE 902.1(25)
MISSISSIPPI[a]

a. Counties identified with (H) shall be considered "hot and humid climate areas" for purposes of the application of Section 502.1.1.

CLIMATE MAPS

County	Zone	County	Zone	County	Zone	County	Zone	County	Zone	County	Zone	County	Zone
Adair	12B	Bates	11B	Caldwell	12B	Cass	11B	Clinton	12B	Gentry	13B	McDonald	9B	Reynolds	10B
Andrew	12B	Benton	11B	Callaway	11B	Cedar	11B	Cole	11B	Greene	10B	Mercer	13B	Ripley	9B
Atchison	13B	Bollinger	10B	Camden	11B	Chariton	12B	Cooper	11B	Grundy	12B	Miller	11B	Saline	11B
Audrain	12B	Boone	11B	Cape Girardeau	9B	Christian	10B	Crawford	10B	Harrison	13B	Mississippi	9B	Schuyler	13B
Barry	9B	Buchanan	12B	Carroll	12B	Clark	13B	Dade	10B	Henry	11B	Moniteau	11B	Scotland	13B
Barton	10B	Butler	9B	Carter	10B	Clay	11B	Dallas	10B	Hickory	11B	Monroe	12B	Scott	9B
								Daviess	12B	Holt	12B	Montgomery	11B	Shannon	10B
								De Kalb	12B	Howard	11B	Morgan	11B	Shelby	12B
								Dent	10B	Howell	9B	New Madrid	9B	St Charles	10B
								Douglas	10B	Iron	10B	Newton	9B	St Clair	11B
								Dunklin	9B	Jackson	11B	Nodaway	13B	St Francois	10B
								Franklin	10B	Jasper	9B	Oregon	9B	St Louis	10B
								Gasconade	11B	Jefferson	10B	Osage	11B	St Louis City	10B
										Johnson	11B	Ozark	9B	Ste Genevieve	10B
										Knox	12B	Pemiscot	9B	Stoddard	9B
										Laclede	10B	Perry	10B	Stone	9B
										Lafayette	11B	Pettis	11B	Sullivan	12B
										Lawrence	10B	Phelps	10B	Taney	9B
										Lewis	12B	Pike	12B	Texas	10B
										Lincoln	11B	Platte	11B	Vernon	11B
										Linn	12B	Polk	10B	Warren	11B
										Livingston	12B	Pulaski	10B	Washington	10B
										Macon	12B	Putnam	13B	Wayne	10B
										Madison	10B	Ralls	12B	Webster	10B
										Maries	11B	Randolph	12B	Worth	13B
										Marion	12B	Ray	11B	Wright	10B

Zone 9B
Zone 10B
Zone 11B
Zone 12B
Zone 13B

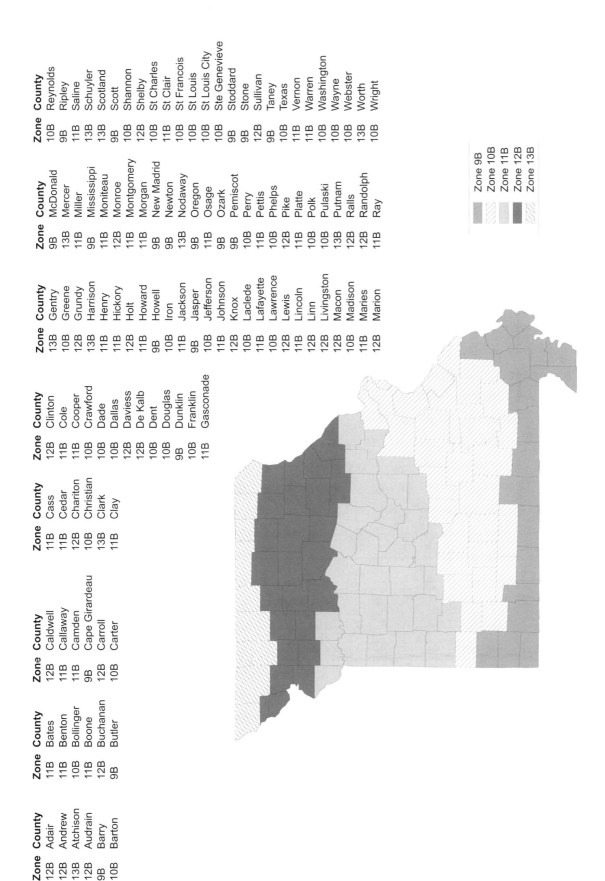

FIGURE 902.1(26)
MISSOURI

CLIMATE MAPS

Zone	County	Zone	County	Zone	County	Zone	County	Zone	County	Zone	County	Zone	County
15	Beaverhead	15	Custer	15	Garfield	15	Lewis And Clark	15	Missoula	16	Powell	16	Sheridan	16	Valley
15	Big Horn	16	Daniels	16	Glacier	16	Liberty	15	Musselshell	15	Prairie	16	Silver Bow	15	Wheatland
16	Blaine	15	Dawson	15	Golden Valley	15	Lincoln	15	Park	15	Ravalli	15	Stillwater	15	Wibaux
15	Broadwater	16	Deer Lodge	16	Granite	15	Madison	15	Petroleum	15	Richland	15	Sweet Grass	15	Yellowstone
15	Carbon	15	Fallon	16	Hill	15	McCone	16	Phillips	16	Roosevelt	15	Teton	15	Yellowstone National Park
15	Carter	15	Fergus	15	Jefferson	15	Meagher	16	Pondera	15	Rosebud	16	Toole		
15	Cascade	16	Flathead	15	Judith Basin	15	Mineral	15	Powder River	15	Sanders	15	Treasure		
15	Chouteau	15	Gallatin	15	Lake										

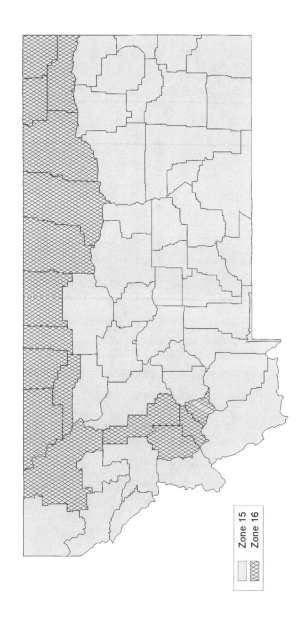

**FIGURE 902.1(27)
MONTANA**

CLIMATE MAPS

Zone	County	Zone	County	Zone	County	Zone	County	Zone	County	Zone	County
13B	Adams	14B	Cass	14B	Deuel	13B	Gosper	13B	Johnson	13B	Merrick	15	Sioux
14B	Antelope	14B	Cedar	14B	Dixon	14B	Grant	13B	Kearney	14B	Morrill	14B	Stanton
14B	Arthur	13B	Chase	13B	Dodge	14B	Greeley	14B	Keith	13B	Nance	13B	Thayer
14B	Banner	14B	Cherry	13B	Douglas	13B	Hall	14B	Keya Paha	13B	Nemaha	14B	Thomas
14B	Blaine	14B	Cheyenne	13B	Dundy	13B	Hamilton	14B	Kimball	13B	Nuckolls	14B	Thurston
14B	Boone	13B	Clay	13B	Fillmore	13B	Harlan	14B	Knox	13B	Otoe	14B	Valley
15	Box Butte	13B	Colfax	13B	Franklin	13B	Hayes	13B	Lancaster	13B	Pawnee	13B	Washington
14B	Boyd	14B	Cuming	13B	Frontier	13B	Hitchcock	14B	Lincoln	13B	Perkins	14B	Wayne
14B	Brown	14B	Custer	13B	Furnas	14B	Holt	14B	Logan	13B	Phelps	13B	Webster
13B	Buffalo	14B	Dakota	13B	Gage	14B	Hooker	14B	Loup	14B	Pierce	14B	Wheeler
14B	Burt	15	Dawes	14B	Garden	14B	Howard	14B	Madison	13B	Platte	13B	York
13B	Butler	13B	Dawson	14B	Garfield	13B	Jefferson	14B	McPherson				

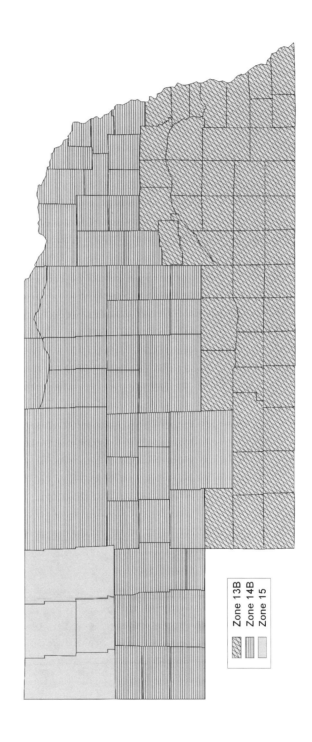

FIGURE 902.1(28)
NEBRASKA

CLIMATE MAPS

Zone	County
12B	Carson City
12B	Churchill
5B	Clark
13B	Douglas
15	Elko
12B	Esmeralda
15	Eureka
13B	Humboldt
13B	Lander
12B	Lincoln
13B	Lyon
12B	Mineral
12B	Nye
12B	Pershing
12B	Storey
12B	Washoe
15	White Pine

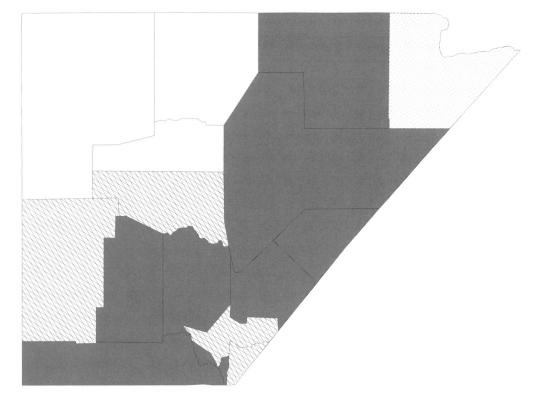

FIGURE 902.1(29)
NEVADA

CLIMATE MAPS

Zone	County
15	Belknap
15	Carroll
15	Cheshire
16	Coos
15	Grafton
15	Hillsborough
15	Merrimack
15	Rockingham
15	Strafford
15	Sullivan

☐ Zone 15
▨ Zone 16

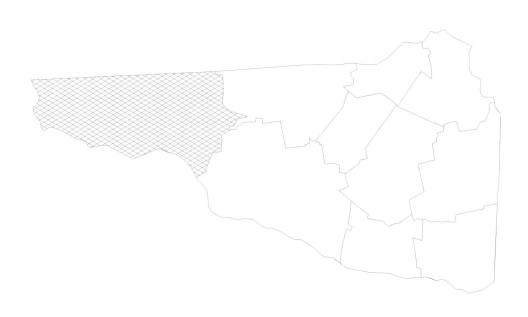

**FIGURE 902.1(30)
NEW HAMPSHIRE**

CLIMATE MAPS

FIGURE 902.1(31)
NEW JERSEY

CLIMATE MAPS

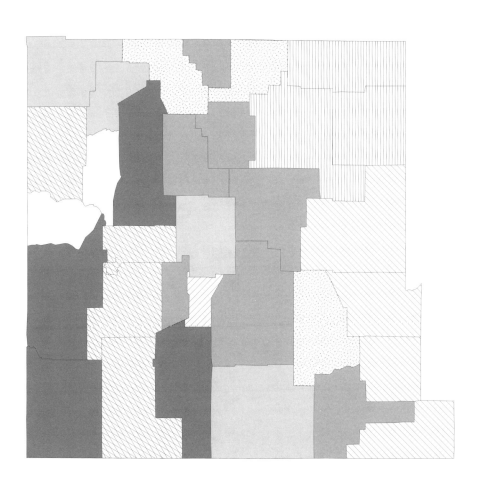

FIGURE 902.1(32)
NEW MEXICO

CLIMATE MAPS

Zone	County	Zone	County
14A	Albany	14A	Niagara
15	Allegany	15	Oneida
11B	Bronx	14A	Onondaga
15	Broome	14A	Ontario
15	Cattaraugus	12B	Orange
14A	Cayuga	14A	Orleans
14A	Chautauqua	14A	Oswego
15	Chemung	15	Otsego
15	Chenango	12B	Putnam
15	Clinton	10B	Queens
14A	Columbia	14A	Rensselaer
15	Cortland	11B	Richmond
15	Delaware	12B	Rockland
13A	Dutchess	14A	Saratoga
14A	Erie	14A	Schenectady
16	Essex	15	Schoharie
16	Franklin	15	Schuyler
15	Fulton	14A	Seneca
14A	Genesee	15	St Lawrence
14A	Greene	15	Steuben
16	Hamilton	11B	Suffolk
15	Herkimer	15	Sullivan
15	Jefferson	15	Tioga
10B	Kings	15	Tompkins
15	Lewis	15	Ulster
14A	Livingston	15	Warren
14A	Madison	15	Washington
14A	Monroe	14A	Wayne
14A	Montgomery	12B	Westchester
11B	Nassau	14A	Wyoming
10B	New York	14A	Yates

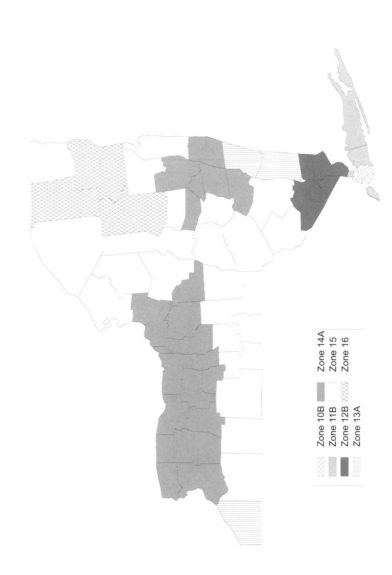

FIGURE 902.1(33)
NEW YORK

CLIMATE MAPS

Zone	County	Zone	County	Zone	County	Zone	County	Zone	County	Zone	County
8	Alamance	8	Caldwell	7A	Currituck	7A	Greene	7A	Lee	6B	New Hanover (H)	7A	Richmond	6B	Tyrrell (H)
8	Alexander	7A	Camden	6B	Dare (H)	8	Guilford	7A	Lenoir	7A	Northampton	7A	Robeson	7A	Union
11A	Alleghany	6B	Carteret (H)	8	Davidson	7A	Halifax	7A	Lincoln	6B	Onslow (H)	8	Rockingham	8	Vance
7A	Anson	8	Caswell	8	Davie	7A	Harnett	9B	Macon	8	Orange	7A	Rowan	7A	Wake
11A	Ashe	8	Catawba	6B	Duplin (H)	9B	Haywood	9B	Madison	6B	Pamlico (H)	7A	Rutherford	8	Warren
11A	Avery	8	Chatham	8	Durham	9B	Henderson	7A	Martin	7A	Pasquotank	6B	Sampson (H)	7A	Washington
6B	Beaufort (H)	9B	Cherokee	7A	Edgecombe	7A	Hertford	8	McDowell	6B	Pender (H)	7A	Scotland	11A	Watauga
7A	Bertie	7A	Chowan	8	Forsyth	7A	Hoke	7A	Mecklenburg	7A	Perquimans	7A	Stanly	7A	Wayne
6B	Bladen (H)	9B	Clay	8	Franklin	6B	Hyde (H)	11A	Mitchell	8	Person	9B	Stokes	9B	Wilkes
6B	Brunswick (H)	7A	Cleveland	8	Gaston	8	Iredell	7A	Montgomery	7A	Pitt	9B	Surry	7A	Wilson
9B	Buncombe	6B	Columbus (H)	7A	Gates	9B	Jackson	7A	Moore	7A	Polk	9B	Swain	8	Yadkin
8	Burke	6B	Craven (H)	9B	Graham	7A	Johnston	7A	Nash	8	Randolph	9B	Transylvania	11A	Yancey
7A	Cabarrus	7A	Cumberland	8	Granville	6B	Jones (H)								

Zone 6B
Zone 7A
Zone 8
Zone 9B
Zone 11A

a. Counties identified with (H) shall be considered "hot and humid climate areas" for purposes of the application of Section 502.1.1.

**FIGURE 902.1(34)
NORTH CAROLINA**[a]

CLIMATE MAPS

Zone	County	Zone	County	Zone	County	Zone	County	Zone	County	Zone	County	Zone	County
16	Adams	16	Burleigh	16	Emmons	17	Kidder	16	Mercer	17	Ramsey	17	Sheridan	17	Towner
17	Barnes	17	Cass	17	Foster	16	La Moure	16	Morton	16	Ransom	16	Sioux	17	Traill
17	Benson	17	Cavalier	16	Golden Valley	16	Logan	17	Mountrail	17	Renville	16	Slope	17	Walsh
16	Billings	16	Dickey	17	Grand Forks	17	McHenry	17	Nelson	16	Richland	16	Stark	17	Ward
17	Bottineau	17	Divide	16	Grant	16	McIntosh	16	Oliver	17	Rolette	17	Steele	17	Wells
16	Bowman	16	Dunn	17	Griggs	16	McKenzie	17	Pembina	16	Sargent	17	Stutsman	17	Williams
17	Burke	17	Eddy	16	Hettinger	17	McLean	17	Pierce						

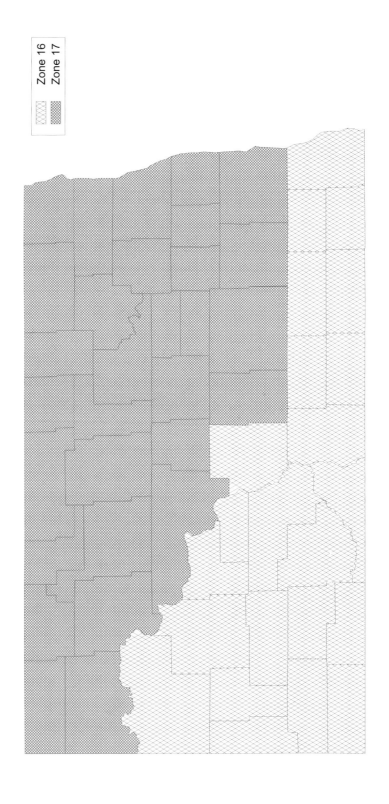

FIGURE 902.1(35)
NORTH DAKOTA

CLIMATE MAPS

Zone	County	Zone	County
11B	Adams	12B	Licking
13B	Allen	13A	Logan
13A	Ashland	13A	Lorain
13A	Ashtabula	14A	Lucas
11B	Athens	12B	Madison
13B	Auglaize	13A	Mahoning
12A	Belmont	13A	Marion
11B	Brown	13A	Medina
12B	Butler	11B	Meigs
13A	Carroll	13B	Mercer
13A	Champaign	13A	Miami
13A	Clark	12A	Monroe
11B	Clermont	12B	Montgomery
12B	Clinton	12A	Morgan
13A	Columbiana	13A	Morrow
12B	Coshocton	12B	Muskingum
13A	Crawford	12A	Noble
13A	Cuyahoga	13A	Ottawa
13A	Darke	14A	Paulding
14A	Defiance	12A	Perry
13A	Delaware	12B	Pickaway
13A	Erie	11B	Pike
12A	Fairfield	13A	Portage
12B	Fayette	12B	Preble
12B	Franklin	13B	Putnam
14A	Fulton	13A	Richland
11B	Gallia	12B	Ross
13A	Geauga	13A	Sandusky
12B	Greene	11B	Scioto
12B	Guernsey	13A	Seneca
11B	Hamilton	13A	Shelby
13A	Hancock	13A	Stark
13A	Hardin	13A	Summit
13A	Harrison	13A	Trumbull
14A	Henry	13A	Tuscarawas
11B	Highland	13B	Union
12A	Hocking	13B	Van Wert
13A	Holmes	11B	Vinton
13A	Huron	12B	Warren
11B	Jackson	11B	Washington
13A	Jefferson	13A	Wayne
13A	Knox	14A	Williams
13A	Lake	14A	Wood
11B	Lawrence	13A	Wyandot

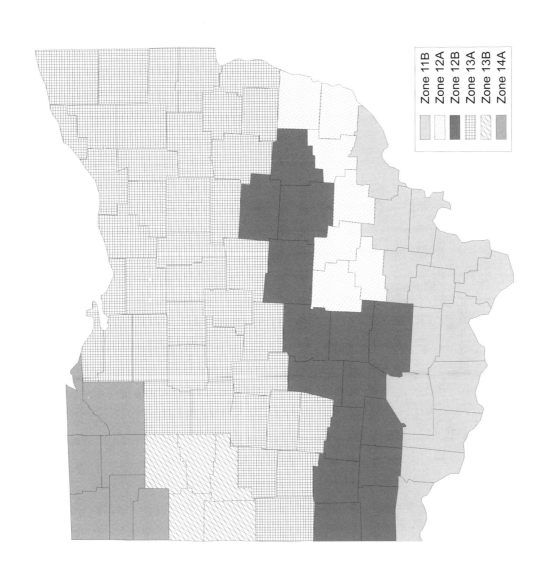

FIGURE 902.1(36)
OHIO

CLIMATE MAPS

Zone	County	Zone	County	Zone	County	Zone	County	Zone	County	Zone	County	Zone	County
8	Adair	8	Cherokee	8	Delaware	7B	Haskell	7B	Lincoln	7B	Muskogee	8	Payne	7B	Stephens
9B	Alfalfa	6B	Choctaw (H)	9B	Dewey	7B	Hughes	8	Logan	8	Noble	7B	Pittsburg	10B	Texas
7B	Atoka	10B	Cimarron	9B	Ellis	7B	Jackson	6B	Love (H)	9B	Nowata	7B	Pontotoc	7B	Tillman
10B	Beaver	7B	Cleveland	8	Garfield	6B	Jefferson (H)	9B	Major	7B	Okfuskee	7B	Pottawatomie	8	Tulsa
8	Beckham	7B	Coal	7B	Garvin	6B	Johnston (H)	6B	Marshall (H)	8	Oklahoma	6B	Pushmataha (H)	8	Wagoner
8	Blaine	7B	Comanche	7B	Grady	8	Kay	8	Mayes	8	Okmulgee	9B	Roger Mills	9B	Washington
7B	Bryan	7B	Cotton	9B	Grant	8	Kingfisher	7B	McClain	8	Osage	9B	Rogers	8	Washita
8	Caddo	9B	Craig	7B	Greer	7B	Kiowa	7B	McCurtain	9B	Ottawa	7B	Seminole	9B	Woods
8	Canadian	8	Creek	7B	Harmon	7B	Latimer	7B	McIntosh	8	Pawnee	7B	Sequoyah	9B	Woodward
6B	Carter (H)	8	Custer	9B	Harper	7B	Le Flore	7B	Murray						

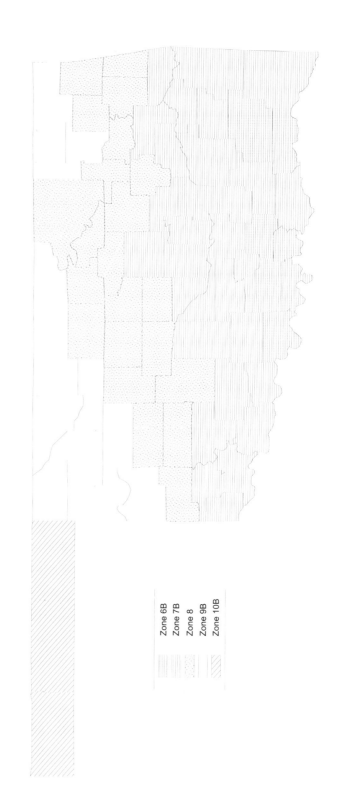

FIGURE 902.1(37)
OKLAHOMA[a]

a. Counties identified with (H) shall be considered "hot and humid climate areas" for purposes of the application of Section 502.1.1.

CLIMATE MAPS

Zone	County	
15	Baker	
10A	Benton	
10A	Clackamas	
11A	Clatsop	
11A	Columbia	
9A	Coos	
14A	Crook	
9A	Curry	
14A	Deschutes	
9A	Douglas	
12A	Gilliam	
15	Grant	
15	Harney	
12A	Hood River	
11A	Jackson	
13B	Jefferson	
9A	Josephine	
14A	Klamath	
15	Lake	
10A	Lane	
11A	Lincoln	
12B	Linn	
10A	Malheur	
12A	Marion	
10A	Morrow	
10A	Multnomah	
13B	Polk	
11A	Sherman	
12A	Tillamook	
13B	Umatilla	
15	Union	
13B	Wallowa	
10A	Wasco	
13B	Washington	
10A	Wheeler	
	Yamhill	

Zone 9A
Zone 10A
Zone 11A
Zone 12A
Zone 12B
Zone 13B
Zone 14A
Zone 15

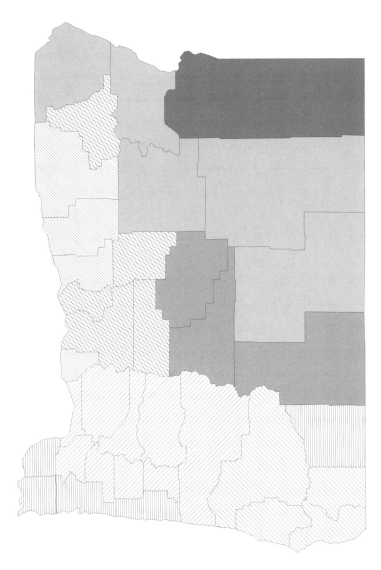

FIGURE 902.1(38)
OREGON

CLIMATE MAPS

Zone	County	Zone	County	Zone	County	Zone	County	Zone	County	Zone	County
11B	Adams	13B	Blair	13B	Carbon	13B	Columbia	14A	Erie	12B	Huntingdon
12A	Allegheny	15	Bradford	13B	Centre	14A	Crawford	12A	Fayette	13B	Indiana
13B	Armstrong	11B	Bucks	11B	Chester	12B	Cumberland	15	Forest	15	Jefferson
12A	Beaver	14A	Butler	14A	Clarion	12B	Dauphin	11B	Franklin	12B	Juniata
13B	Bedford	13B	Cambria	15	Clearfield	10B	Delaware	12B	Fulton	14A	Lackawanna
12B	Berks	15	Cameron	13B	Clinton	15	Elk	12A	Greene	11B	Lancaster

Zone	County	Zone	County
14A	Lawrence	14A	Mercer
12B	Lebanon	12B	Mifflin
12B	Lehigh	13B	Monroe
13B	Luzerne	11B	Montgomery
13B	Lycoming	13B	Montour
15	McKean	12B	Northampton
		13B	Northumberland
		12B	Perry
		10B	Philadelphia
		13B	Pike
		15	Potter
		13B	Schuylkill
		13B	Snyder
		13B	Somerset
		14A	Sullivan
		15	Susquehanna
		15	Tioga
		13B	Union
		14A	Venango
		14A	Warren
		12A	Washington
		15	Wayne
		13B	Westmoreland
		14A	Wyoming
		11B	York

Zone 10B
Zone 11B
Zone 12A
Zone 12B
Zone 13B
Zone 14A
Zone 15

FIGURE 902.1(39)
PENNSYLVANIA

CLIMATE MAPS

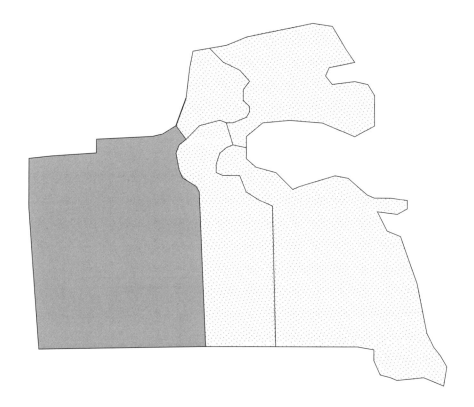

FIGURE 902.1(40)
RHODE ISLAND

CLIMATE MAPS

Zone	County	Zone	County
7A	Abbeville	7A	Greenwood
6B	Aiken (H)	5A	Hampton (H)
5A	Allendale (H)	5A	Horry (H)
7A	Anderson	5A	Jasper (H)
5A	Bamberg (H)	7A	Kershaw
5A	Barnwell (H)	7A	Lancaster
5A	Beaufort (H)	7A	Laurens
5A	Berkeley (H)	6B	Lee (H)
6B	Calhoun (H)	6B	Lexington (H)
5A	Charleston (H)	6B	Marion (H)
7A	Cherokee	6B	Marlboro (H)
7A	Chester	6B	McCormick (H)
7A	Chesterfield	6B	Newberry (H)
6B	Clarendon (H)	7A	Oconee
5A	Colleton (H)	6B	Orangeburg (H)
6B	Darlington (H)	7A	Pickens
6B	Dillon (H)	6B	Richland (H)
5A	Dorchester (H)	6B	Saluda (H)
6B	Edgefield (H)	7A	Spartanburg
7A	Fairfield	6B	Sumter (H)
6B	Florence (H)	7A	Union
5A	Georgetown (H)	6B	Williamsburg (H)
7A	Greenville	7A	York

Zone 5A
Zone 6B
Zone 7A

**FIGURE 902.1(41)
SOUTH CAROLINA**[a]

a. Counties identified with (H) shall be considered "hot and humid climate areas" for purposes of the application of Section 502.1.1.

CLIMATE MAPS

Zone	County
15	Aurora
15	Beadle
14B	Bennett
14B	Bon Homme
16	Brookings
16	Brown

Zone	County
15	Brule
15	Buffalo
15	Butte
15	Campbell
14B	Charles Mix
16	Clark

Zone	County
14B	Clay
16	Codington
15	Corson
15	Custer
15	Davison
16	Day

Zone	County
16	Deuel
15	Dewey
14B	Douglas
15	Edmunds
15	Fall River
15	Faulk

Zone	County
16	Grant
14B	Gregory
15	Haakon
16	Hamlin
15	Hand
15	Hanson

Zone	County
15	Harding
15	Hughes
14B	Hutchinson
15	Hyde
14B	Jackson
15	Jerauld

Zone	County
15	Jones
15	Kingsbury
15	Lake
15	Lawrence
15	Lincoln
15	Lyman

Zone	County
16	Marshall
15	McCook
16	McPherson
15	Meade
14B	Mellette
15	Miner
15	Minnehaha
15	Moody
15	Pennington
15	Perkins
15	Potter
16	Roberts
15	Sanborn
15	Shannon
15	Spink
15	Stanley
15	Sully
14B	Todd
14B	Tripp
15	Turner
14B	Union
15	Walworth
14B	Yankton
15	Ziebach

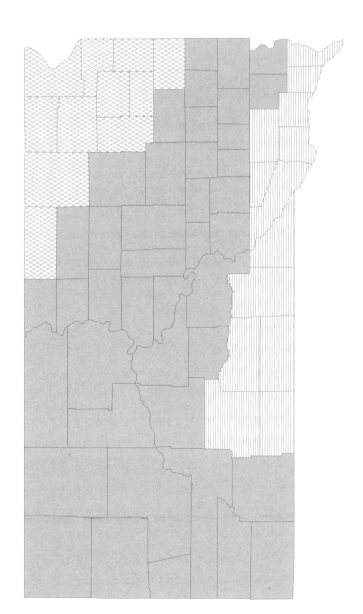

FIGURE 902.1(42)
SOUTH DAKOTA

CLIMATE MAPS

Zone	County	Zone	County	Zone	County	Zone	County	Zone	County	Zone	County	Zone	County
9B	Anderson	10B	Claiborne	10B	Fentress	9B	Hawkins	8	Lauderdale	8	Meigs	9B	Roane	9B	Trousdale
8	Bedford	9B	Clay	8	Franklin	8	Haywood	8	Lawrence	8	Monroe	9B	Robertson	10B	Unicoi
9B	Benton	9B	Cocke	9B	Gibson	8	Henderson	8	Lewis	9B	Montgomery	8	Rutherford	9B	Union
8	Bledsoe	8	Coffee	8	Giles	9B	Henry	8	Lincoln	8	Moore	10B	Scott	9B	Van Buren
8	Blount	9B	Crockett	9B	Grainger	9B	Hickman	8	Loudon	10B	Morgan	8	Sequatchie	9B	Warren
8	Bradley	9B	Cumberland	9B	Greene	9B	Houston	9B	Macon	9B	Obion	9B	Sevier	9B	Washington
10B	Campbell	8	Davidson	9B	Grundy	9B	Humphreys	8	Madison	9B	Overton	7B	Shelby (H)	8	Wayne
9B	Cannon	9B	De Kalb	9B	Hamblen	9B	Jackson	8	Marion	8	Perry	9B	Smith	9B	Weakley
9B	Carroll	8	Decatur	8	Hamilton	9B	Jefferson	8	Marshall	10B	Pickett	9B	Stewart	9B	White
10B	Carter	9B	Dickson	10B	Hancock	10B	Johnson	8	Maury	8	Polk	9B	Sullivan	8	Williamson
9B	Cheatham	8	Dyer	8	Hardeman	8	Knox	8	McMinn	9B	Putnam	9B	Sumner	9B	Wilson
8	Chester	7B	Fayette (H)	8	Hardin	9B	Lake	8	McNairy	8	Rhea	8	Tipton		

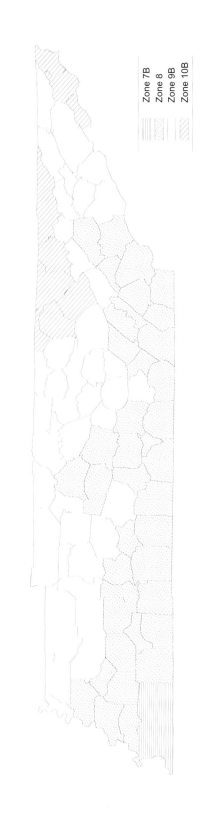

a. Counties identified with (H) shall be considered "hot and humid climate areas" for purposes of the application of Section 502.1.1.

FIGURE 902.1(43)
TENNESSEE[a]

CLIMATE MAPS

Zone	County	Zone	County	Zone	County	Zone	County	Zone	County	Zone	County
5A	Anderson (H)	6B	Callahan	4B	Comal (H)	6B	Denton	7B	Gaines	7B	Henderson (H)	5B	Mason (H)	5B	San Saba (H)
6B	Andrews	2B	Cameron (H)	5B	Comanche (H)	7B	Dickens	3B	Galveston (H)	2B	Hidalgo (H)	3B	Matagorda (H)	5B	Schleicher (H)
5A	Angelina (H)	6B	Camp	5B	Concho (H)	3C	Dimmit (H)	7B	Garza	5B	Hill (H)	3C	Maverick (H)	7B	Scurry
3B	Aransas (H)	9B	Carson	6B	Cooke	8	Donley	5A	Gillespie (H)	8	Hockley	5B	McCulloch (H)	6B	Shackelford
7B	Archer	6B	Cass	5B	Coryell (H)	3C	Duval (H)	6B	Glasscock	5B	Hood (H)	5B	McLennan (H)	5A	Shelby (H)
9B	Armstrong	9B	Castro	7B	Cottle	6B	Eastland	3B	Goliad (H)	6B	Hopkins	3C	McMullen (H)	9B	Sherman
3C	Atascosa (H)	4B	Chambers (H)	5B	Crane (H)	6B	Ector	4B	Gonzales (H)	5A	Houston (H)	4B	Medina (H)	5B	Smith (H)
4B	Austin (H)	5A	Cherokee (H)	5B	Crockett (H)	5A	Edwards (H)	9B	Gray	6B	Howard	5B	Menard (H)	5B	Somervell (H)
9B	Bailey	7B	Childress	7B	Crosby	6B	El Paso	6B	Grayson	6B	Hudspeth	6B	Midland (H)	2B	Starr (H)
5A	Bandera (H)	7B	Clay	6B	Culberson	5B	Ellis (H)	6B	Gregg	6B	Hunt	4B	Milam (H)	6B	Stephens
4B	Bastrop (H)	8	Cochran	5B	Dallam	6B	Erath	4B	Grimes (H)	9B	Hutchinson	5B	Mills (H)	6B	Sterling
7B	Baylor	6B	Coke	5B	Dallas (H)	5B	Falls (H)	4B	Guadalupe (H)	5B	Irion (H)	6B	Mitchell	7B	Stonewall
3B	Bee (H)	5B	Coleman (H)	7B	Dawson	6B	Fannin	8	Hale	6B	Jack	4B	Montague	5A	Sutton (H)
5B	Bell (H)	6B	Collin	3C	De Witt (H)	4B	Fayette (H)	8	Hall	3B	Jackson (H)	9B	Montgomery (H)	8	Swisher
4B	Bexar (H)	7B	Collingsworth	9B	Deaf Smith	6B	Fisher	5B	Hamilton (H)	5A	Jasper (H)	6B	Moore	5B	Tarrant (H)
5A	Blanco (H)	4B	Colorado (H)	6B	Delta	9B	Floyd	9B	Hansford	6B	Jeff Davis	6B	Morris	6B	Taylor
7B	Borden					7B	Foard	7B	Hardeman	4B	Jefferson (H)	7B	Motley	5A	Terrell (H)
5B	Bosque (H)					4B	Fort Bend (H)	4B	Hardin (H)	2B	Jim Hogg (H)	5A	Nacogdoches (H)	7B	Terry
6B	Bowie					6B	Franklin	4B	Harris (H)	3C	Jim Wells (H)	5B	Navarro (H)	6B	Throckmorton
3B	Brazoria (H)					5B	Freestone (H)	6B	Harrison	5B	Johnson (H)	5A	Newton (H)	6B	Titus
4B	Brazos (H)					3C	Frio (H)	9B	Hartley	6B	Jones	6B	Nolan	5B	Tom Green (H)
5A	Brewster (H)							6B	Haskell	3C	Karnes (H)	3B	Nueces (H)	5A	Travis (H)
8	Briscoe							5B	Hays (H)	6B	Kaufman	9B	Ochiltree	5A	Trinity (H)
2B	Brooks (H)							9B	Hemphill	5A	Kendall (H)	9B	Oldham	5A	Tyler (H)
5B	Brown (H)									2B	Kenedy (H)	4B	Orange (H)	6B	Upshur
4B	Burleson (H)									7B	Kent	5B	Palo Pinto	5B	Upton (H)
5A	Burnet (H)									5A	Kerr (H)	5A	Panola (H)	4B	Uvalde (H)
4B	Caldwell (H)									5A	Kimble (H)	6B	Parker	4B	Val Verde (H)
3B	Calhoun (H)									7B	King	9B	Parmer	6B	Van Zandt
										4B	Kinney (H)	5A	Pecos (H)	3B	Victoria (H)
										2B	Kleberg (H)	5A	Polk (H)	4B	Walker (H)
										7B	Knox	9B	Potter	4B	Waller (H)
										3C	La Salle (H)	5A	Presidio (H)	6B	Ward
										6B	Lamar	6B	Rains	4B	Washington (H)
										8	Lamb	9B	Randall	3C	Webb (H)
										5B	Lampasas (H)	5B	Reagan (H)	3B	Wharton (H)
										4B	Lavaca (H)	5A	Real (H)	9B	Wheeler
										4B	Lee (H)	6B	Red River	7B	Wichita
										5B	Leon (H)	6B	Reeves	7B	Wilbarger
										4B	Liberty (H)	3B	Refugio (H)	2B	Willacy (H)
										5B	Limestone (H)	9B	Roberts	5B	Williamson (H)
										9B	Lipscomb	4B	Robertson (H)	4B	Wilson (H)
										3C	Live Oak (H)	6B	Rockwall	6B	Winkler
										5B	Llano (H)	5B	Runnels (H)	6B	Wise
										6B	Loving	5B	Rusk (H)	6B	Wood
										7B	Lubbock	5A	Sabine (H)	8	Yoakum
										7B	Lynn	5A	San Augustine (H)	6B	Young
										4B	Madison (H)	4B	San Jacinto (H)	2B	Zapata (H)
										6B	Marion	3C	San Patricio (H)	3C	Zavala (H)
										6B	Martin				

Zone 2B
Zone 3B
Zone 3C
Zone 4B
Zone 5A
Zone 5B
Zone 6B
Zone 7B
Zone 8
Zone 9B

**FIGURE 902.1(44)
TEXAS[a]**

a. Counties identified with (H) shall be considered "hot and humid climate areas" for purposes of the application of Section 502.1.1.

CLIMATE MAPS

Zone	County
14B	Beaver
12B	Box Elder
15	Cache
14B	Carbon
15	Daggett
12B	Davis
15	Duchesne
14B	Emery
14B	Garfield
10B	Grand
12B	Iron
12B	Juab
10B	Kane
13B	Millard
15	Morgan
13B	Piute
15	Rich
12B	Salt Lake
13B	San Juan
14B	Sanpete
13B	Sevier
15	Summit
12B	Tooele
15	Uintah
12B	Utah
15	Wasatch
10B	Washington
14B	Wayne
12B	Weber

Zone 10B
Zone 12B
Zone 13B
Zone 14B
Zone 15

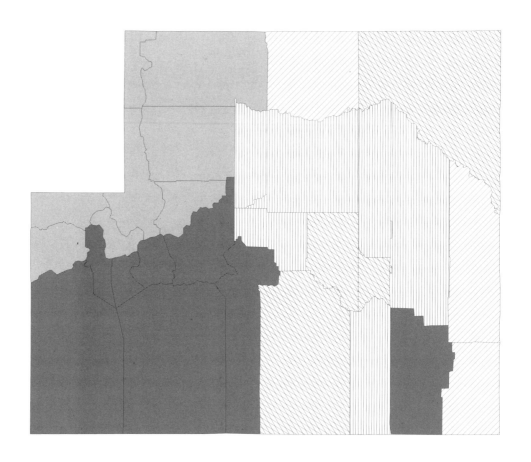

FIGURE 902.1(45)
UTAH

CLIMATE MAPS

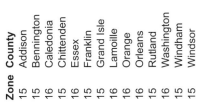

Zone	County
15	Addison
15	Bennington
16	Caledonia
15	Chittenden
16	Essex
15	Franklin
15	Grand Isle
16	Lamoille
16	Orange
16	Orleans
15	Rutland
16	Washington
15	Windham
15	Windsor

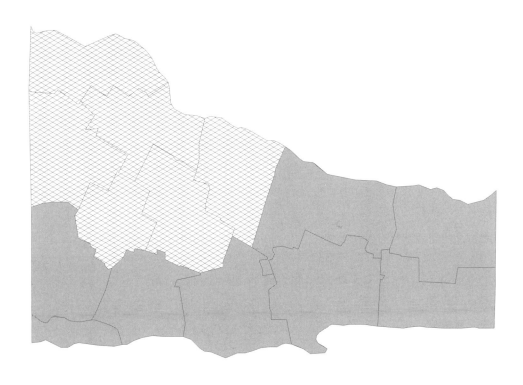

FIGURE 902.1(46)
VERMONT

CLIMATE MAPS

Zone	County	Zone	County	Zone	County	Zone	County	Zone	County	Zone	County
8	Accomack	9B	Campbell	11B	Floyd	8	Isle Of Wight	10B	Montgomery	10B	Orange
9B	Albemarie	9B	Caroline	9B	Fluvanna	8	James City	11B	Nansemond	11B	Page
10B	Alleghany	11B	Carroll	10B	Franklin	9B	King And Queen	10B	Nelson	10B	Patrick
9B	Amelia	8	Charles City	11B	Frederick	9B	King George	8	New Kent	9B	Pittsylvania
9B	Amherst	9B	Charlotte	10B	Giles	9B	King William	9B	Northampton	9B	Powhatan
9B	Appomattox	8	Chesterfield	8	Gloucester	8	Lancaster	9B	Northumberland	9B	Prince Edward
10B	Arlington	11B	Clarke	9B	Goochland	10B	Lee	9B	Nottoway	9B	Prince George
11B	Augusta	10B	Craig	11B	Grayson	10B	Loudoun			8	
11B	Bath	10B	Culpeper	10B	Greene	9B	Louisa				
9B	Bedford	8	Cumberland	8	Greensville	9B	Lunenburg				
11B	Bland	10B	Dickenson	9B	Halifax	11B	Madison				
9B	Botetourt	8	Dinwiddie	9B	Hanover	8	Mathews				
8	Brunswick	9B	Essex	8	Henrico	9B	Mecklenburg				
10B	Buchanan	10B	Fairfax	9B	Henry	8	Middlesex				
9B	Buckingham	10B	Fauquier	11B	Highland						

Zone	County
10B	Prince William
11B	Pulaski
11B	Rappahannock
8	Richmond
9B	Roanoke
11B	Rockbridge
10B	Rockingham
8	Russell

Zone	County
10B	Scott
11B	Shenandoah
11B	Smyth
8	Southampton
10B	Spotsylvania
10B	Stafford
8	Surry
8	Sussex
11B	Tazewell
11B	Warren
11B	Washington
8	Westmoreland
10B	Wise
11B	Wythe
8	York

Independent Cities

Zone	City	Zone	City	Zone	City	Zone	City	Zone	City	Zone	City
10B	Alexandria	10B	Clifton Forge	10B	Fairfax	8	Hampton	10B	Manassas	10B	Norton
9B	Bedford	8	Colonial Heights	10B	Falls Church	11B	Harrisonburg	10B	Manassas Park	8	Petersburg
11B	Bristol	10B	Covington	8	Franklin	8	Hopewell	10B	Martinsville	8	Poquoson
9B	Buena Vista	9B	Danville	10B	Fredericksburg	9B	Lexington	8	Newport News	8	Portsmouth
9B	Charlottesville	8	Emporia	11B	Galax	9B	Lynchburg	8	Norfolk	11B	Radford
8	Chesapeake										

Zone	City
8	Richmond
9B	Roanoke
9B	Salem
9B	South Boston
11B	Staunton

Zone	City
8	Suffolk
8	Virginia Beach
11B	Waynesboro
8	Williamsburg
11B	Winchester

**FIGURE 902.1(47)
VIRGINIA**

CLIMATE MAPS

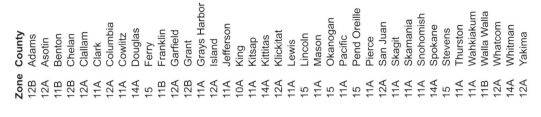

Zone	County
12B	Adams
12A	Asotin
11B	Benton
12B	Chelan
12A	Clallam
11A	Clark
12A	Columbia
11A	Cowlitz
14A	Douglas
15	Ferry
11B	Franklin
12A	Garfield
12B	Grant
11A	Grays Harbor
12A	Island
11A	Jefferson
10A	King
11A	Kitsap
14A	Kittitas
12A	Klickitat
11A	Lewis
15	Lincoln
11A	Mason
15	Okanogan
11A	Pacific
15	Pend Oreille
11A	Pierce
12A	San Juan
11A	Skagit
11A	Skamania
11A	Snohomish
14A	Spokane
15	Stevens
11A	Thurston
11A	Wahkiakum
11B	Walla Walla
12A	Whatcom
14A	Whitman
12A	Yakima

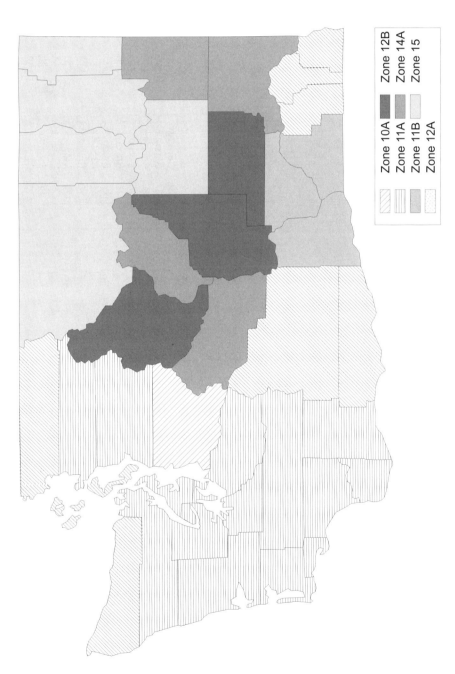

FIGURE 902.1(48)
WASHINGTON

CLIMATE MAPS

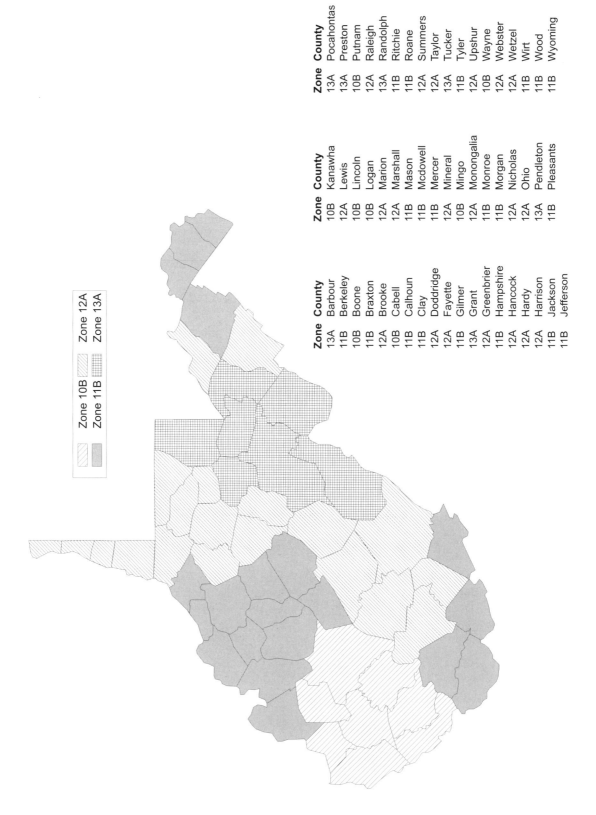

FIGURE 902.1(49)
WEST VIRGINIA

CLIMATE MAPS

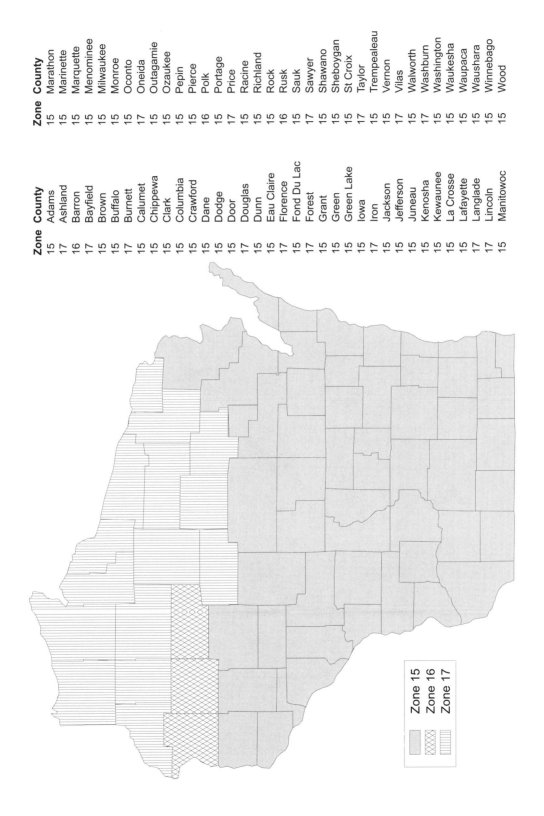

FIGURE 902.1(50)
WISCONSIN

Zone	County	Zone	County
15	Adams	15	Marathon
17	Ashland	15	Marinette
16	Barron	15	Marquette
17	Bayfield	15	Menominee
15	Brown	15	Milwaukee
15	Buffalo	15	Monroe
17	Burnett	15	Oconto
15	Calumet	17	Oneida
15	Chippewa	15	Outagamie
15	Clark	15	Ozaukee
15	Columbia	15	Pepin
15	Crawford	15	Pierce
15	Dane	16	Polk
15	Dodge	15	Portage
15	Door	17	Price
17	Douglas	15	Racine
15	Dunn	15	Richland
15	Eau Claire	15	Rock
17	Florence	16	Rusk
15	Fond Du Lac	15	Sauk
17	Forest	17	Sawyer
15	Grant	15	Shawano
15	Green	15	Sheboygan
15	Green Lake	15	St Croix
15	Iowa	17	Taylor
17	Iron	15	Trempealeau
15	Jackson	15	Vernon
15	Jefferson	17	Vilas
15	Juneau	15	Walworth
15	Kenosha	17	Washburn
15	Kewaunee	15	Washington
15	La Crosse	15	Waukesha
15	Lafayette	15	Waupaca
17	Langlade	15	Waushara
17	Lincoln	15	Winnebago
15	Manitowoc	15	Wood

CLIMATE MAPS

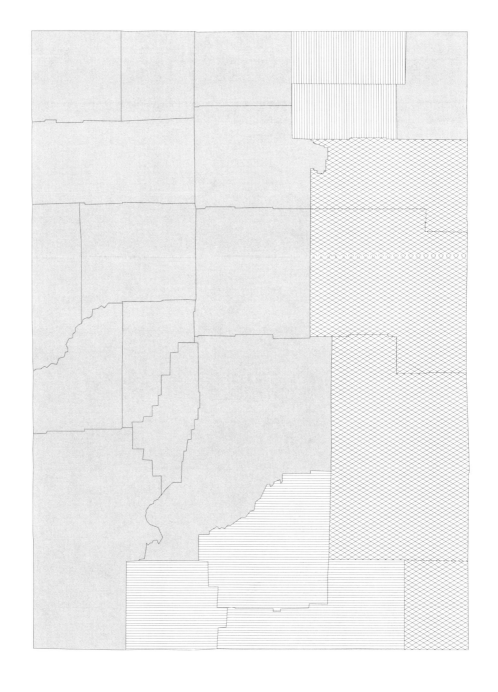

**FIGURE 902.1(51)
WYOMING**

CHAPTER 10
REFERENCED STANDARDS

This chapter lists the standards that are referenced in various sections of this document. The standards are listed herein by the promulgating agency of the standard, the standard identification, the effective date and title, and the section or sections of this document that reference the standard. The application of the referenced standards shall be as specified in Section 107.

AAMA

American Architectural Manufacturers Association
1827 Walden Office Square
Suite 104
Schaumburg, IL 60173-4268

Standard reference number	Title	Referenced in code section number
101/I.S.2—97	Voluntary Specifications for Aluminum, Vinyl (PVC) and Wood Windows and Glass Doors	Table 502.1.4.1, 601.3.2.2, 802.3.1
101/I.S.2/NAFS—02	Voluntary Performance Specification for Windows, Skylights and Glass Doors	802.3.1

AMCA

Air Movement and Control Association International
30 West University Drive
Arlington Heights, IL 60004-1806

Standard reference number	Title	Referenced in code section number
500—89	Test Methods for Louvers, Dampers, and Shutters	802.3.4

ANSI

American National Standards Institute
25 West 43rd Street
Fourth Floor
New York, NY 10036

Standard reference number	Title	Referenced in code section number
Z21.10.3—98	Gas Water Heaters, Volume III - Storage Water Heaters with Input Ratings Above 75,000 Btu per Hour, Circulating Tank and Instantaneous—with Addenda Z21.10.3a-99	Table 504.2.1, Table 804.2
Z21.13—99	Gas-Fired Low-Pressure Steam and Hot Water Boilers	Table 803.2.2(5)
Z21.47—00	Gas-Fired Central Furnaces—with Addenda Z21.47a-2000	Table 803.2.2(4)
Z21.56—98	Gas-Fired Pool Heaters—with Z21.56a—with Addenda-1999	Table 504.2.1
Z83.8—96	Gas Unit Heaters—with Addendum Z83.8a-1997	Table 803.2.2(4)
Z83.9—96	Gas-fired Duct Furnaces	Table 803.2.2(4)

ARI

Air Conditioning and Refrigeration Institute
4301 North Fairfax Drive
Suite 200
Arlington, VA 22203

Standard reference number	Title	Referenced in code section number
210/240—94	Unitary Air-Conditioning and Air-Source Heat Pump Equipment	Table 503.2, Table 803.2.2(1), Table 803.2.2(2)
310/380—93	Standard for Packaged Terminal Air-Conditioning and Heat Pumps	202, Table 803.2.2(3)
325—98	Ground Water-Source Heat Pumps	Table 803.3.2(2)
340/360—2000	Commercial and Industrial Unitary Air-Conditioning and Heat Pump Equipment	Table 803.2.2(1), Table 803.2.2(2)
365—94	Commercial and Industrial Unitary Air-Conditioning Condensing Units	Table 803.3.2(1)
460-94	Remote Mechanical-Draft Air-Cooled Refrigerant Condensers	Table 803.3.2(6)
550/590—98	Water Chilling Packages Using the Vapor Compression Cycle	Table 803.3.2(2)

REFERENCED STANDARDS

ARI—continued

Standard reference number	Title	Referenced in code section number
560—92	Absorption Water Chilling and Water Heating Packages	Table 803.3.2(2)
13256-1 (1998)	Water-source Heat Pumps - Testing and Rating for Performance - Part 1: Water-to-Air and Brine-to-Air Heat Pumps	Table 803.2.2(2)

ASHRAE

American Society of Heating, Refrigerating and Air-Conditioning Engineers, Inc.
1791 Tullie Circle, NE
Atlanta, GA 30329-2305

Standard reference number	Title	Referenced in code section number
136—1993 (RA 2001)	A Method of Determining Air Change Rates in Detached Dwellings	402.2.3.9
146-1998	Testing and Rating Pool Heaters	Table 804.2
13256-1 (1998)	Water-source Heat Pumps - Testing and Rating for Performance - Part 1: Water-to-Air and Brine-to-Air Heat Pumps	Table 803.2.2(2)
55—1992	Thermal Environmental Conditions for Human Occupancy	202
90.1—2001	Energy Standard for Buildings Except Low-Rise Residential Buildings	701.1, 801.2, 802.1, 802.2
ASHRAE—1999	ASHRAE HVAC Applications Handbook-1999	504.2.2
ASHRAE—2001	ASHRAE Fundamentals Handbook- 2001	Table 302.1, 402.4.2, 502.2.1.1.2, 502.2.2, 503.3.1, 803.2.1
ASHRAE—2000	ASHRAE HVAC Systems and Equipment Handbook-2000	503.3.1, 803.2.1

ASME

American Society of Mechanical Engineers
Three Park Avenue
New York, NY 10016-5990

Standard reference number	Title	Referenced in code section number
A112.18.1-2000	Plumbing Fixture Fittings	504.6.1
PTC 4.1 - 1964	Steam Generating Units	Table 803.2.2(5)

ASTM

ASTM International
100 Barr Harbor Drive
West Conshohocken, PA 19428-2859

Standard reference number	Title	Referenced in code section number
C 236—93e1	Standard Test Method for Steady-State Thermal Performance of Building Assemblies by Means of a Guarded Hot Box	602.1.1.1
C 518—98	Standard Test Method for Steady-State Thermal Transmission Properties by Means of the Heat Flow Meter Apparatus	Table 503.3.3.3
C 976 - 90(1996)96e1	Standard Test Method for Thermal Performance of Building Assemblies by Means of a Calibrated Hot Box	602.1.1.1
E 96—00	Standard Test Methods for Water Vapor Transmission of Materials	502.1.1, 602.1.7, 802.1.2
E 283—99	Test Method for Determining the Rate of Air Leakage Through Exterior Windows, Curtain Walls and Doors Under Specified Pressure Differences Across the Specimen	502.1.3, Table 502.1.4.1, 802.3.2, 802.3.7
E 779—99	Standard Test Method for Determining Air Leakage Rate by Fan Pressurization	402.2.3.9

CTI

Cooling Technology Institute
2611 FM 1960 West, Suite H-200
Houston, TX 77068-3730

Standard reference number	Title	Referenced in code section number
STD-201 (1996)	Certification Standard for Commercial Water Cooling Towers	Table 803.3.2(6)
ATC-105 (1997)	Acceptance Test Code for Water Cooling Towers	Table 803.3.2(6)

REFERENCED STANDARDS

DOE

U.S. Department of Energy
c/o Superintendent of Documents
U.S. Government Printing Office
Washington, DC 20402-9325

Standard reference number	Title	Referenced in code section number
10 CFR Part 430, Subpart B, Appendix E (1998)	Uniform Test Method for Measuring the Energy Consumption of Water Heaters	Table 504.2.1, Table 803.2.2(4), Table 804.2
10 CFR Part 430, Subpart B, Appendix N (1998)	Uniform Test Method for Measuring the Energy Consumption of Furnaces and Boilers	Table 503.2, Table 803.2.2(5), Table 804.2
10 CFR Part 430, Subpart B, Test Procedures (1998)	Energy Conservation Program for Consumer Products	202
ORNL/Sub-86-72143/1-88	Building Foundation Design Handbook	Table 502.2, 502.2.1.5, 502.2.3.5
DOE/EIA—0376 (Current Edition)	State Energy Prices and Expenditure Report	806.2.3

HI

Hydronics Institute, Division of the Gas Appliance Manufacturers Association
P.O. Box 218
Berkeley Heights, NJ 07054

Standard reference number	Title	Referenced in code section number
HBS	I=B=R - Testing and Rating Standard for Heating Boilers, 1989 Ed.	Table 803.2.2(5)

ICC

International Code Council, Inc.
5203 Leesburg Pike, Suite 600
Falls Church, VA 22041-3401

Standard reference number	Title	Referenced in code section number
IBC—03	International Building Code®	201.3
ICC EC—03	ICC Electrical Code™	201.3
IEBC—03	International Existing Building Code™	101.2
IFC—03	International Fire Code®	201.3
IFGC—03	International Fuel Gas Code®	201.3
IMC—03	International Mechanical Code®	201.3, 202, 503.3.3.4, 503.3.3.4.1, 503.3.3.4.2, 803.2.5, 803.2.6, 803.2.8.1, 803.2.8.1.1, 803.2.8.1.2, 803.3.4, 803.3.8.1
IPC—03	International Plumbing Code®	201.3

IESNA

Illuminating Engineering Society of North America
120 Wall Street, 17th Floor
New York, NY 10005-4001

Standard reference number	Title	Referenced in code section number
90.1-2001	Energy Standard for Buildings Except Low-Rise Residential Buildings	701.1, 801.2, 802.1, 802.2

NFRC

National Fenestration Rating Council, Inc.
8484 Georgia Avenue
Suite 320
Silver Spring, MD 20910

Standard reference number	Title	Referenced in code section number
100—01	Procedure for Determining Fenestration Product U-Factors	102.5.2, 601.3.2, 601.3.2.1

REFERENCED STANDARDS

NFRC—continued

Standard reference number	Title	Referenced in code section number
200—01	Procedure for Determining Fenestration Product Solar Heat Gain Coefficients and Visible Transmittance at Normal Incidence	102.5.2, 601.3.2, 601.3.2.1
400-01	Procedure for Determining Fenestration Product Air Leakage	Table 502.1.4.1, 601.3.2.2, 802.3.1

NOAA

National Oceanic and Atmospheric Administration
U.S. Department of Commerce
c/o Superintendent of Documents
U.S. Government Printing Office
Washington, DC 20402-9325

Standard reference number	Title	Referenced in code section number
CLIM 81-2	Annual Degree Days To Selected Bases 1961-1990 Normals	Table 302.1, 402.2.3.7

SMACNA

Sheet Metal and Air Conditioning Contractors National Association, Inc.
4021 Lafayette Center Drive
Chantilly, VA 20151-1209

Standard reference number	Title	Referenced in code section number
SMACNA—85	HVAC Air Duct Leakage Test Manual	402.2.3.8, 803.3.6

UL

Underwriters Laboratories Inc.
333 Pfingsten Road
Northbrook, IL 60062-2096

Standard reference number	Title	Referenced in code section number
181A—98	Closure Systems for Use with Rigid Air Ducts and Air Connectors — with Revisions through December 1998	503.3.3.4.3, 803.2.8
181B—95	Closure Systems for Use with Flexible Air Ducts and Air Connectors —with Revisions through December 1998	503.3.3.4.3, 803.2.8
727—98	Oil-Fired Central Furnaces—with Revisions through January 1999	Table 803.2.2(4)
731—95	Oil-Fired Unit Heaters—with Revisions through January 1999	Table 803.2.2(4),

WDMA

Window and Door Manufacturers Association
1400 East Touhy Avenue, Suite 470
Des Plaines, IL 60018

Standard reference number	Title	Referenced in code section number
101/I.S.2—97	Voluntary Specifications for Aluminum, Vinyl (PVC) and Wood Windows and Glass Doors	Table 502.1.4.1, 601.3.2.2, 802.3.1
101/I.S.2/NAFS—02	Voluntary Performance Specification for Windows, Skylights and Glass Doors	802.3.1

APPENDIX

The sections and construction details in Details 502.2.1.5(1) and 502.2.1.5(2), and Tables 502.2.3.1(1), 502.2.3.1(2), 502.2.3.1(3), 502.2.3.2, 502.2.3.3, 502.2.3.5 and 502.2.3.6 are intended to be representative and not all-inclusive. Adopting agencies are encouraged to add construction details and sections appropriate to their specific areas. Utilization of these tables should be correlated with local industry group practices and model code research recommendations.

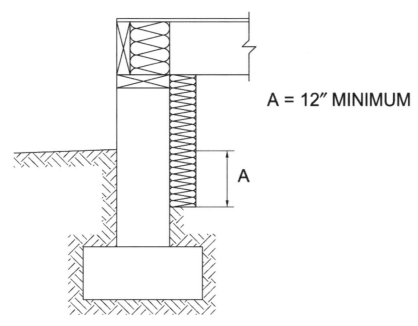

For SI: 1 inch = 25.4 mm.

DETAIL 502.2.1.5(1)
CRAWL SPACE WALL INSULATION—INSTALLATION NO. 1

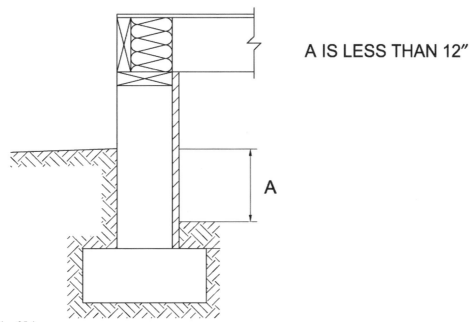

For SI: 1 inch = 25.4 mm.

DETAIL 502.2.1.5(2)
CRAWL SPACE WALL INSULATION—INSTALLATION NO. 2

APPENDIX

TABLE 502.2.3.1(1)
WALL ASSEMBLIES
(U_w selected shall not exceed the U_o determined by Section 502.2.3.1 for any wall section)

WALL DETAILS[a]		TYPE AND SPACING OF FRAMING (nominal)	R-VALUE OF CAVITY INSULATION	R-VALUE OF SHEATHING	U_w[b]
Typical schedules:					
Typical interior finish— 1. Gypsum wallboard; 2. Lath and plaster; or 3. $^3/_8$″ minimum wood paneling	Typical exterior finish— 1. Stucco; 2. Wood or plywood siding; or 3. Brick veneer				
WOOD STUD CONSTRUCTION		4″ Studs @ 16″ o.c.	11	noninsulating	0.085
			13	noninsulating	0.076
			13	3	0.064
			13	5	0.056
			13	7	0.051
			15	noninsulating	0.070
			15	3	0.059
			15	5	0.053
			15	7	0.048
		6″ Studs @ 16″ o.c.	19	noninsulating	0.058
			19	3	0.050
			19	5	0.046
			19	7	0.041
			21	noninsulating	0.052
			21	3	0.046
			21	5	0.042
			21	7	0.038
		6″ Studs @ 24″ o.c.	21	noninsulating	0.050
STEEL STUD CONSTRUCTION		4″ Studs @ 16″ o.c.	11	noninsulating	0.14
			13	noninsulating	0.13
		6″ Studs @ 16″ o.c.	19	noninsulating	0.11
		4″ Studs @ 24″ o.c.	11	noninsulating	0.12
			13	noninsulating	0.11
		6″ Studs @ 24″ o.c.	19	noninsulating	0.10

For SI: 1 inch = 25.4 mm.

a. Details shown are for insulation and are not complete construction details.
b. U_w calculated based on the ASHRAE *Fundamentals Handbook* 2001.

APPENDIX

TABLE 502.2.3.1(2)
WALL ASSEMBLIES
(U_w selected shall not exceed the U_o determined by Section 502.2.3.1 for any wall section)

WALL DETAILS[f]	R-VALUE OR TYPE		U_w AND R_o FOR WALL THICKNESS LISTED[a]			
PLAIN CONCRETE MASONRY BLOCK CONSTRUCTION			6″	8″	10″	12″
Plain block wythe	No insulation, no interior finish	U_w R_o	0.37 2.70	0.33 3.03	0.31 3.23	0.30 3.33
	Loose fill in cores, no interior finish	U_w R_o	0.18 5.56	0.13 7.69	0.11 9.09	0.09 11.11
Interior finish: 1/2″ gypsum board on furring strips	No insulation, interior finish	U_w R_o	0.24 4.17	0.23 4.35	0.22 4.55	0.21 4.76
	No insulation, foil-backed gypsum board interior finish	U_w R_o	0.18 5.56	0.17 5.88	0.16 6.25	0.16 6.25
Cavity insulation and interior finish: 1/2″ gypsum board on furring strips	1″ extruded polystyrene, interior finish	U_w R_o	0.13 7.69	0.13 7.69	0.12 8.33	0.12 8.33
	2″ expanded polystyrene, interior finish	U_w R_o	0.09 11.11	0.09 11.11	0.09 11.11	0.09 11.11
	2″ extruded polystyrene, interior finish	U_w R_o	0.08 12.50	0.08 12.50	0.08 12.50	0.08 12.50
	2″ polyisocyanurate, interior finish	U_w R_o	0.06 16.67	0.06 16.67	0.06 16.67	0.06 16.67
Interior finish: 1/2″ gypsum board over fibrous batt or loose fill between studs out from wall	R-11, 2 × 3 studs, interior finish	U_w R_o	0.07 14.29	0.07 14.29	0.07 14.29	0.07 14.29
	R-13, 2 × 3 studs, interior finish	U_w R_o	0.06 16.67	0.06 16.67	0.06 16.67	0.06 16.67
	R-19, 2 × 4 studs, interior finish	U_w R_o	0.05 20.00	0.05 20.00	0.05 20.00	0.05 20.00
MULTI-WYTHE WALLS			U_w AND R_o FOR WALL THICKNESS LISTED[b,c,d,e]			
			8″	10″	12″	14″
Plain block and clay wythes	No insulation, no interior finish	U_w R_o	0.32 3.13	0.26 3.85	0.24 4.17	0.22 4.55
	Loose fill in cavity, no interior finish	U_w R_o	NA NA	0.12 8.33	0.12 8.33	0.11 9.09
Cavity insulation and interior finish: 1/2″ gypsum board on furring strips	Loose fill, interior finish	U_w R_o	0.11 9.03	0.10 10.00	0.10 10.00	0.10 10.00
	Loose fill foil-backed gypsum board, interior finish	U_w R_o	0.10 10.00	0.09 11.11	0.09 11.11	0.09 11.11
	1″ expanded polystyrene in cavity, interior finish	U_w R_o	NA NA	0.13 7.69	0.12 8.33	0.12 8.33
	2″ expanded polystyrene in cavity, interior finish	U_w R_o	NA NA	0.08 12.50	0.08 12.50	0.08 12.50
	1″ extruded polystyrene in cavity, interior finish	U_w R_o	NA NA	0.11 9.09	0.11 9.09	0.11 9.09
	2″ extruded polystyrene in cavity, interior finish	U_w R_o	NA NA	0.07 14.29	0.07 14.29	0.07 14.29
	1″ polyisocyanurate in cavity, interior finish	U_w R_o	NA NA	0.08 12.50	0.08 12.50	0.08 12.50
	2″ polyisocyanurate in cavity, interior finish	U_w R_o	NA NA	0.05 20.00	0.05 20.00	0.05 20.00
	1″ expanded polystyrene in cavity foil-backed gypsum board, interior finish	U_w R_o	NA NA	0.09 11.11	0.09 11.11	0.09 11.11
	1″ extruded polystyrene in cavity foil-backed gypsum board, interior finish	U_w R_o	NA NA	0.08 12.50	0.08 12.50	0.08 12.50

For SI: 1 inch = 25.4 mm, 1 pound per cubic foot = 0.1572 kg/m^3.

a. The U_w values are for blocks made with concrete having a density of 80 pounds per cubic foot; for other densities, the U_w must be calculated based on the R-values provided in NCMA *TEK* 6-1A or the ASHRAE *Fundamentals Handbook* 2001.
b. 8″ composite wall: 4″ dense outer wythe and hollow-unit inner wythe.
c. 10″ cavity wall: 4″ dense outer wythe, 2″ air space and 4″ hollow-unit inner wythe.
d. 12″ cavity wall: 4″ dense outer wythe, 2″ air space and 6″ hollow-unit inner wythe.
e. 14″ cavity wall: 4″ dense outer wythe, 2″ air space and 8″ hollow-unit inner wythe.
f. Refer to drawings in Tables 502.2.3.1(1) and 502.2.3.1(3).
NA = Not Applicable.

APPENDIX

TABLE 502.2.3.1(3)
WALL ASSEMBLIES
(U_w selected shall not exceed the U_o determined by Section 502.2.3.1 for any wall section)

WALL DETAILS[d]	R-VALUE OF INSULATION	U_w	R_o
Interior finish 1/4" gypsum board applied on furring strips			
BRICK MASONRY CONSTRUCTION WITH LOOSE FILL	Solid grout in space	0.38	2.63
	2" space with loose fill R-4	0.16	6.25
	4" space with loose fill R-8	0.10	10.00
BRICK MASONRY CONSTRUCTION WITH INSULATION	4	0.12	8.33
	6	0.09	11.11
	11	0.07	14.29
NORMAL-WEIGHT CONCRETE CONSTRUCTION	4	0.18	5.56
	6	0.13	7.69
	7	0.12	8.33
	11	0.08	12.50
LIGHTWEIGHT CONCRETE CONSTRUCTION	4	0.17	5.88
	6	0.12	8.33
	7	0.11	9.09
	11	0.08	12.50
INSULATING CONCRETE FORM SYSTEM (ICF)[c]	12	0.07	13.55
	15	0.06	16.55
	16	0.06	17.55
	17	0.05	18.55
	20	0.05	21.55
	22	0.04	23.55

For SI: 1 inch = 25.4 mm.

a. The R-value listed is the sum of the values for the exterior and interior insulation layers.
b. The manufacturer shall be consulted for the U_w and R_o values if the insulated concrete form system (ICF) uses metal form ties to connect the interior and exterior insulation layers.
c. These values shall be permitted to be used for concrete masonry wall assemblies with exterior and interior insulation layers.
d. Details shown are for insulation and are not complete construction details.

APPENDIX

TABLE 502.2.3.2
ROOF/CEILING ASSEMBLIES
(U_r selected shall not exceed the value specified in Section 502.2.3.2)

ROOF DETAILS[a, b, c]	R-VALUE OF INSULATION[c]	U_r	R_o
Typical interior finish— 1. Gypsum wallboard; or 2. Lath & plaster *(Diagram: rafters, ceiling joist, insulation, interior finish)*	19	0.050	20.00
	22	0.040	25.00
	30	0.030	33.33
	38	0.025	40.00
(Diagram: air space and ventilation desirable, built-up roof, sheathing, insulation, interior finish, ceiling joist or rafters)	19	0.050	20.00
	22	0.040	25.00
	30	0.030	33.33
	38	0.025	40.00
(Diagram: cathedral-type ceiling, built-up roof, rigid insulation, wood or plywood sheathing, beam)	**Wood decking**		
	9	0.080	12.50
	Plywood		
	10	0.080	12.50
	19	0.050	20.00
	30	0.030	33.33

a. Details shown are for insulation and are not complete construction details.
b. Skylights not exceeding one percent of the roof are permitted.
c. Insulation installed between joints.

TABLE 502.2.3.3
FLOOR ASSEMBLIES
(U_r selected shall not exceed the U_o specified in Section 502.2.3.3)

FLOOR DETAILS[a]	R-VALUE OF INSULATION[c]	U_r	R_o
(Diagram: insulation, sub-floor, floor joist, girder)	No insulation	0.32	3.13
	7	0.11	9.09
	11	0.08	12.50
	19	0.05	20.00

a. Details shown are for insulation and are not complete construction details.

APPENDIX

TABLE 502.2.3.5
CRAWL SPACE FOUNDATION WALL ASSEMBLIES
(*U*-factor selected shall not exceed the *U*-factor determined by Section 502.2.3.5)

WALL DETAILS[a]	*R*-VALUE OF INSULATION	*U*-FACTOR
WOOD FOUNDATION	11	0.10
	13	0.09
	19	0.06
CONCRETE/MASONRY FOUNDATION—INTERIOR INSULATION	5	0.15
	10	0.08
	11	0.08
	13	0.07
	19	0.05
CONCRETE/MASONRY FOUNDATION—EXTERIOR INSULATION	3	0.20
	5	0.15
	10	0.08
	15	0.06
INSULATING CONCRETE FORM SYSTEM (ICF)[b, c, d]	12	0.08
	15	0.06
	16	0.06
	17	0.06
	20	0.05
	22	0.04

a. Details shown are for insulation and are not complete construction details.
b. The *R*-value listed is the sum of the values for the exterior and interior insulation layers.
c. The manufacturer shall be consulted for the *U*-factor if the insulated concrete form system (ICF) uses metal form ties to connect the interior and exterior insulation layers.
d. These values shall be permitted to be used for concrete masonry wall assemblies with exterior and interior insulation layers.

**TABLE 502.2.3.6
BASEMENT FOUNDATION WALL ASSEMBLIES**
(*U*-factor selected shall not exceed the *U*-factor determined by Section 502.2.3.6)

WALL DETAILS[a]	R-VALUE OF INSULATION	U-FACTOR
WOOD FOUNDATION	11	0.08
	13	0.08
	19	0.06
CONCRETE/MASONRY FOUNDATION—INTERIOR INSULATION	5	0.15
	6.5	0.12
	10	0.08
	11	0.08
	19	0.06
CONCRETE/MASONRY FOUNDATION—EXTERIOR INSULATION	3	0.20
	5	0.15
	10	0.09
	15	0.06
INSULATING CONCRETE FORM SYSTEM (ICF)[b, c, d]	12	0.07
	15	0.06
	16	0.06
	17	0.05
	20	0.05
	22	0.04

a. Details shown are for insulation and are not complete construction details.
b. The *R*-value listed is the sum of the values for the exterior and interior insulation layers.
c. The manufacturer shall be consulted for the *U*-factor if the insulated concrete form system (ICF) uses metal form ties to connect the interior and exterior insulation layers.
d. These values shall be permitted to be used for concrete masonry wall assemblies with exterior and interior insulation layers.

APPENDIX REFERENCED STANDARDS

ASHRAE-2001	ASHRAE *Fundamentals Handbook*	Tables 502.2.3.1(1) and 502.2.3.1(2)
NCMA TEK 6-1A	R-Values of Multi-Wythe Concrete Masonry Walls	Table 502.2.3.1(2)

INDEX

A

ADDITIONS AND ALTERATIONS
   Defined 202
   Requirements 101.2.2.2, 502.2.5
**ADMINISTRATION AND
   ENFORCEMENT** . . . . Chapter 1
AIR ECONOMIZERS 803.2.6, 803.3.3.5
AIR INFILTRATION
   Defined 202
   Requirements 402.1.3.10
AIR LEAKAGE 502.1.4, 601.3.2.2, 802.3
AIR SYSTEM BALANCING 503.3.3.7, 803.3.8.1
AIR TRANSPORT FACTOR
   Defined 202
   Requirements 503.3.3.6
ALTERNATE MATERIALS 103
ANNUAL FUEL UTILIZATION EFFICIENCY
   Defined 202
APPROVED
   Defined 202
AUTOMATIC
   Defined 202

B

BALANCING 503.3.3.7, 803.3.8
BALLASTS 805.3
BASEMENT WALLS
   Defined 202
   Requirements . . . . 402.2.3.10, Table 502.2,
    Figure 502.2(6), 502.2.1.6, 502.2.3.6,
    Tables 502.2.4(1-9), 502.2.4.9, Table 602.1,
    602.1.5, 802.1.1.2, Tables 802.2(1-37), 802.2.8
BELOW-GRADE WALLS (see BASEMENT WALLS)
BI-LEVEL SWITCHING 805.2.2.1
BOILERS . . . . Table 503.2, Table 803.2.2(5), 804.2
BUILDING ENVELOPE
   Compliance documentation 104, 402.5, 806.5
   Defined 202
   Requirements 402, 502, 602, 802
   System performance criteria 402.2.1, 402.2.3.3,
    806.4
   System performance method 402, 806

C

CAULKING AND WEATHERSTRIPPING . . . 502.1.4.2,
   602.1.10, 802.3.2
CIRCULATING PUMPS 504.4, 804.6
CIRCULATING SYSTEMS 504.3, 504.4,
   504.7, 804.4
CLIMATE DATA 302

CLIMATE ZONES . . . . Table 302.1
   By state . . . . Figures 902.1(1-51)
   Envelope requirements . . . . Tables 802.2(1-37)
COEFFICIENT OF PERFORMANCE (COP)
   Defined 202
**COMBINED SERVICE WATER HEATING AND SPACE
   HEATING** 504.2.2
COMMERCIAL BUILDINGS
   Compliance 101.3.2
   Defined 202
   Design by acceptable practice . . . . Chapter 8
   Design by referenced standard . . . . Chapter 7
   Design by total building performance 806
**COMMISSIONING OF HVAC
   SYSTEMS** 503.3.3.7, 803.3.8
   Manuals 803.3.8.3
COMPLIANCE AND ENFORCEMENT 101.4
CONDITIONED FLOOR AREA
   Defined 202
CONDITIONED SPACE
   Defined 202
CONSERVATION OF WATER 509.6
CONTROLS
   Capabilities 503.3.2.2, 803.3.3
   Economizers 803.2.6, 803.3.3.5
   Heat pump 803.3.3.1.1
   Heating and cooling 402.2.3.4,
    Table 402.2.3.4, 803.3.3
   Humidity 503.3.2.4, 803.2.3.3
   Hydronic systems 803.2.4, 803.3.3.7
   Lighting 805.2
   Off hour 803.3.3.3
   Service water heating . . . 504.3, 504.4, 804.2, 804.6
   Shutoff dampers 803.2.7, 803.3.4
   Temperature 503.3.2.1, 803.2.3.1
   Variable air volume systems 803.3.3.6, 803.3.4
   Ventilation 803.2.5, 803.3.5
COOLING WITH OUTDOOR AIR . . 803.2.6, 803.3.3.5
CRAWL SPACE WALLS
   Defined 202
   Requirements . . . . 502.1.5, 502.2, Table 502.2,
    Figure 502.2(5), 502.2.3.5,
    Tables 502.2.4(1-9), 502.2.4.12,
    Table 602.1, 602.1.7

D

DEADBAND 202, 503.3.2.2, 803.3.3.2
DEFINITIONS . . . . Chapter 2
DEGREE DAY COOLING
   Defined 202
DEGREE DAY HEATING
   Defined 202

INDEX

DESIGN CONDITIONS .... Chapter 3
DETAILS............................. Appendix
DUAL DUCT VAV 803.3.4.2
DUCTS
   Defined................................ 202
   Insulation................... 503.3.3.3, 803.2.8
   Sealing........ 503.3.3.4, 803.2.8, 803.3.6, 805.1,
DWELLING UNIT
   Defined................................ 202

E

ECONOMIZER
   Air..................... 803.2.6, 803.3.3.5
   Defined................................ 202
   Requirements....... 803.2.6, 803.3.4.3, 803.3.3.5
   Water............................. 803.3.3.5
ELECTRICAL POWER AND LIGHTING..... 505, 805
ENERGY ANALYSIS, ANNUAL
   Defined................................ 202
   Documentation............. 402.5, 403.2, 806.5
   Requirements...................... 402.1, 806
ENERGY EFFICIENCY RATIO (EER)
   Defined................................ 202
ENVELOPE, BUILDING
   Defined................................ 202
**ENVELOPE DESIGN
   PROCEDURES**............... 502.2.1, 502.2.2,
                     502.2.3, 502.2.4, 602, 802
EQUIPMENT EFFICIENCIES 503.2, 504.2, 603, 803.2,
                 803.3.2, 804.2, 806.3.1, 806.4.1
EQUIPMENT PERFORMANCE REQUIREMENTS
   Boilers..................... Table 803.2.2(5)
   Chillers, nonstandard......... Tables 803.3.2(1-2)
   Condensing units .... Table 803.3.2(1)
   Economizer exception............ Table 803.2.6
   General, residential .... Table 503.2
   Heat rejection equipment .... Table 803.3.2(6)
   Packaged terminal air conditioners
      and heat pump .... Table 803.2.2(3)
   Unitary air conditioners and
      condensing units............ Table 803.2.2(1)
   Unitary and applied heat pumps..... Table 803.2.2
   Warm air duct furnaces and
      unit heaters .... Table 803.2.2(4)
   Warm air furnaces .... Table 803.2.2(4)
   Warm air furnaces/air-conditioning
      units...................... Table 803.2.2(4)
   Water chilling packages,
      standard .... Table 803.3.2(2)
   Water heating.......... 504.2, Table 504.2, 804.2
EXEMPT BUILDINGS 101.2.1
EXISTING BUILDINGS 101.2.2
EXTERIOR LIGHTING 805.5
EXTERIOR SHADING.......... 402.1.3.1.3, 802.2.3

EXTERIOR WALLS
   Defined................................ 202
   Thermal performance 402.2.1.1, 502, 802

F

FENESTRATION 102.5, 402.2.1.2, 601.3.2
   Defined................................ 202
   Rating and labeling 102.5, 601.3.2
FURNACE EFFICIENCY .... Table 503.2,
                         Table 803.2.2(4)
   Defined................................ 202

G

GUESTROOMS 805.2.2.1
GLAZING AREA
   Defined................................ 202
   Requirements 402.2.1.3, 502.2, 502.2.1,
           502.2.2, 502.2.3, 502.2.4, 502.2.5

H

HEAT CAPACITY
   Defined................................ 202
   Requirements 502.2.1.1.2,
           Tables 502.2.1.1.2(1-3), 502.2.4.17,
                 Tables 502.2.4.17(1-2)
HEAT PUMP......... Table 503.2, Tables 803.2.2(2-3)
   Defined................................ 202
HEAT TRAPS 504.7, 804.4
   Defined................................ 202
HEATING AND COOLING CRITERIA .... Table 502.2
**HEATING AND COOLING
   LOADS** 402.4.1, 402.5.6, 503.3.1,
                       803.2.1, 803.3.1
HOT WATER 504.4, 804.2
   Annual energy performance 402.2.3.6, 402.4
   Piping insulation 504.5, 804.5
   System controls 504.3, 504.4, 804.2, 804.6
HUMIDISTAT
   Defined................................ 202
   Requirements 503.3.2.4, 803.2.3.2, 803.3.3.1
HYDRONIC SYSTEM BALANCING......... 803.3.8.2

I

**IDENTIFICATION (MATERIALS, EQUIPMENT
   AND SYSTEM)** 102.1
**INDIRECTLY CONDITIONED SPACE
   (see CONDITIONED SPACE)**
**INFILTRATION, AIR
   LEAKAGE** 402.2.3.9, 502.1.4, 802.3
   Defined................................ 202
INSPECTIONS............................ 105

INDEX

INSULATION
　Identification 102.5
　Installation 102.4, 601.3.1
INSULATING SHEATHING
　Defined 202
　Requirements 502.2.4.1, 502.2.4.16
INTEGRATED PART LOAD VALUE
　Defined 202
INTERIOR LIGHTING POWER 805.5

L

LABELED
　Defined 202
　Requirements 102.3, 102.5.2, 601.3.2, 601.3.3
LIGHTING POWER
　Design procedures 805.5.2
　Exterior connected 805.6, 806.2.8
　Interior connected ... 805.5, Table 805.5.2, 806.4.7
　Manufacturer's information 102.3, 601.3.3
LIGHTING SYSTEMS 505, 605, 805
　Controls, additional 805.2.2
　Controls, exterior 805.2.3
　Controls, interior 805.2.1
　Decorative .... Table 805.5.2
　For visual display terminals .... Table 805.5.2
　Guestrooms 805.2.2.3
　Line voltage 805.5.1.4
　Merchandise .... Table 805.5.2
　Plug-in busway 805.5.1.4
　Track 805.5.1.4
LISTED
　Defined 202
LOW-VOLTAGE LIGHTING
　Defined 202
　Requirements 805.4.1.2
LUMINAIRE
　Defined 202

M

MANUALS 803.3.8.3
MATERIALS AND EQUIPMENT 102, 601.3
**MECHANICAL SYSTEMS AND
　EQUIPMENT** 503, 603.3, 803
MECHANICAL VENTILATION 503.3.3.5, 803.5
　Defined 202
MOISTURE CONTROL 502.1.1, 802.1.2
MULTIFAMILY DWELLING
　Defined 202
**MULTIPLE SINGLE-FAMILY DWELLING
　(TOWNHOUSE)**
　Defined 202
MULTIPLE ZONE SYSTEMS 803.3.4

N

NONCIRCULATING SYSTEMS 804.4
**NONDEPLETABLE/RENEWABLE ENERGY
　SOURCES**
　Defined 202
　Requirements 403, 806.2.4

O

OCCUPANCY
　Defined 202
　Requirements 101.2.2.4, 101.2.3
OCCUPANCY SENSORS 805.2.1.1
OFF-HOUR, CONTROLS 803.3.3.3
OPAQUE AREAS
　Defined 202
ORIENTATION 402.2.2.1, 402.2.3.1.1, 402.5.3
OVERHANG, PROJECTION FACTOR 802.2.3
OZONE DEPLETION FACTOR
　Defined 202

P

PACKAGED TERMINAL AIR CONDITIONER (PTAC)
　Defined 202
　Requirements .... Table 803.2.2(3)
PACKAGED TERMINAL HEAT PUMP
　Defined 202
　Requirements .... Table 803.2.2(3)
**PARALLEL PATH CORRECTION
　FACTORS** .... 502.1.1.1, Table 502.2.1.1.1
PHOTOCELL 805.2.2
PIPE INSULATION 503.3.3.1, 503.3.3.2,
　　　　　　　　　　　　　504.5, 803.2.9, 803.3.7, 804.5
PLANS AND SPECIFICATIONS 104
POOL COVERS 504.3.2
PROJECTION FACTOR 802.2.3
PROPOSED DESIGN
　Defined 202
　Requirements 402, 502.2.2, 806
PUMPING SYSTEMS 504.3, 504.4,
　　　　　　　　　　　　　　　　　803.3.3.7.1, 803.3.8.2

R

***R*-VALUE (see THERMAL RESISTANCE)**
RECOOLING 803.3.4
REFERENCED STANDARDS .... 107, Chapter 10
REHEATING 803.3.4
**RENEWABLE/NONDEPLETABLE ENERGY
　SOURCES** 403, 806.2.4
　Defined 202

INDEX

REPAIR
  Requirements 101.2.2.2
  Defined 202
RESET CONTROL 803.3.3.7.4
RESIDENTIAL BUILDINGS
  Compliance 101.4.1
  Defined 202
  Design by component performance . . . . . Chapter 5
  Design by system analysis and
    renewable sources . . . . Chapter 4
  Design using simplified prescriptive
    requirements . . . . Chapter 6
ROOF ASSEMBLY
  Defined 202
  Requirements . . . . Figure 502.2(2), 502.2.1.2,
    502.2.3.2, Tables 502.2.4(1-9),
    502.2.4.7, Table 602.1, 602.1.2

S

SCOPE 101.2
SCREW LAMP HOLDERS
  Defined 202
  Requirements 805.5.1.1
SEASONAL ENERGY EFFICIENCY RATIO
  Defined 202
SERVICE WATER HEATING
  Defined 202
  Requirements 504, 604, 804, 806.2.8, 806.3.1
SHADING 402.2.3.1.2, 402.2.3.1.3, 402.2.3.1.4
  Projection factor 802.2.3
SHOWER HEADS 504.6.1
SHUTOFF DAMPERS 503.3.3.5, 803.3.3.4
SIMULATION TOOL
  Defined 202
  Requirements 402.4.7, 806.2.1
SINGLE ZONE 503.1, 803.2
SIZING
  Equipment and system 803.2.1, 803.3.1.1
SKYLIGHTS 402.2.1.4, 502.2.1.22.1,
    602.4, 502.2.5
  Defined 202
  Maximum exempt area 502.2.4.4, 802.2.5
SLAB-EDGE INSULATION 102.4.1,
    Figure 502.2(3), 502.2.1.4, 502.2.3.4,
    Tables 502.2.4(1-9), 502.2.4.10, 502.2.4.11,
    Table 602.1, 602.1.6, Tables 802.2(1-37), 802.2.7
SOLAR HEAT GAIN COEFFICIENT (SHGC) . . . 102.3
STANDARD DESIGN
  Defined 202
STANDARDS, REFERENCED . . . . . . 107, Chapter 10
STEEL FRAMING 502.2.1.1.1, 502.2.1.2,
    502.2.1.3, 502.2.3.1, 502.2.3.2, 502.2.3.3,
    502.2.4.16, Tables 502.9.16(1-2), 502.2.4.18,
    Tables 502.2.4.18(1-2), 502.2.4.19, Table 502.2.4.19

STANDARD TRUSS 502.2.4.5, 602.1.2
  Defined 202
SUNROOM ADDITION 502.2.2.5
  Defined 202
SWIMMING POOLS 504.3
SYSTEMS ANALYSIS 402, 806

T

TANDEM WIRING 805.3
TERMITE
  INFESTATION . . . . Table 502.2, Figure 502.2(7)
THERMAL ISOLATION . . . . 502.2.5, Table 502.5
  Defined 202
THERMAL
  MASS . . . . 502.2.1.1.2, Tables 502.2.1.1.2(1-3),
    502.2.4.17, Tables 502.2.4.17(1-2)
THERMAL RESISTANCE (R)
  Defined 202
THERMAL TRANSMITTANCE (U)
  Defined 202
TOTAL BUILDING PERFORMANCE
  Commercial 806
  Residential 402
**TOWNHOUSE (see MULTIPLE SINGLE-FAMILY
  DWELLING)**
**TYPE A-1 AND A-2 RESIDENTIAL BUILDINGS
  (see RESIDENTIAL BUILDINGS)**

U

U-FACTOR (see THERMAL TRANSMITTANCE)

V

VAPOR RETARDER 502.1.1, 802.1.2
VARIABLE AIR VOLUME
  SYSTEMS (VAV) 803.3.3.6, 803.3.4
VENTILATION 503.3.3.5, 803.2.5, 803.3.5
  Defined 202

W

**WALLS (see EXTERIOR WALLS AND ENVELOPE,
  BUILDING)**
**WALLS ADJACENT TO UNCONDITIONED
  SPACE** . . . . . . 802.1.1.3, Tables 802.2(1-37), 802.2.9
WATER CONSERVATION 504.6
WATER ECONOMIZER 803.3.3.5
WATER HEATING 504, 604, 804, 806.2.8, 806.3.1
WINDOW AREA (see GLAZING AREA)
WINDOW PROJECTION FACTOR
  Defined 202
  Requirements 802.2.3
**WINDOW-TO-WALL RATIO
  (see GLAZING AREA)**

WIRING, TANDEM 805.3

Z

ZONE
    Defined 202
    Requirements . . . . Table 402.1.3.5, 503.3.2.1,
                                      803.3.3, 803.3.4
ZONE ISOLATION 803.3.4